TELEPEN
80 7302028 2
WITHDRAWN FROM STOCK
CPC UCW
LLYFRGELL
LIBRARY
ABERYSTWYTH

AF616357

LAKE SIBAYA

MONOGRAPHIAE BIOLOGICAE

Editor

J. ILLIES

Schlitz

VOLUME 36

Dr W Junk bv Publishers The Hague–Boston–London 1979

LAKE SIBAYA

Edited by

B. R. ALLANSON

Dr W Junk bv Publishers The Hague–Boston–London 1979

The distribution of this book is handled by the following team of publishers:

for the United States and Canada

Kluwer Boston, Inc.
160 Old Derby Street
Hingham, MA 02043
USA

for all other countries

Kluwer Academic Publishers Group
Distribution Center
P.O. Box 322
3300 AH Dordrecht
The Netherlands

Library of Congress Cataloging in Publication Data CIP
Main entry under title:

Lake Sibaya.

(Monographiae biologicae; v. 36)
Bibliography: p.
Includes index.
1. Limnology – South Africa – Sibayi, Lake. 2. Lake ecology – South Africa – Sibayi, Lake. 3. Sibayi, Lake, South Africa. I. Allanson, Brian R. II. Series.

QP1.P37 [QH195.S6] 574′.08s [574.5′2632] 79-18805
ISBN 90-6193-088-X

Table of Contents

Addresses of authors

Prof. B. R. Allanson, Institute for Freshwater Studies, Rhodes University, Grahamstown 6140, South Africa.

Dr C. M. Breen, Department of Botany, University of Natal, P.O. Box 375, Pietermaritzburg 3200, South Africa.

Dr M. N. Bruton, J. L. B. Smith Institute of Ichthyology, Rhodes University, Grahamstown 6140, South Africa.

Mr A. Bruton, Electron Microscopy Unit, Rhodes University, Grahamstown 6140, South Africa.

Mrs P. M. Eva, Institute for Freshwater Studies, Rhodes University, Grahamstown 6140, South Africa.

Dr R. C. Hart, Institute for Freshwater Studies, Rhodes University, Grahamstown 6140, South Africa.

Dr B. J. Hill, Queensland Fisheries, P.O. Box 36, North Quay, Brisbane, Australia 4000.

Dr D. K. Hobday, Bureau of Economic Geology, University of Texas, University Station, Box X, Austin, Texas 78712, U.S.A.

Dr C. Howard-Williams, D.S.I.R. (Freshwater Section) P.O. Box 415, Taupo, New Zealand, (Previously of Institute for Freshwater Studies, Rhodes University, Grahamstown.)

Preface

This is the story of a lake; somewhat fragmented but each chapter tells of the specific and sometimes peripheral interests of each author. Some of us have lived at the lake for long periods of time, others for short periods, and each has attempted to contribute to an overall understanding of the lake as an ecosystem.

Apart from some earlier unpublished studies of the lake during a series of expeditions under the leadership of the late Dr G. G. Campbell of Durban, serious work started in July 1964 with a team of staff and students from the Department of Zoology & Entomology at Rhodes University, Grahamstown. In the following year we managed to attract funds for the continuation of our work from the South African IBP committee. This provided a strong measure of continuity to our planning, and by degrees – indeed slowly – we gathered up the interest and skills available to us and with what simple equipment we could acquire, settled down to what in effect was the first reasonably integrated study of the limnology of a lake in the Republic of South Africa.

This volume is an assemblage and review of the significant work which has been published or, because of its somewhat fragmentary nature, remained unpublished until an opportunity arose to place it into perspective. The request from Professor Dr Joachim Illies in 1974 to assemble our work into a volume for the Monographiae Biologicae series provided just this, and the incentive and spur needed to take a closer and more critical look at what we were achieving in our study of Lake Sibaya and other lakes along the coastal rimland of Southern Africa.

South Africa does not possess any striking limnological feature – indeed to some it might, with good cause, be considered a limnological desert! Nevertheless a group of determined biologists and chemists have during the past three decades established unequivocally a set of limnological parameters for the sub-continent.

Much of what has been done was in direct response to problems of river and reservoir eutrophication and the toxic pollution of our meagre surface waters. As emigration from the metropolitan area of the Witwatersrand increased, the coastal rimland began to feel the pressure of human demands for new water resources. The coastal lakes are obviously a prime target of this demand and until quite recently we were ignorant of their properties. This volume and the studies of Howard-Williams and Allanson (1978) on a southern coastal lake, Swartvlei, will, I hope, go a long way to providing the requisite information upon which their conservative utilization can be based.

Acknowledgements

I have been aided in the preparation and editing of this volume by a variety of people with special skills. Of particular note is the skill of my secretary,

Mrs P. Eva. Without her understanding help the editing of the contributions would have been much more difficult. I should like to acknowledge the sterling work of my wife in the preparation of the indexes. My colleagues who contributed to the content of the volume have been at considerable pains to present manuscripts of quality, and were indulgent of my editorial suggestions and demands. I have attempted to maintain a constant figure format in each chapter. This required redrawing many of the text figures. This labour was the responsibility of our artist, Miss Robin Allanson. Likewise many of the photographic plates required specialized attention, and Mr Philip Fischer of the Department of Zoology & Entomology provided much needed professional advice and execution. Our collective thanks are due to these collaborators.

Much of the editorial work was done in the library of the Freshwater Biological Association, Windermere, while I was on sabbatical leave. The assistance of the librarian, J. E. M. Horne, O.B.E. and his colleagues is greatly appreciated as is wise help of Mr. Michael Henderson of the publishers.

Our thanks are due to the Total Oil Company of South Africa and BP South Africa who served the objects of our research so generously through the provision of fuel. The Chairman's Fund of the Anglo American Company awarded grants to assist the continuation of our work during 1976, 1977 and 1978, which were greatly appreciated. The Department of Bantu Affairs in Natal and in particular the Chief Bantu Affairs Commissioner, Mr J. Kotze, and later Mr C. Hardwick of the Department of Agriculture and Conservation of the KwaZulu government, provided the buildings of the research station and their maintenance. Without this material assistance the continuity of the programme would have suffered.

Finally, I wish to record my deep appreciation of the generous financial framework provided by the South African Council for Scientific and Industrial Research and the Council of Rhodes University. In this regard the President, Dr C. van der Merwe Brink and the then Vice-Chancellor of the university, Dr J. Hyslop, deserve special mention for their faith in a group of nascent limnologists.

DEDICATION

To the life and work of
Dr R. E. Boltt

Palman qui meruit, ferat
Horatio, Viscount Nelson

1 Geological evolution and geomorphology of the Zululand coastal plain

D. K. Hobday

Introduction

Lake Sibaya is situated on the seaward margin of the Zululand Coastal Plain. The coastal plain (Fig. 1) commences in the south at Mtunzini, broadening out northward to 80 km at the national border, beyond which it occupies almost half the width of Mozambique. Rocks underlying the coastal plain are mainly Cretaceous to Palaeocene sediments, which abut landward against late Karoo volcanics of the Lebombo Range. Relatively thin, discontinuous Tertiary shallow marine and beach deposits overlie the Cretaceous with erosive unconformity. Postdating these are complex Quaternary sediments representing a variety of depositional environments, including shorezone, aeolian dune, fluviatile, lagoonal and paludal, which constitute the major physiographic elements of the coastal plain.

The coastline of southern Africa was initiated by Mesozoic fragmentation of the protocontinent Gondwanaland. The first evidence of rifting was the generation of rifts roughly parallel to the present outline of the subcontinent. Along the east coast of Africa marine incursions extended gradually southward along these rifted grabens, and are recorded by mid-Jurassic limestones and shales in Tanzania (King, 1962, p. 60) and by Neocomian (Early Cretaceous) deposits in Mozambique and Zululand. Offshore drilling on the Agulhas Bank by SOEKOR has recently established an Early Cretaceous rifting stage, during which marine conditions were established in the south, followed by continental drifting (Du Toit, 1977). The continental outline was established by the end of the Cretaceous Period.

Following separation from the other southern hemisphere continents and India, southern Africa was subjected to a series of epeirogenic uplifts with concomitant down-flexuring of the margins (King and King, 1959). Depending upon the position of the axis of no relative movement, these tectonic events were responsible for either emergence of the coastal margins (axis seaward of the coastline) or marine transgression (axis inland). Thus King and King (1959) have postulated marginal tilting during the mid-Cretaceous, early Miocene, early Pliocene, and end Pliocene–early Pleistocene, leading to a series of transgressions and regressions across the coastal plain.

In addition to the tectonic changes in the relative levels of land and sea there were world-wide sea-level changes consequent upon lithospheric plate generation and interaction (Hallam, 1971; Flemming and Roberts, 1973) and glacio-eustatism (Trowbridge, 1954). Late Cretaceous, Eocene and Miocene transgressions have been ascribed to accelerated sea floor spreading (Flemming and Roberts, op. cit.). According to Tankard (1976) plate

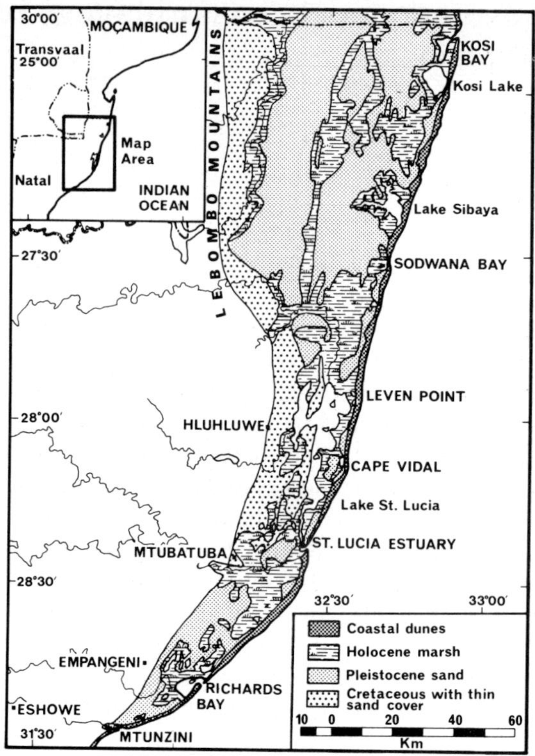

Fig. 1. The main geological elements of the coastal plain.

tectonic processes may account for the Miocene transgressive marine deposits of southern Africa. McCarthy (1967), from a study of coastal sediments in Natal, favours a general Late Tertiary regression, which may be reconciled with Hallam's (1963) postulate of continued subsidence of the ocean floors from the Miocene. The generally descending chronological order of shorelines would suggest a series of glacio-eustatic sea-level oscillations of approximately equal magnitude, superimposed on an overall regression (Tankard, 1976). In Zululand this would appear to hold true for the Pleistocene, where the highest (and probably oldest) shorelines are of an elevation that cannot be accounted for by glacio-eustatism alone.

Mesozoic and Cenozoic History of the Coastal Plain

Cretaceous

On the downwarped margins produced by the breakup of Gondwanaland were deposited early Cretaceous terrigenous conglomerates now exposed at isolated localities along the inner margin of the Zululand Coastal Plain (Frankel, 1960). Channeled and truncated Stormberg (end Karoo) lavas, which outcrop in the Lebombos, dip steeply beneath the Cretaceous (Fig. 2).

The basal Cretaceous conglomerate exposed in the Mfolozi Valley attains 26 m in thickness, and contains boulders almost a metre in diameter (Frankel, 1960). These deposits probably originated in the beds of high-gradient, anastomosing torrential streams of alluvial fans. Most of the rock types of the eroded hinterland, including basalt, dolerite, rhyolite, granite, banded ironstone and quartzite are represented. There is an upward and seaward gradation from a predominance of boulders into pebbles and then cross-bedded sands, marls and silts. This records a rapid reduction of flow energy into distal fan braided channels. Fossil plant material, all gymnosperms, are scattered throughout and indicate a Neocomian-Wealden or earlier age (Frankel, 1960).

In northern Zululand up to 250 m of Aptian and Albian conglomerates, limestones, siltstones and sandstones are developed. A short distance north

Table 1. Generalized stratigraphic relationships of the Zululand Coastal Plain.

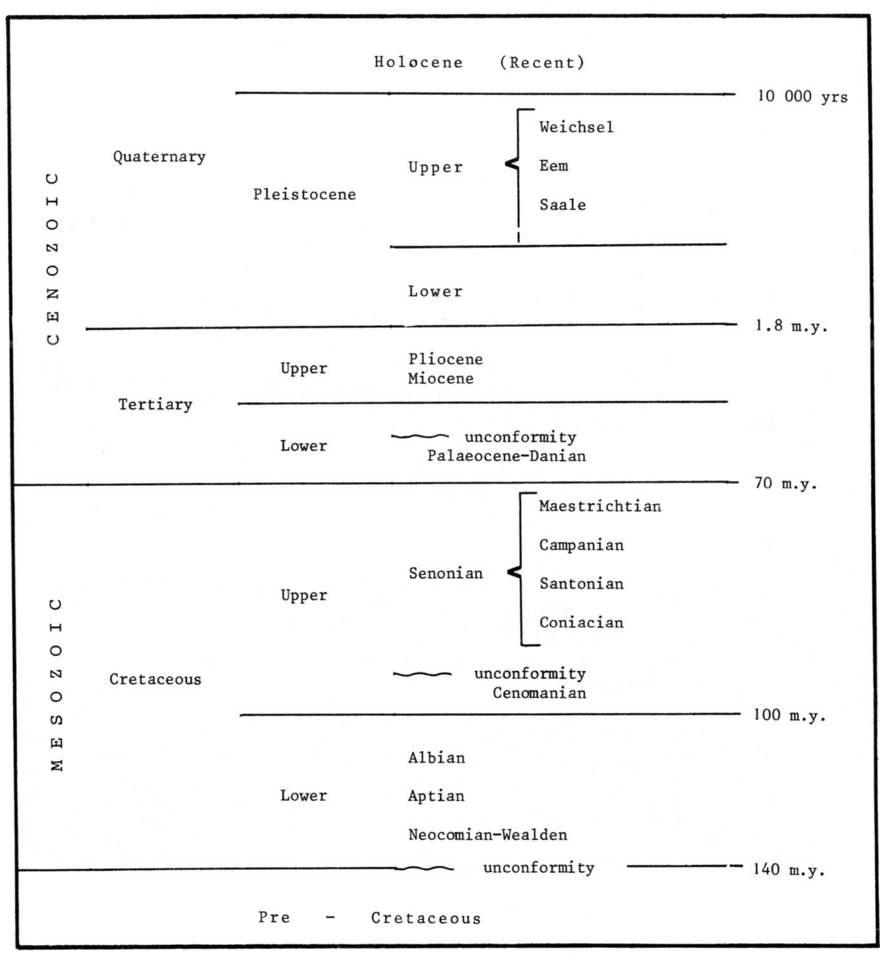

		Holocene (Recent)			
					10 000 yrs
CENOZOIC	Quaternary	Pleistocene	Upper	Weichsel Eem Saale	
			Lower		
					1.8 m.y.
	Tertiary	Upper	Pliocene Miocene		
		Lower	unconformity Palaeocene-Danian		
					70 m.y.
MESOZOIC	Cretaceous	Upper	Senonian	Maestrichtian Campanian Santonian Coniacian	
			unconformity Cenomanian		
					100 m.y.
		Lower	Albian Aptian Neocomian-Wealden		
			unconformity		140 m.y.
		Pre - Cretaceous			

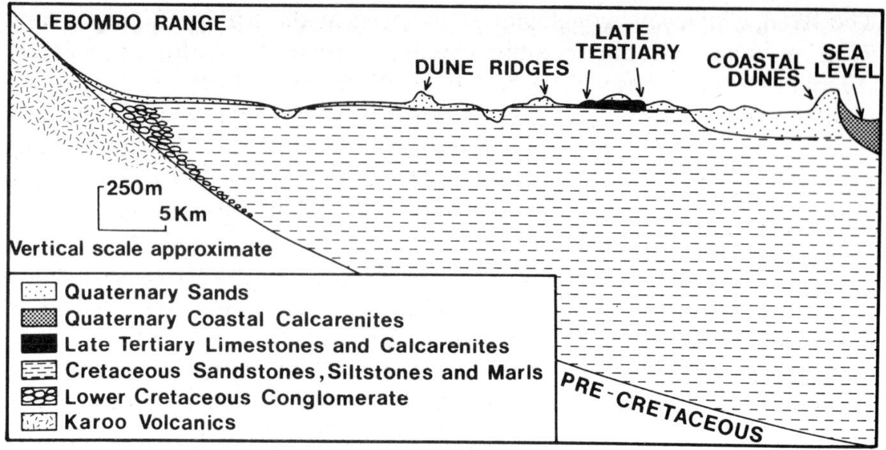

Fig. 2. Cross-section illustrating gross stratigraphic relationships and major topographic features of the coastal plain.

of the Zululand border Middle Aptian ammonites indicate that the Early Cretaceous sea lapped the base of the Lebombo (Du Toit, 1954, p. 393).

The upper Cretaceous follows conformably (Table 1), and shows progressive onlap. There is, however, a major intra-Cretaceous unconformity. The uppermost Cenomanian and Turonian stages are absent and the Coniacian (lower Senonian) oversteps Middle Cenomanian to Aptian and pre-Cretaceous rocks (Kennedy and Klinger, 1971).

Plate 1. Coarse bioclastic Upper Cretaceous shoreline deposits 50 cm above wave-eroded basement rocks.

Generally there is an increase in open marine character upward through the Upper Cretaceous succession. This presumably resulted both from progressive widening of the proto-ocean gulf and the worldwide transgressive character of the Senonian and Maestrichtian. Maximum thicknesses of accumulation occurred in northern Zululand where almost 2000 m of Cretaceous sediments have been revealed by subsurface exploration (Truswell, 1970, p. 145). A short distance beyond the southern limit of the coastal plain 733 m of Cretaceous were penetrated in an offshore borehole (Du Toit and Leith, 1974).

Early Tertiary

The early-Tertiary record of Zululand is sparse, mainly representing a time of non-deposition. Records from the continental shelf (Du Toit and Leith, 1974) indicate that substantial sediment accumulation occurred farther seaward during this generally regressive episode.

Boreholes and excavations at Richards Bay reveal Palaeocene (Danian) deposits of a lithology identical to the underlying Upper Cretaceous (Orr and Chapman, 1974). Geotechnical tests indicated a marked change in consolidation values at the Cretaceous-Palaeocene faunal boundary, suggesting a disconformity (Maud and Orr, 1975). The preserved Palaeocene succession attains a thickness of 50 m. On the basis of the foraminiferal content Orr and Chapman (1974) postulate an open shelf origin of intermediate depth. Palaeocene marine inundation of the coastal plain must therefore have been extensive, although not positively recorded elsewhere in Zululand.

In Zululand the only evidence of Eocene deposition comes from reworked sediments of this age at Uloa (Frankel, 1968). Middle Eocene marine sediments are, however, known from Mozambique (Du Toit, 1954, p. 431). It is possible that the record further south was obliterated by coastal erosion associated with the Late Tertiary transgression.

Late Tertiary

Separated from the older rocks by a marked Oligocene hiatus are sporadic developments of shelly limestones and calcarenites of Late Tertiary age. There is some dispute concerning the precise dating of these rocks. King (1953, 1969, 1970) envisages two distinct depositional events, Lower Miocene and Pliocene, between which occurred an unconformity. On the other hand, Frankel (1968) and Maud and Orr (1975) view all of these deposits as being of Late Miocene age. McCarthy (1967) and Stapleton (in press) question the presence of an unconformity between the two formations, which they relate to a single depositional episode.

Shelly limestone ('pecten bed')

The highly fossiliferous, locally conglomeratic 'pecten bed' occurs at a number of localities in the Mfolozi Valley (King, 1953; Frankel, 1966) and at the base of a ridge which extends intermittently from Richards Bay to Lake St. Lucia (King, 1970), and beyond. Isolated exposures occur inland of Lake Sibaya and in Mozambique (Soares and Da Silva, 1970).

At the base of the formation, resting on a smooth, marine-planed surface, is a thin ferruginous layer containing phosphatic, pyritic and iron oxide concretions, septarian nodules, pyritized Cretaceous ammonites, Tertiary fossils, and ironstone fragments described by King (1953) as reworked Early Tertiary laterite. According to King, (1953, 1970) mid Tertiary uplift led to erosion of the Early Tertiary or 'African' planation, thereby providing Miocene sediment which included fragments of the Early Tertiary lateritic cover.

Overlying this coarse lag, which probably represents a concentration during marine transgression of the coarsest detritus derived from the hinterland and by *in situ* winnowing, is a massive coquina-like limestone from which over 100 species of macrofossils have been recovered (King, 1953), along with 45 species of foraminifera. Characteristic of the formation is *Aequipecten uloa* (King, 1953) along with a variety of other molluscs and smaller numbers of corals, brachiopods, echinoderms and crustaceans.

Pebbles and boulders occur scattered throughout the formation, particularly at Uloa, where it terminates against a low cliff to the west (Frankel, 1968). The 'pecten bed' seldom exceeds 5 m in thickness, lacks stratification and shows little vertical or lateral change in character (King, 1970).

The abundance of thick-shelled molluscs, many of which are broken, together with the coarseness of the extrabasinal detritus, suggests a high-energy, nearshore marine environment. According to Stapleton (in press) deposition probably took place in shallow, temperate, normal marine conditions.

Calcarenite formation

The contact of the overlying calcarenites on the 'pecten bed' varies from abrupt, as at Uloa, to the gradational relationship evident in some outcrops further to the north and east. Thin section analysis reveals that the rock is composed almost exclusively of quartz with foraminifera and a sparry calcite cement. Oysters, abraded shells and spherical or discoidal pebbles are locally present.

In contrast to the underlying formation, bedding is well developed. Subhorizontal dips predominate, with seaward inclinations alternating with shallow scour channels and occasional high-angle cross-bedding which becomes more abundant upward. Exposures of these calcarenites on the Nebela Peninsula of Lake St. Lucia (Hobday, 1965) clearly represent a beach foreshore deposit, overlain by aeolianites. A boulder bed, representing a transgressive shoreline accumulation is sometimes present at the base of the formation, particularly where the 'pecten bed' is not developed.

According to King and King (1959) and King (1966, 1972) end Miocene-early Pliocene uplift, accompanied by marginal downwarping, caused the sea to cover the entire area of the coastal plain, bevelling it to low relief in the process, and depositing calcareous sands. Deposition of these transgressive marine sediments was followed by a Quaternary expulsion of the sea and the accumulation of regressive sands which were redistributed by slumping and wind action, and remained unconsolidated. A contrasting view, held by McCarthy (1967), is that there was a single Late Tertiary phase of gradual emergence of the coastal plain. This is supported by Stapleton's (in press) study of the planktonic foraminifera, which in his opinion indicate that the age difference between the two formations cannot be more than 3 m.y. The Nebela Peninsula exposures record shore-zone regression, with prograding beach deposits conformably above the 'pecten bed' gradationally overlain by wind-deposited sandstones, presumably once coastal dunes. In the Uloa area the record is more complex, and vertically-sided channels several metres deep, are cut into the 'pecten bed' and filled with sands of the calcarenite formation (King, 1970). Frankel (1968) too records an unconformity at Uloa, which he ascribes to scour along an emergent shore-line.

To summarize, at some localities the sequence from the truncated Cretaceous surface through the ferruginous layer and 'pecten bed' into the overlying calcarenites represents a classical transgressive–regressive couplet such as is commonly preserved in coastal margins of low relief elsewhere in the world. In other areas there was apparently a break in sedimentation, with possible subaerial exposure intervening between the deposition of the lower and upper formations.

Pleistocene Geology and Geomorphology

Boulder beds, old red sands, younger coversands, coastal dunes and calcarenites together indicate a complex Quaternary history of erosion and sedimentation. In the absence of accurate dating techniques it is impossible to unravel these developments in any detail except to recognize a sequence of events, which during the later Pleistocene at least, can be related to sea-level fluctuations of known extent and duration. Archaeological evidence (Davies, 1970, 1976) provides a relative chronology for some deposits.

Relict Dune Ridges

Prominent north-south trending dune ridges are a conspicuous topographic feature of the coastal plain. They are generally composed of red or yellowish sand and are distinguished from younger accumulations of light-coloured sand by their higher relief. A boulder bed is commonly present at the base of the western cordon, and gives way eastward to scattered pebble horizons. Davies (1976) has mapped a series of dune cordons (Fig. 3) which decrease in age from west to east. Davies assumes that each cordon was constructed in the vicinity of the contemporary shore. The oldest shoreline, of Plio-Pleistocene age, is nearly 100 km inland near the Mozambique border.

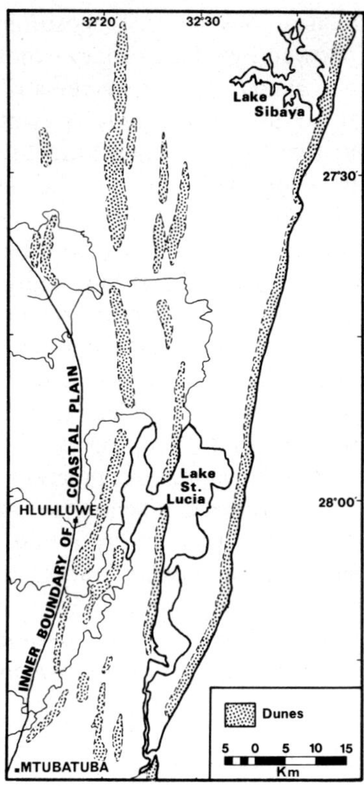

Fig. 3. Distribution of prominent dune cordons of the coastal plain (modified after Davies, 1976).

The innermost dune cordon occurs adjacent to Lebombo volcanics in the Mkuze Game Reserve, where it is 30 m high. Its base is at 95 m. This ridge, together with an adjacent ridge to seaward, is poorly preserved. These are paralleled to the east by a more continuous ridge, which is traced by Davies from Mtunzini to Mozambique. Its base is between 60 and 75 m, and it attains a maximum elevation of 130 m. A basal boulder bed is of variable thickness. The boulders, mainly of quartzite, are up to 30 cm in diameter.

Another high and relatively persistent ridge extends along the western shores of Lake St. Lucia. It rests on an irregular surface at elevations which range between 5 and 30 m. At its base are pebbly, water-lain deposits with herringbone cross-bedding suggestive of tidal flow reversals (Hobday, 1965, p. 48). Above this is well-sorted red sand and very fine-grained loessic sand with ventifacts above, clearly of aeolian origin. These upper sands are locally incised by coarser pebbly sands, probably the deposits of ephemeral streams.

There are several differences between the eastern and western cordons, which in conjunction with the decreasing elevation of their bases, confirm a progressively younger age seaward. The eastern ridges are higher, steeper sided and less eroded. Those in the west have better developed soil profiles, and are commonly a deeper red, displaying advanced mineral diagenesis. The red colour is entirely epigenetic and is due to a coating of iron oxides around the grains. These iron oxides were probably derived from the breakdown of ferro-magnesian minerals. In general the sands of the low western ridges contain feldspar which has become partly kaolinized, whereas this mineral breakdown is less pronounced seaward. Indurated ferruginous hardpans, a product of advanced podsolization, occur at certain localities in all dune ridges, but are generally thickest and most mature in the inland ridges.

The coarser basal sediments are probably of estuarine or beach origin. Wave action would have been subdued by frictional degradation on the shallow platform to seaward. Tidal processes were therefore more important than they are along the present coast. The bulk of the dune ridges were constructed by shoreline and aeolian processes during marine regression.

The relationship of these dune ridges to discrete Pleistocene sea-level oscillations is not clear. Certainly they display sequential decrease in the elevation of their bases such that the oldest ridges are probably Early Pleistocene (Davies, 1976; Hobday, 1976) whereas the youngest are probably Late Pleistocene. Davies (1976) provisionally proposes that the youngest cordon represents the last interglacial high sea-level (Eemian) and last glacial (Weichsel). This estimate is based on the low elevation of the floor of the deposits and the associated artifacts. Davies identified Late Acheulian and earlier artifacts beneath the ridge, and Tugela and later industries on the upper dune surface. The two next older cordons are assigned by Davies to the preceding cycle of the penultimate high interglacial level (Holstein) and the Saalian glacial regression. These bear Acheulian artifacts. Older ridges which overlie pre-Acheulian artifacts are viewed by Davies as relating to the antepenultimate cycle (Cromer and Elster).

Coastal Calcarenites

Outcropping intermittently along the Zululand coast are calcarenites displaying a series of marine-cut platforms, some of which are overlain by younger deposits. These rocks are exposed only on the seaward side of the high coastal dunes (Fig. 2) where they are being reduced by wave attack and solution weathering. Lithification of these deposits was probably a consequence of their coastal situation (*cf.* the beachrock cementation processes of Russell, 1959), whereas sediments of similar age inland of the coastal dunes remained uncemented.

The oldest calcarenites are aeolian sandstones which crop out at isolated points along the coast. Occurrences at Black Rock have been described by Coetzee (1975) and in the Lake St. Lucia area by Hobday (1976). The calcarenites extend from an unknown depth below sea-level to elevations of 25 m. The upper parts have been subjected to advanced karst weathering, with the development of solution pipes up to 20 m deep (Coetzee, 1975). Grain size ranges from fine to medium sand, which is very well sorted. Rounded quartz grains constitute up to 95 per cent of the rock, with zircon, rutile, ilmenite, feldspar, garnet and shell fragments as accessories. The cement is sparry calcite. Large-scale cross-bedding is the predominant structure, with northerly and westerly azimuths most common. Concurrent palaeowinds thus differed somewhat from the prevailing directions of the present day, which are mainly subparallel to the coastline. There is evidence in the Lake St. Lucia that these deposits originated during the second last glacial regression (Saalian).

Overlying the aeolianites with an irregular erosive contact, which ranges from below present sea-level to 6 m above, are coarser, less well-sorted beach deposits (Plate 2) which are related to the last interglacial Eemian high (Hobday, 1976). Beach foreshore sandstones represented by plane beds with heavy mineral laminae, dip gently seaward, where they overlie and merge laterally with predominantly trough cross-bedded upper shoreface facies. Two distinct phases of deposition with intervening erosion and pothole development are presented. An intraformational marine erosion surface occurs between 3.5 and 5.5 m, and is often accentuated by discoidal pebbles and oyster beds.

Post-dating these beach sandstones is an aeolian calcarenite which strongly resembles the earlier calcarenite. It is far less widely distributed, however, and has a maximum observed thickness of only 10 m. This unit is thought to have been deposited by renewed wind dominance during a brief intra-Eemian regression.

Renewed rise of the sea to above its present level is evidenced by a third generation of coarse shingle beach deposits with a maximum elevation of 4.5 m. This unit is also dated as Eemian on the basis of its geomorphological characteristics (Hobday, 1976).

Plate 2. Late Pleistocene (Eemian) raised beach deposits overlying aeolianites at the level of the modern wave-cut terrace, Mission Rocks. Note high coastal dunes to landward and planed surface of beach sandstones.

Coastal Terraces

Wave cut terraces which relate to the Eemian calcarenites (Plate 2) occur at a number of localities along the Zululand coast. Because of their original seaward slope, subsequent covering by sand and fallen debris, and their partial obliteration by karsting and undermining by wave attack, absolute elevations are difficult to obtain. Nevertheless, levels at between +1.5 and 4.5 m, 5 m and 8 m are most common. All three terraces are seldom represented together at any one locality, and often only one is recognizable along a particular stretch of coast. Thus at Black Rock, Coetzee (1975) finds a single, poorly-defined terrace at 4 m, and in southern Mozambique elevations of just over one metre and about 5 m are most common.

These three terraces and their associated sandstone and conglomerate deposits are ascribed to three Eemian sea-level peaks (Hobday, 1976) which occurred over a time range of 128,000 to 73,000 years ago (Suggate, 1974). The sea-level estimates are based on palaeotemperature determinations from deep-sea cores (Emiliani, 1970; Shackleton and Opdyke, 1973, amongst others), and are corroborated by radiometric dating of terraces in various parts of the world (*e.g.* Broecker *et. al.*, 1968). During the brief intra-Eemian regressions there was rapid lithification of the sands, such as is characteristic of carbonate-rich sediments. Sea-level probably did not fall much below the present, so that limited amounts of sand were available for dune formation. Renewed advance of the sea concentrated pebbles brought down by rivers during their brief period of steepened gradient and planed the calcarenites, developing new shoreline deposits on their surface.

The platform corresponding to the modern sea-level has been cut during the past 5000 years. In places this erosion coincides with a late Eemian terrace which is evidenced by a well-cemented pebble veneer. This re-occupation of an earlier level possibly accounts for the well-developed planation, which is commonly too broad to have been cut during the past few thousand years alone. For example, the intertidal platform at Black Rock has a maximum width of 80 m (Coetzee, 1975).

The modern platform, which is described in detail by Coetzee (op. cit). has slightly raised rims at its seaward edge, behind which are shallow pools. According to Coetzee the platform surface is lowered by solution, while the outer rim is protected by algal encrustation. Cliffs at the inner margin are sometimes undercut at the base.

Other Deposits of the Last Interglacial

Clays, sands, lignites and coral limestones, all tentatively attributed to the last (Eemian) interglacial, provide an interesting record of environmental and climatic changes during the late history of the coastal plain.

Port Durnford Formation

The Port Durnford Formation (Anderson, 1907; Hobday and Orme, 1974) is intermittently exposed in sea cliffs along the southern Zululand coast. A lower clayey unit contains a variety of fossil remains including the extinct elephant *Paleoloxodon zulu*, buffalo *Bubalus andersoni*, and other mammalian fauna, fish, fossil wood, and various invertebrates of fresh or brackish water environments. This essentially lagoonal facies is overlain by up to 2.5 m of lignite or peat, which accumulated both by *in situ* growth and decay of swamp vegetation, and by inrafting of floating debris. The swamp surface was subsequently covered by wave and tidal current deposited sands, followed by aeolian accumulation. This sequence therefore records a back-barrier lagoon, which filled, developed into a swamp, and was covered by the sands of retreating barrier to seaward. The lignite, which must have accumulated very close to sea-level has its base some 1 to 3 m above modern sea-level, except where it has slumped.

Coral limestones

Highly fossiliferous limestones (Plate 3) occur in False Bay, the western arm of the Lake St. Lucia system. They are unique to Zululand in that they contain thermophile corals typical of tropical conditions, along with large reef molluscs, brachiopods, echinoderms, crustaceans and open marine microfossils. Radiometric dating indicates an age in excess of 50,000 years (Orme, 1973). The maximum elevation of these deposits is 3.5 m above sea level, which together with the faunal content, suggests that they originated in the warm waters of an Eemian interglacial. False Bay was at that time a marine embayment opening directly to the sea (Hobday, 1976).

Plate 3. Late Pleistocene corals fossilized in growth position, False Bay, Lake St. Lucia.

Features Associated with Marine Regression of the Last Glaciation

The last glaciation was accompanied by lowered sea-levels which attained several distinct minima of over 100 m below the present. Climatic amelioration after 17,000 years B.P. was accompanied by rapid transgression until about 10,000 B.P., when there was a decrease in the rate of sea-level rise. The Pleistocene-Holocene (Recent) boundary is conventionally taken as 10,000 years. The sea probably attained its present level some 5,000 years ago (Curray, 1969), although some estimates are younger.

Valley incision

Rejuvenation of the lower courses of coastal rivers, which resulted from the lowering of base level, caused deep valley incision. The rivers would have scoured mainly into sediments filling the palaeo-valleys of earlier Pleistocene regressions. Radiometric dating (Maud, 1968; Hobday, 1976) indicates that most of the material filling the valleys relates to the last glaciation, suggesting that practically all of the older valley-fill was eroded. The maximum depths of valley incision would have depended largely upon the size of the river. Thus, in the case of the Mfolozi River, 16 km from the sea >67 m of alluvium were penetrated without meeting bedrock (Pullin, 1966, cited in Maud, 1968). In Lake St. Lucia incision in the Hells Gate area by the combined discharge of three minor rivers, the Nyalazi, Hluhluwe and Mzenene, was 33 m (Hobday, 1976). By contrast, small, intermittent streams scoured to depths of less than 5 m.

In addition to vertical degradation the rivers extended their courses across the emergent continental shelf. Their former courses are probably recorded by submarine canyons, which terminate close inshore. The extreme narrowness of the shelf (in places as little as 4 km) would have provided steep gradients of up to 1 in 30 (Hobday, 1965) and consequently, deep entrenchment occurred. Near Sodwana Bay the canyons commence at a depth of 36 m and plunge to 650 m a distance of only 6 km offshore (Maud, 1968). As suggested by Maud (1968), Hill (1975) and Hobday (1976) the canyons may be related to earlier connections to the sea of rivers in the Lake Sibaya and St. Lucia areas. While this origin would appear to be indisputable, Flemming (1976) has recently shown that canyons off the Natal coast currently trap large quantities of sediment being transported alongshore. Gravity flow of this material to the canyon base probably causes present-day erosion and increase in canyon dimension, a process documented in California by Dill (1964).

Submerged platforms and sand ridges

Other features which originated during episodes of lowered sea-level are submarine platforms at depths of 11, 18, 36, 47, 70, 80 and 95 m (Maud, 1968), and lithified dune ridges, locally referred to as ‘reefs’. It is not clear whether these features are a product of the last glaciation alone, or whether some are relicts from earlier glaciations. Very possibly only the larger, best-preserved features of maximum relief are referable to the final regression, recording wave abrasion and beach ridge accumulation during halting stages in the overall retreat. The low preservation potential of features associated with transgressing shorelines makes it unlikely that they relate to stillstands during the rise of the sea to its present level.

High coastal dunes

The coastal dunes (Plate 4) which extend from Mtunzini northward well into Mozambique constitute one of the largest persistent coastal cordons in the world. The dunes, with crest elevations attaining a maximum of almost 200 m, are interupted only by rare estuaries, major river outlets, and oblique aeolian blowouts. Elsewhere the dunes seal off former river outlets, for example at Leven Point (Hill, 1975) and Mission Rocks (Davies, 1976) in the St. Lucia area, and at Lake Sibaya (Davies, op. cit.).

A core of older deposits such as the Port Durnford Formation or red sand is present in places, and the dune sands abut landward against red sand overlain by Holocene coversands and marsh. Sands of the high dunes are medium grained and very well sorted. Quartz grains with frosted surfaces are dominant, with concentrations of opaque heavy minerals increasing with depth. In places the sands are weakly cemented by rainwater percolation or lateral seepage. The latter phenomenon sometimes produces a ferruginous, lithified sandstone at the base of the dunes, for example at Ochre Hill.

Steep, large-scale cross-bedding of aeolian character is the main internal structure. Occasionally this bedding is evidenced by vegetation patterns on the dune flanks. Foreset inclinations suggest that winds from the northeast and southwest were equally effective in fashioning the dunes, a conclusion

Plate 4. The high coastal dunes at the Zululand–Mozambique border, with parallel low dunes to landward. (Photo C. J. Ward).

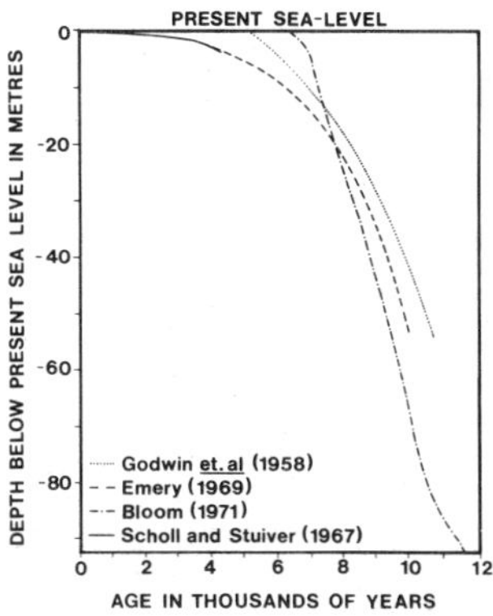

Fig. 4. Sea-level curves for the postglacial transgression, modified after Tankard (1976).

which is supported by their topography. Along certain sections of the coast, such as the Lake St. Lucia area, the steep relict slip faces are mainly towards the south. Elsewhere the slip faces to the north. There has been some modification by present day winds.

Derivation of the dune sand was clearly related to an episode of lowered sea-level. The present narrow beaches provide little sand to the dunes, and the prevailing wind regime is largely destructive. With retreat of the sea exposing a broad expanse of loose sand to seaward, and with the strengthening of onshore winds by greater sea-land temperature contrasts during the glaciation, dune construction would have occurred on the inner margin of the sand belt. Further landward migration was retarded by calcarenites and older topographic features, upon which the dunes in places became superimposed. During the subsequent phase of climatic amelioration the dunes were stabilized by vegetation.

The Postglacial Transgression

End Pleistocene and Holocene advance of the shoreline to its present position decreased river gradients and reduced their capability of maintaining an outlet to the sea. Estimates of the rate of sea-level rise are presented in Fig. 4. Smaller river mouths became blocked by littoral and aeolian processes, and were permanently sealed by vegetation. Only the largest rivers and estuaries, such as the Mfolozi and Kosi Bay, maintained a semi-permanent outlet.

Waters of the blocked rivers became dammed up behind the dune barrier. Where discharge was relatively small, *e.g.* rivers of the Sibaya system, a freshwater lake formed with a relict fauna of marine affinities (Allanson, *et al.*, 1966). In the opinion of Maud (1968) Lake Sibaya represents the betrunked outlet of the Pongola River which now discharges into Delagoa Bay 160 km to the north. Lake St. Lucia, with larger freshwater inflow, has had a more complex history. Borehole cores indicate that lagoonal confinement occurred early during the transgression Hobday (1976), *i.e.* the former outlet at Leven Point (Hill, 1975) became blocked. Nevertheless, foraminiferal evidence (Phleger, 1976) shows that intermittent marine incursions have

Plate 5. The northern part of the Kosi Lake system showing the outlet on the left and drowned tributary valley centre right. (Photo C. J. Ward).

been common during the infilling of the lake. The impounded waters of the Lake St. Lucia rose well above sea-level, as has occurred in Lake Sibaya, and flooded low-lying swamps and inter-dune depressions both north and south of the present Lake. Eventually the southward-spreading waters met up with the Mfolozi system, possibly via a small meandering tributary system, cut-off meanders of which are still evident on air photographs. A further variation is exemplified by the Kosi lake system (Plate 5) where the position of a former outlet in the south is overlain by low dunes, and the only surviving outlet in the north is maintained mainly by tidal scour rather than by river discharge (Orme, 1973).

Holocene Sea-levels and Sedimentation

A postglacial Holocene sea-level high with a maximum of 4 to 7 m above the present has been postulated by a number of workers in Natal and Zululand (Krige, 1932; Hobday, 1965; McCarthy, 1967; Maud, 1968; Davies, 1970). These views accord with the sea-level curves of Zeuner (1959) and Fairbridge (1961). It would appear, however, that many of the low emergent features of the Natal–Zululand coast are either older than Holocene or may be accounted for by processes which do not involve a relative change in sea-level. For example the raised limestones of False Bay attributed by Hobday (1965) to a Holocene high sea-level were radiometrically dated and shown by Orme (1973) to be almost certainly Eemian. Raised shoreline terraces are probably all Eemian or older. Many of the depositional terraces around the coastal lakes and lagoons are due to back-barrier accumulation of floodwaters above sea-level. Prior to the development of flood controls in the lower Mfolozi Valley, the Mfolozi Flats were commonly inundated by floodwaters to elevations of 5 m and more.

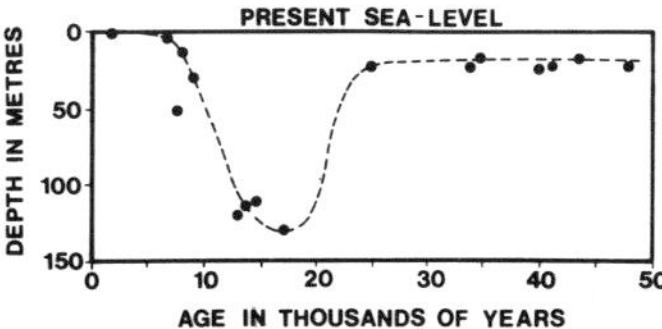

Fig. 5. Estimates of sea-levels based on South African data, after Tankard (1976).

High beach ridges develop around the margins of Lake St. Lucia at the present day, particularly during flood periods of raised lake level when onshore winds cause additional mounding of water against the shore.

Data from stable parts of the world (Chappell, 1974) and coast-lines where the shelf is very shallow and therefore not subject to hydro-isostatic downwarping (Scholl and Stuiver, 1967; Bloom, 1971) suggest that no substantial changes have occurred since the sea attained its present Holocene level. Tankard (1976) has constructed a sea-level curve (Fig. 5) based on meagre South African data which suggests that following the end Pleistocene eustatic low 17,000 years B.P. the sea rose very rapidly until 9,000 B.P. when it was at −25 m, and was followed by a more gradual rise to a metre or so below present sea-level between 5000 and 6000 B.P.

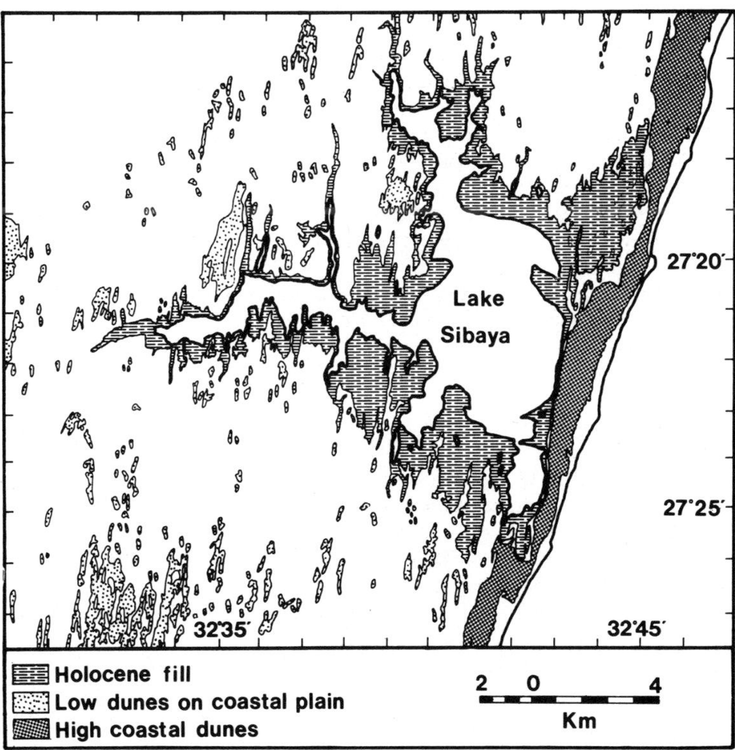

Fig. 6. Geomorphology of the Lake Sibaya area showing high coastal dunes, low parallel dune ridges and areas of Holocene infilling. Modified after Orme (1973).

Since the end of the postglacial transgression there has been local redistribution of dunesands as indicated by parabolic blowouts and the development of smaller, poorly vegetated dune trains (Plate 4). More important has been the reduction in extent of coastal lagoons, which once dominated the outer margin of the coastal plain. The main processes have involved segmentation, infilling and reedswamp encroachment (Hobday, 1965; Orme, 1973). Orme (op. cit.) has estimated that all Zululand lagoons have experienced at least a 60 percent reduction in area during the Holocene. The Lake St. Lucia system once extended 112 km parallel to the coastal dunes. Its present length is 40 km. Lake Sibaya had an initial area of almost 150 km^2. The pattern of infilling is illustrated in Fig. 6. Reduction in depth has in some cases been even more marked. Thirty metres of sediment has accumulated in Hells Gate, Lake St. Lucia, in the past 6000 years (Vogel and van Urk, 1975). Elsewhere, where freshwater inflow is limited such as Lake Sibaya, a lesser amount of infilling has occurred (Hill, 1975).

Acknowledgements

Mr C. J. Ward kindly contributed air photographs of the Zululand coastal areas.

References

Allanson, B. R., B. J. Hill, R. E. Boltt and V. Schultz. 1966. An estuarine fauna in a freshwater lake in South Africa. Nature, Lond. 209: 532–533.

Anderson, W. 1907. On the discovery in Zululand of marine fossiliferous rocks of Tertiary age containing mammalian remains. Third and Final Report of the Geological Survey of Natal and Zululand, London. 121–127.

Bloom, A. L. 1971. Glacial-eustatic and isostatic controls of sea level since the last glaciation. In K. K. Turekian (editor) Late Cenozoic Glacial Ages, Yale University Press, New Haven. 355–379.

Broecker, W. S., D. L. Thurler, J. Goddard, J. T-L. Ku, R. K. Matthews and K. J. Mesolella. 1968. Milankovitch hypothesis supported by precise dating of coral reefs and deep sea sediments. Science, N.Y. 159, 293–300.

Chappell, J. 1974. Late Quaternary glacio – and hydro-isostasy on a layered earth. Quaternary Res. 4: 405–428.

Coetzee, F. 1975. Coastal aeolianites at Black Rock, northern Zululand. Trans. Geol. Soc. S. Afr., 78: 313–322.

Curray, J. R. 1969. History of continental shelves. In D. J. Stanley (editor). The New Concepts of Continental Margin Sedimentation. Am. Geol. Inst., Washington D.C., JC61–JC618.

Davies, O. 1970. Pleistocene beaches of Natal. Ann. Natal Mus. 20: 403–442.

Davies, O. 1976. The Older Coastal Dunes in Natal and Zululand and their relation to former shorelines. Ann. S. Afr. Mus. 71: 19–32.

Dill, R. F. 1964. Sedimentation and erosion in Scripps submarine canyon head. In R. L. Miller (editor) Marine Geology, Macmillan, London. 23–41.

Du Toit, A. L. 1954. Geology of South Africa. 3rd ed., Oliver and Boyd, London. 611 pp.

Du Toit, S. R. 1977. The Mesozoic history of the Agulhas Bank in terms of plate-tectonic theory. Geokongres 77 Abstracts. 17th Congress of the Geological Society of South Africa, Rand Afrikaans University, Johannesburg. 32 pp.

Du Toit, S. R. and M. J. Leith. 1974. The J(c)-1 bore-hole on the continental shelf near Stanger, Natal. Trans. Geol. Soc. S. Afr. 77: 247–252.

Emiliani, C. 1970. Pleistocene paleotemperatures. Science, N.Y. 168: 822–825.

Fairbridge, R. W. 1961. Eustatic changes in sealevel. In L. H. Ahrens (editor). Physics and Chemistry of the Earth. Pergamon, London. 99–185.

Flemming, B. W. 1976. Underwater sand dunes along the southeast African continental margin – observations and implications. Technical Report 8, Joint Geological Survey/University of Cape Town Marine Geology Programme. 35–46.

Flemming, N. C. and D. G. Roberts. 1973. Tectono-eustatic changes in sea level and seafloor spreading. Nature, Lond. 243: 19–22.

Frankel, J. J. 1960. The geology along the Umfolosi River south of Mtubatuba, Zululand. Trans. Geol. Soc. S. Afr., 63: 231–264.

Frankel, J. J. 1966. The basal rocks of the Tertiary at Uloa, Zululand, South Africa. Geol. Mag. 103: 214–230.

Frankel, J. J. 1968. Tertiary sediments in the lower Umfolozi River valley, Zululand. Trans. Geol. Soc. S. Afr. 71: 135–145.

Hallam, A. 1963. Major epeirogenic and eustatic changes since the Cretaceous and their possible relationship to crustal structure. Am. J. Sci. 261: 397–423.

Hallam, A. 1971. Mesozoic geology and the opening of the North Atlantic. J. Geol. 79: 129–157.

Hill, B. J. 1975. The origin of Southern African coastal lakes. Trans. R. Soc. S. Afr. 41: 225–240.

Hobday, D. K. 1965. The Geomorphology of the Lake St. Lucia Area. Unpub. M.Sc. thesis, Univ. Natal. 130 pp.

Hobday, D. K. 1976. Quaternary sedimentation and development of the lagoonal complex, Lake St. Lucia, Zululand. Ann. S. Afr. Mus. 71: 93–113.

Hobday, D. K. and A. R. Orme. 1974. The Port Durnford Formation: a major Pleistocene barrier-lagoon complex along the Zululand coast. Trans. Geol. Soc. S. Afr. 77: 141–149.

Kennedy, W. J. and H. Klinger. 1971. A major intra-Cretaceous unconformity in eastern South Africa. J. Geol. Soc. Lond. 127: 183–186.

King, L. C. 1953. A Miocene marine fauna from Zululand. Trans. Geol. Soc. S. Afr. 56: 59–91.

King, L. C. 1962. The Morphology of the Earth. Oliver and Boyd, London. 699 pp.

King, L. C. 1966. An extensive marine Pliocene formation in Natal and Mozambique: its geomorphological implications. Trans. Geol. Soc. S. Afr. 69: 201–210.

King, L. C. 1969. On the ages of Tertiary rocks at Uloa and Umkwelane, Zululand and their geomorphological significance. Geol. Mag. 106: 206–208.

King, L. C. 1970. Uloa revisited. A review of the essential data from this classic locality, and a statement of Tertiary history of the Zululand coastal area. Trans. Geol. Soc. S. Afr. 73: 151–157.

King, L. C. 1972. The coastal plain of Southeast Africa: its form, deposits and development. Z. Geomorph. 16: 239–251.

King, L. C. and L. A. King. 1959. A reappraisal of the Natal Monocline. S. Afr. Geog. J. 41: 15–30.

Krige, L. J. 1932. The geological history of Durban. S. Afr. J. Sci. 29, 25–40.

Maud, R. R. 1968. Quaternary geomorphology and soil formation in coastal Natal. Z. Geomorph. 7: 155–199.

Maud, R. R. and W. N. Orr. 1975. Aspects of post-Karroo geology in the Richards Bay area. Trans. Geol. Soc. S. Afr. 78: 101–109.

McCarthy, M. J. 1967. Stratigraphical and sedimentological evidence from the Durban region of major sea-level movements since late Tertiary. Trans. Geol. Soc. S. Afr. 70: 135–165.

Orme, A. R. 1973. Barrier and lagoon systems along the Zululand coast, South Africa. In D. R. Coates (editor) Coastal Geomorphology, State Univ. of New York. 181–217.

Orr, W. N. and J. Chapman. 1974. Danian (Palaeocene) planktonic foraminifera from Zululand, South Africa. S. Afr. J. Sci. 70: 247–249.

Phleger, F. B. 1976. Holocene ecology of Lake St. Lucia Lagoon, Zululand, based on foraminifera. In Report of the St. Lucia Scientific Advisory Council Meeting, Charters Creek, 1976.

Russell, R. J. 1959. Caribbean beach rock observations. Z. Geomorph. 3: 227–236.

Scholl, D. W. and M. Stuiver. 1967. Recent submergence of southern Florida. A comparison with adjacent coasts and other eustatic data. Bull. Geol. Soc. Am. 78: 437–454.

Shackleton, N. J. and N. D. Opdyke. 1973. Oxygen isotope and palaeo-magnetic stratigraphy of equatorial Pacific core V28–238: oxygen isotope temperatures and ice volumes on a 10^5 year and 10^6 year scale. Quaternary Res. 2: 1–14.

Soares, A. F. and G. H. Da Silva. 1970. Contribuicão para o estudo da geologia do Mapuló. Rev. Cienc. Geol., Univ. Lourenco Marques 3: 1–85.

Stapleton, R. P. (in press). Planktonic foraminifera and the age of the Uloa pecten bed. Ann. Geol. Surv. S. Afr.

Suggate, R. P. 1974. When did the last interglacial end? Quaternary Res. 4: 246–252.

Tankard, A. J. 1976. Cenozoic sea-level changes: a discussion. Ann. S. Afr. Mus. 71: 1–17.

Trowbridge, A. C. 1954. Mississippi River and Gulf Coast terraces and sediments as related to Pleistocene history – a problem. Bull. Geol. Soc. Am. 65, 793–812.

Truswell, J. F. 1970. An introduction to the historical geology of South Africa. Purnell, Cape Town. 167 pp.

Vogel, J. C. and H. van Urk. 1975. Isotopic investigation of Lake St. Lucia. National Physical Research Laboratory, Pretoria. 12 pp.

Zeuner, F. E. 1959. The Pleistocene Period, its climate, chronology and faunal successions. Hutchinson. 300 pp.

2 The dune forest: its structure and maintenance

C. M. Breen

Structure

The forested dunes are one of the most striking features of the coastal plain of Zululand. This is because, in spite of being confined to a narrow belt seldom exceeding 1 km in width, they attain considerable height. At Lake Sibaya the highest peak is 134 m, but further south in the vicinity of St. Lucia they are considerably higher (183 m). It seems probable that the southerly winds (land-breezes), at times when sea level was much lower and there was a wide expanse of sand, were the major factor in fashioning the dune crests in this area (Hobday 1965). This has resulted in them having a N.N.E.–S.S.W. orientation. However, it is the north-easterly winds, the sea-breezes, that have the greatest influence on the vegetation of the dune system, because they pick up salt spray from the breaking waves, particularly those along the rocky outcrop at Mabibi (Fig. 1), and carry it inland, depositing it on the vegetation.

The combined effects of wind and salt on the coastal vegetation of Natal and Zululand have been well documented (Bayer 1938; Edwards 1967; Moll 1967; Moll 1970; Ward 1971 and Tinley 1976), and are detectable in that they result in a dense low-growing scrub comprised of deformed bushes which have a uniform or 'hedged' canopy. The more common species contributing to this scrub are *Ficus burtt-davyi* Hutch., *Grewia occidentalis* L., *Mimusops caffra* E. Mey. ex A. DC. *Acokanthera oblongifolia* (Hochst.) Codd, *Carissa bispinosa* (L.) Desf. ex Brenan, *Brachylaena discolor* A. DC. and in slightly more protected areas *Colpoon compressum* Berg., *Sophora inhambanensis* Klotzsch, *Euclea natalensis* A. DC., *Sideroxylon inerme* L. and *Canthium obovatum* Klotzsch. The influence of wind and salt are progressively diminished with increasing altitude and distance from the sea, so that the physiognomy of the vegetation gradually changes from closed dune scrub to dune forest where there is protection from the saltspray bearing winds.

The forest comprises three strata: canopy, subcanopy and a herb and shrub layer. The canopy is continuous but forms an uneven layer with some of the larger trees *e.g. Chrysophyllum viridifolium* Wood and Franks, *Ficus polita* Vahl, *Celtis africana* N. L. Burm. *Balanites maughamii* Sprague, *Ziziphus mucronata* Willd. and *Mimusops caffra* occurring as emergents. The general canopy layer is formed mainly by *Drypetes natalensis* Hutch., *Acacia karroo* Hayne, *Mimusops caffra, Croton gratissimus* Burch, *Ptaeroxylon obliquum* (Thunb.) Radlk. and *Euclea natalensis* A. DC. The height of the canopy is variable. On the fairly level areas and in the valleys, it may be up to 15 m high, but on the ridges, particularly nearer the coast, it is

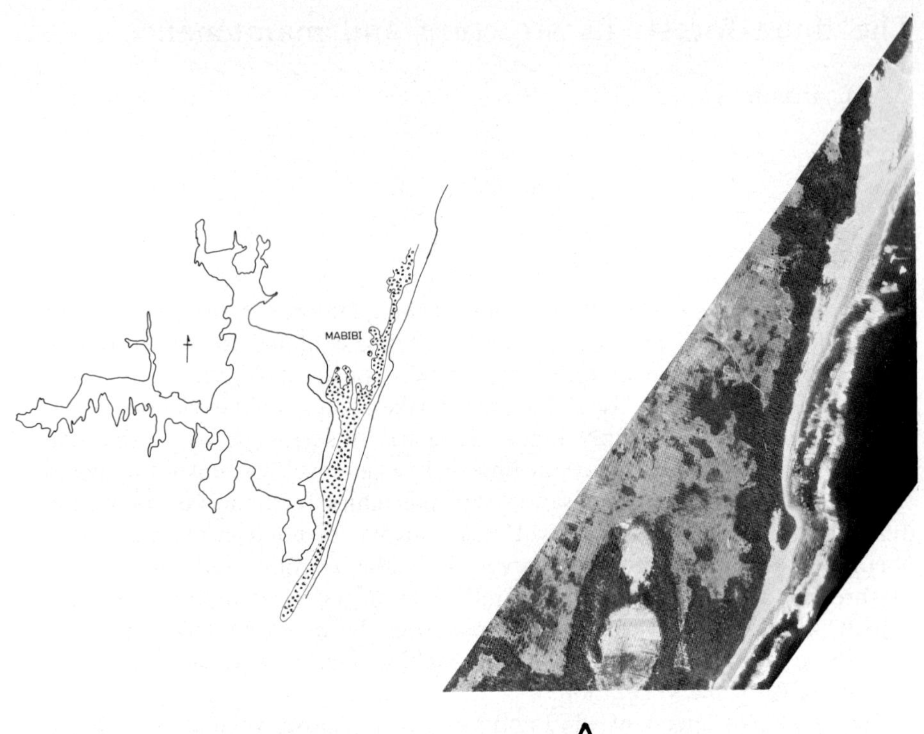

Fig. 1. Portion of the dune forest east of Lake Sibaya in relation to the settlement of Mabibi. A is portion of an aerial photograph showing wave action over the Mabibi reefs.

considerably lower, often being less than 4 m. In these areas canopy and subcanopy merge so that only canopy and herb and shrub layers are present. Where a subcanopy is developed it is discontinuous and comprised of many of the canopy species together with *Cassipourea gerrardii* Alston, *Plectroniella armata* (K. Schum.) Robyns, *Deinbollia oblongifolia* (Sond.) Radlk., *Drypetes gerrardii* Hutch., *Teclea natalensis* (Sond.) Engl., *Clerodendrum glabrum* E. Mey. and *Croton gratissimus.*

Where the canopy is continuous, the herbaceous stratum is not well developed, a feature in common with the dune forest elsewhere in Natal (Moll 1968). However, in areas where the canopy has been broken, dense undergrowth develops. Thickets of *Isoglossa woodii* C.B. Cl. more than 2 m tall are common. *Tricalysia lanceolata* (Sond.) Burtt Davy, *Clerodendrum glabrum, Grewia occidentalis* and saplings of *Croton gratissimus* and *Deinbollia oblongifolia* develop where there is sufficient light.

Lianes are reported as being common in dune forest (Bayer 1938; Moll 1968). However at Lake Sibaya they are perhaps less common than elsewhere except in disturbed areas where *Acacia kraussiana* Meisn. ex Benth. forms an impenetrable thicket. Other lianes are *Dalbergia armata* E. Mey. *Monanthotaxis caffra* (Sond.) Benth. *Secamone frutescens* Decne., *Smilax kraussiana* Meisn., *Dioscorea sylvatica* (Kunth) Eckl. and species of *Grewia* and *Rhoicissus.*

On the inland margin of the system, the dunes drop steeply onto the coastal plain, at places having a slope of almost 30° (Breen 1971), which is not much less than the 35° recorded as the angle of repose for similar sands at St. Lucia (Hobday 1965). This combination of loose unconsolidated substrate and steep slope, has the effect of reducing the size of the plants, but not to the extent observed on the seaward side. In spite of the canopy being discontinuous, herbs and shrubs do not form dense stands possibly because of the extreme drainage on these slopes. The canopy comprises *Acacia karroo, Ziziphus mucronata, Balanites maughamii, Croton gratissimus, Ptaeroxylon obliquum* and *Euclea natalensis* and seldom exceeds 5 m in height. However once the slope decreases and the dune merges with the coastal plain the trees become very large including *Celtis africana, Cola microcarpa* Brenan, *Croton gratissimus* and, particularly along the lake shore, *Acacia karroo.*

In addition to the three intergrading physiognomic forms of the dune forest *viz.* the dune scrub, the dune forest and the dune forest of the steep western slopes, there is another, the strand vegetation which plays an important rôle in the stabilization of the dune forest. It occupies a narrow belt of relatively unstable sand between the high tide level and the dune scrub. Although this association of species has been referred to as the Pioneer Strand Community (Bews 1912; Bayer 1938; Edwards 1967; Ward 1971), it seems unlikely that they represent stages in the succession, as there is no evidence that they are giving way to more complex or stable communities. It may therefore, be more realistic to regard them, as Walker (1970) has done, the classical hydrarch successions *i.e.* that they are plants simply colonizing sites as and when they become suitable and that, in consequence, almost any sequence would be possible in the succession to dune forest. Most of the species occupying this zone are widespread (Bews 1912; Bayer 1938), and do not change much between Natal and Zululand. They are salt tolerant and effectively bind the sand, particularly those which have horizontal stems and tolerate burying, such as *Scaevola thunbergii* Eckl. and Zeyh. and *Ipomoea brasiliensis* (L.) Sweet. Further away from the sea *Carpobrotus dimidiatus* (Haw.) L. Bol. and *Gazania rigens* (L.) Gaertn. grow on the more stable sand. Where this zone of vegetation has been destroyed and the stability of the sand reduced, the effects on dune vegetation have been disastrous (Ward 1971).

In spite of being 4° south of the tropic of Capricorn, northern Zululand is a transitional area where tropical conditions give way to subtropical (Bews in Aitken and Gale 1921). The southerly extension of tropical conditions results partly from the ameliorating effect that the warm Mozambique current, which flows down the east coast, has on the climate. These conditions have therefore provided a migration route by which tropical species have entered South Africa. Depending on their individual tolerance, species extended their ranges to varying degrees. Some (Table 1) have become extremely widespread *e.g. Acacia karroo* whereas others, although extending down to the Cape are confined more to the eastern parts of South Africa *e.g. Maytenus acuminata.* The transitional nature of the dune vegetation

Table 1. Tree species occurring in the dune forest at Lake Sibaya (Breen 1971) which are considered to have had a northern origin, and grouped according to their southern distribution limit (Coates Palgrave 1977).

	Species having their southern limit in:			
Widespread	Zululand	Natal	Transkei	Eastern Cape
Celtis africana N. L. Burm.	*Morus mesozygia* Stapf.	*Ximenia caffra* Sond.	*Trema orientalis* (L.) Blume	*Chaetachme aristata* Planch.
Acacia karroo Hayne	*Sophora inhambanensis* Klotzsch	*Acacia schweinfurthii* Bren. & Exell	*Albizia adianthifolia* (Schum.) W. F. Wight	*Ficus natalensis* Hochst.
Maytenus acuminata (L.f.) Loes	*Dalbergia nitidula* Welw. Ex. Bak.	*Bauhinia tomentosa* L.	*Erythroxylum emarginatum* Thonn.	*Capparis sepiaria* L.
Apodytes dimidiata E. Mey. ex Arn.	*Balanites maughamii* Sprague	*Trichilia emetica* Vahl	*Drypetes gerrardii* Hutch.	*Clausena anisata* (Willd.) J. D. Hook ex Benth.
Dodonaea viscosa Jacq.	*Tapura fischeri* Engl.	*Cleistanthus schlechteri* (Pax) Hutch.	*Antidesma venosum* E. Mey. ex Tul.	*Ptaeroxylon obliquum* (Thunb.) Radlk.
Ziziphus mucronata Willd.	*Bridelia cathartica* Bertol	*Croton gratissimus* Burch.	*Acalypha glabrata* Thunb.	*Ekebergia capensis* Sparrm.
Carissa bispinosa (L.) Desf. ex Brenan	*Blighia unijugata* Bak.	*Spirostachys africana* Sond.	*Allophylus africanus* Beauv.	*Cassine aethiopica* Thunb.
	Tabernaemontana elegans Stapf.	*Xeromphis obovata* (Hochst.) Keay	*Strychnos henningsii* Gilg	*Scolopia zeyheri* (Nees) Harv.
			Vangueria infausta Burch.	*Syzygium cordatum* Hochst.
				Diospyros natalensis (Harv.) Brenan
				Euclea natalensis A. DC.
				Strychnos spinosa Lam.
				Clerodendrum glabrum E. Mey.

may be illustrated by grouping the trees according to the geographical areas in which they find their southern limit. Considering the species which are thought to have had a northern origin (Table 1) and expressing these as a percentage of the number of tree species recorded by Breen (1971), it appears that 7% do not extend south of Zululand and might therefore be lost completely from the South African flora if there was large-scale disturbance of the dune forest. An additional 7% extend further south into Natal, and only about 75% enter the Eastern Cape. When the species of uncertain origin are considered 22% are found to have either their northern or southern limits in the areas of Zululand and southern Mozambique (Table 2). These species should therefore also be included in any analysis of the transitional character of the dune forest in the vicinity of Lake Sibaya. Thus nearly 30% of the species appear to have their limit in the northern Zululand and southern Mozambique area. No doubt as vegetation survey increases, the ranges of many of these species will probably be shown to be more extensive. Even were this to be the case, they would be unlikely to occur in significant numbers within the areas that are added to their presently known ranges. Palmer and Pitman (1972) have recorded wider distribution in some species but it is not clear whether these records are based only on visual sitings or whether they are substantiated by herbarium specimens. However, even if this number was halved, there can be little doubt that the dune forest in the vicinity of Lake Sibaya must be of particular botanical interest.

Table 2. Tree species occurring in the dune forest at Lake Sibaya (Breen 1971) that have uncertain origin but with either northern or southern limits in the northern Zululand/southern Mozambique area (Coates Palgrave 1977).

Ficus burtt-davyi Hutch.	*Inhambanella henriquesii* (Engl. & Warb.) Dubard
Colpoon compressum Berg.	*Manilkara concolor* (Harv. ex C.H. Wr.) Gerstner
Monanthotaxis caffra (Sond.) Verdc.	*Diospyros inhacaensis* F. White
Craibia zimmermanii (Harms) Harms ex Dunn	*Diospyros scabrida* (Harv. ex Hiern) De Winter
Erythroxylum pictum E. Mey.	*Linociera peglerae* (C.H. Wr.) Gilg & Schellenb.
Erythrococca berberidea Prain	*Acokanthera oblongifolia* (Hochst.) Codd
Suregada africana (Sond.) Kuntze	*Acokanthera schimperi* (A. DC.) Schweinf.
Allophylus natalensis (Sond.) de Winter	*Ephippiocarpa orientalis* (S. Moore) Markgraf
Pancovia golungensis (Hiern) Exell & Mendonca	*Tarenna junodii* (Schinz.) Brem.
Cola natalensis Oliv.	*Enterospermum littorale* Hiern
Xylotheca kraussiana Hochst.	*Canthium setiflorum* Hiern
Dovyalis longispina (Harv.) Warb.	*Coffea racemosa* Lour.

Maintenance

(a) *The Effect of Human Activity*

The African people occupying the eastern zone of the coastal plain are probably derived from two sources: the Tonga-Bantu who migrated southwards along the coastal plain from present-day Mozambique, and that portion of the Tekela-Ngunis who moved seawards from modern Swaziland (Bryant 1929). The latter group subsequently turned south and eventually moved into Natal. It is difficult to establish the impact these people had on the vegetation of the sandy coastal plain in general and of the dune forest in particular. It seems probable that they would have been attracted to the Lake Sibaya area because of the lack of water and suitable arable land elsewhere (Tinley 1976). In addition, since they were derived from a people who had a fishing tradition (Torres pers. comm.*), the lake would have provided a source of food.

Tinley (1976) noted that African people were sparsely distributed throughout the whole area but that one of the more important concentrations occurs in the Mabibi pans area (Fig. 1). This group is closest to the dune forest and has probably had most impact on the system.

These people make use of the dune system in two ways: as a source of food, building materials, medicines *etc.*; and as an area for the cultivation of crops, particularly maize. In the former, the impact on the vegetation is dispersed because the requirements are precise in that they are defined in terms of particular species or even the suitability of an individual plant or part thereof. As a result, although many of the species are actively sought after for various purposes (Watt and Breyer-Brandwijk 1962; Palmer and Pitman 1972), the impact is not confined to a small area. Breen (1971) concluded that because the canopy species showed high survival rates, these activities were on a scale that had little effect on the stability of the dune forest.

In contrast, crop planting in the dune forest results in almost total destruction of the vegetation, initially in localized areas but as each field becomes unproductive it is either extended or new ones started. Eventually the fields coalesce and very little if any of the dune forest remains. This is clearly evident in Plate 1, where encroachment from the north has greatly reduced the area covered by dune forest. As pressure on the land in the vicinity of the Mabibi pans increased and productivity decreased, there was a tendency to seek suitable sites within the high dune system east of the main basin of the lake. The extent to which this has happened may be judged from Plate 1a.

The dunes are comprised of white unconsolidated sand with little or no clay fraction. Maud (1968) mentions a clay content of 1% for similar sands and it would seem as low, if not lower, at Lake Sibaya. As a result, the substrate is easily eroded by both wind and water and is also readily leached. Not surprisingly therefore these fields soon become unproductive and are

* J. Torres, Department of African Studies, University of Natal.

Plate 1a.

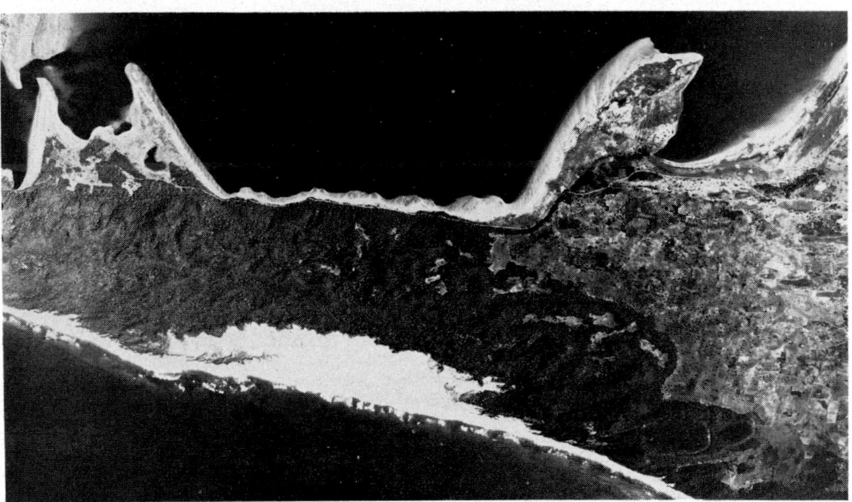

Plate 1b.

Plate 1. a. Aerial photograph taken in 1942 of the dune forest east of the lake. Note the considerable areas of unstable sand, distributed mainly in hollows. b. Aerial photograph taken in 1964 in the dune forest east of the lake. Note colonization of the exposed areas shown in Plate 1a including introduction of *Casuarina equisetifolia* in the right foreground. Aerial photographs reproduced under Government Printer's copyright authority no 4084 of 21.11.68.

Table 3. The influences of cultivation and subsequent regeneration on selected aspects of dune soils. Each value is the mean from three sites, except in the abandoned sites, where only two areas were sampled. 95% confidence limits of the mean are given in parenthesis. Data from Furness (1972).

Parameter		Site			
		Undisturbed	Recently disturbed	Abandoned	Regenerating
N (ppm)	upper 5 cm	52.33 (4.32)	43.83 (8.24)	14.00	41.8 (2.9)
	at 15 cm	20.17 (1.73)	28.83 (11.93)	7.00	24.7 (13.1)
C (%)	upper 5 cm	0.62 (0.02)	0.13 (0.05)	0.05	0.24 (0.18)
	at 15 cm	0.14 (0.05)	0.08 (0.04)	0.04	0.11 (0.07)
pH (in H_2O)					
	upper 5 cm	7.33 (0.36)	7.27 (0.28)	6.10	7.17 (0.35)
	at 15 cm	7.73 (0.24)	7.50 (0.20)	6.25	7.63 (0.17)

abandoned. The extent to which soil conditions deteriorate during cultivation is evident in the data from Table 3. Sites were grouped as undisturbed dune forest, recently disturbed (*i.e.* being prepared for planting), abandoned and regenerating in which tree-cover had developed. The nitrogen content of the soils was not immediately reduced by clearing but, by the time the sites were abandoned, nitrogen levels had dropped considerably, a factor which would contribute to the decreased productivity. During regeneration however, considerable amounts of nitrogen were added to the soil so that by the time trees had re-established themselves, there was little difference between undisturbed and regenerating sites. Carbon levels decreased with disturbance, and thus this potential source of nutrients was also utilized during crop production. Although carbon levels rose during regeneration they had not been fully reinstated at the time of the study. Since carbon influences both water-holding and base exchange capacity, the reduction in carbon content could reduce productivity. A decrease in pH during cultivation was also evident, but this returned to the level observed in the undisturbed forest during regeneration.

It is clear that fields are abandoned when the soils have been impoverished. This must also influence the rate at which regeneration takes place. The aerial photograph taken in 1942 (Plate 1a) provides what is probably a conservative idea of the extent of cultivation in the dune forest when control was effected approximately six years later (Breen 1971). Seventeen years passed between institution of control and the aerial survey of 1964 (Plate 1b). A comparison of these photographs illustrates the extent to which cover was re-established. It is noteworthy that even after this period some areas had not been re-colonized by trees.

Investigation of a typical abandoned field illustrates that *A. karroo* is the major tree species to colonize these sites (Table 4). Although it may occur as a fairly large tree along the margins of cleared areas there are very few seedlings. Progressing further into the cleared area however, there is a dramatic increase in the number of small *A. karroo* with little change in the number of larger ones. In the centre of the old field, although there were fewer seedlings, they were still abundant. The larger specimens of *A. karroo* may represent either the earliest colonizers or trees which were conserved during felling. Other species of trees show the opposite trends, their numbers decrease in the open areas and the few colonizers tend to be of species such as *Clerodendrum glabrum, Deinbollia oblongifolia, Croton gratissimus* and *Grewia* spp.

Where *A. karroo* occurs in the dune forest as tall trees, it tends to form groves. Because of its prominent role in succession, these groves have been interpreted as representing abandoned fields (Breen 1971). During 1969 cultivation in the dune system had started once again and, at a site where clearing had just been completed, it was confirmed that they were re-clearing a site that had not been used for a number of years. It was not possible to establish the intervening period, but it was likely to be at least the period during which control was effective (17 years). Unfortunately, growth ring counts were also not possible due to the equitable climate in this area making growth rings indistinct. Since most of the trees had been felled it was only possible to estimate tree size by the girth of the stump and to divide them into *A. karroo* and 'other species'. Considering the ratio of *A. karroo* to other species in various girth classes (Table 5) it is apparent that only in the smaller girth classes (less than 30 cms), are the 'other' species dominant. Above 30 cms the ratio tends to be considerably greater than unity, reading a maximum of thirteen for trees with girths of 60–69 cms. *A. karroo* is therefore dominant among the larger trees and it is only some time after it has formed a grove that other tree species enter the succession to any degree. Because this is such a slow process, these groves, which may already have been protected for nearly twenty years, would still require many years before the succession proceeded to climax dune forest.

Table 4. Data illustrating the colonization of abandoned fields, principally by *A. karroo*. Numbers of trees were recorded in two height classes, in each of three plots (15×20 m). Plots were positioned with their long axes parallel to the forest margin and at increasing distances from the dune forest.

	Plot position		
Criterion	Border: un-disturbed forest	Margin of cleared area	Middle of cleared area
A. karroo			
25–200 cm	0	125	98
>200 cm	14	23	14
Total	14	158	112
Other species			
25–200 cm	43	21	15
>200 cm	73	11	8
Total	116	36	23

Table 5. The numbers of stumps in various girth classes in an old abandoned field in which the trees had recently been felled.

Species	Girth Class (cm) <20	20–29	30–39	40–49	50–59	60–69	70–80	>80
A. karroo	9	25	20	16	19	13	8	36
Other species	37	30	20	7	3	1	0	9
Ratio *A. karroo*/ other species	0.2	0.8	1.0	2.3	6.3	13	8	4

The habit of the local Africans to return to previously abandoned sites, might also explain the presence of the fringe of large *A. karroo* observed elsewhere (Table 4). It may be that the boundary of the new field does not follow the original boundary and consequently part of the grove remains standing. Undoubtedly this presence provides abundant seed to aid colonization when the field is abandoned again.

A. karroo has been implicated in the succession elsewhere in the dune system (Edwards 1967) and in other parts of South Africa. It is interesting to speculate on why it should have such competitive advantages in the impoverished soils of the dune system. A possibility is that belonging to the Leguminosae, and having root nodules, it is capable of fixing atmospheric nitrogen. Although the marked increase in nitrogen beneath groves of *A. karroo* may offer tentative support for this hypothesis it requires further investigation.

The dune forest of Zululand in general, but in the vicinity of Lake Sibaya in particular, is a very sensitive ecosystem, not only because of its geographical location which results in an unique species composition, but also because it is under constant stress. The stress along the coastal margin results from the salt-laden sea-breezes which, upon disruption of the dune scrub, enter beneath the canopy and deposit salt on plants not adapted to these conditions. Once these plants are killed the unstable sand is rapidly eroded and frequently drastic action, such as the planting of exotic species *e.g. Casuarina* sp. is implemented in order to halt this process. *Casuarina* has been planted at many places along the coast, including Lake Sibaya (Plate 1b). The maintenance of the dune forest therefore requires the conservation of the dune scrub in an undisturbed condition.

Inland the forest is more stable, but it is developed on impoverished soils and as a result, any area that is cultivated takes at least twenty years to attain sufficient tree cover to create conditions that are suitable for further invasion by climax species. For these to achieve the status of forest trees would no doubt require many more years. Perhaps about a hundred years of preservation might allow these sites to resemble climax dune forest. The value of two or three crops, produced on nutrient deficient soils which are readily leached, must be weighed not only against the long period required for recovery, but also against the expenses that might be incurred if cultivation of the dune system developed on a scale that created erosion problems. The cost of rehabilitation, even with exotic species, would be very high and rehabilitation would be essential in order to prevent sand moving inland and into Lake Sibaya.

The dune forest is thus a system with very low productivity which, taken with other stresses, makes it very susceptible to destruction. It cannot therefore be exploited other than by conservation orientated activities, without potentially disastrous ecological implications that would be difficult and expensive to control.

Plate 2. Destruction of the littoral fringe of *Acacia karroo* during elevation in lake level. a The dune forest viewed from the lake in 1968 at low lake level. *Acacia karroo* is clearly differentiated from the climax of the steep tree clad slopes.

Plate 2b. Inundation of the *A. karroo* fringe and collapse of canopy.

Plate 2c. Final phase of destruction in January 1979.

(b) *The Effect of Lake Inundation upon the Marginal Fringe*

B. R. Allanson

Plate 2 demonstrates very clearly the remarkable effect of rising lake level upon the integrity of the marginal fringe of *Acacia karroo* at the base of the dunes. In 1968 when a site for the research station was chosen, a mature fringe of *A. karroo* existed along the eastern shore of the lake, separating the lake from the climax community of the dune forest. Unfortunately we did not understand the ecological significance of this particular marginal dominance of this leguminous tree. The lake rose slowly during the early years of the subsequent decade 1968–1978 until a rapid rise in lake level (Allanson, Chapter 4, this volume) inundated the roots and trunks of the fringing trees as shown in Plate 2b. The trees rapidly died and the trunks and branches under water were swiftly colonized by the burrowing mayfly *Povilla adusta* which caused wholesale destruction of the fringe canopy, leaving only, as Plate 2c shows, the larger trunks standing. Wave action brings about the erosion of the banks and new elevated terraces are built between and covering the *A. karroo* stumps. The canopy material moves rapidly under the influence of wave action into the lake, contributing in some degree to its nutrient input during this inundation phase.

The successional implications are now a good deal clearer than when we first visited the lake. Now that the lake is falling these devasted terraces will, as they dry out, become colonized by seedlings of *A. karroo* and a pioneer community will develop. It is not unreasonable to suggest that the development of such a community may take between 17 and 20 years keeping pace with the hydrological events which B. J. Hill has described in Chapter 3, and the meteorological cycles described by Tyson (1978). By so doing, this pioneer community never quite reaches the stage of successional change to mature dune climax before it will become inundated again during the next phase of lake elevation.

The moral of the story is clear: do not build field research stations in *A. karroo* communities on the shores of lakes!

References

Aitken, R. D. and G. W. Gale. 1921. Botanical survey of Natal and Zululand. Bot. Surv.Mem. S. Afr. No. 2.

Bayer, A. W. 1938. An account of the plant ecology of the coastbelt and midlands of Zululand. Ann. Natal Mus., 8: 371–455.

Bews, J. W. 1912. The vegetation of Natal. Ann. Natal Mus. 2: 253–331.

Breen, C. M. 1971. An account of the plant ecology of the dune forest at Lake Sibaya. Trans. R. Soc. S. Afr. 39: 223–233.

Bryant, A. T. 1929. Olden times in Zululand and Natal. Longmans Green and Co., London.

Coates Palgrave, K. 1977. Trees of Southern Africa. C. Struik, Cape Town.

Edwards, D. 1967. A plant ecological survey of the Tugela river basin. Mem. Bot. Surv. S. Afr. No. 36.

Furness, H. D. 1972. An investigation of the regeneration of the coastal dune forest of the Lake Sibaya area, Zululand. Unpublished report, Department of Botany, University of Natal, Pietermaritzburg.

Hobday, D. K. 1965. The geomorphology of the Lake St. Lucia area. M.Sc. thesis, Department of Geology, University of Natal, Durban.

Maud, R. R. 1968. Quarternary geomorphology and soil formation in coastal Natal. Z. Geomorph., 7: 155–199.

Moll, E. J. 1967. An account of the plant ecology of the Hawaan Forest, Natal. Jl S. Afr. Bot., 34: 61–76.

Moll, E. J. 1968. An investigation of the plant ecology of the Hawaan Forest, Natal, using an ordination technique. Bothalia, 10: 121–128.

Moll, E. J. 1970. Vegetation studies in the Three Rivers Region, Natal. Ph.D. thesis, Department of Botany, University of Natal, Pietermaritzburg.

Palmer, E. and N. Pitman. 1972. Trees of Southern Africa. Balkema. Cape Town.

Tinley, K. L. 1976. The ecology of Tongaland. Wildlife Soc., Natal Branch, Durban.

Tyson, P. D. 1978. Rainfall changes over South Africa during the period of meteorological record. In: Biogeography & Ecology of Southern Africa. Ed. M. S. A. Werger. pp. 53–69. Dr W. Junk bv. Publishers.

Walker, D. 1970. Direction and rate in some British postglacial hydroseres. In: The vegetational history of the British Isles. Ed. D. Walker and R. West, Cambridge University Press.

Ward, C. J. 1971. The plant ecology of the Isipingo Beach Area, Natal, South Africa. MSc. thesis, Department of Botany, University of Natal, Pietermaritzburg.

Watt, J. M. and M. G. Breyer-Brandwijk. 1962. The medicinal and poisonous plants of southern and eastern Africa. Livingstone, Edinburgh and London.

3 Bathymetry, morphometry and hydrology of Lake Sibaya

B. J. Hill

Introduction

Lake Sibaya approaches to within 1 km of the sea (Plate 1) but is cut off from the sea by a series of high forested sand dunes. The surface of the lake is about 20 m above sea level but the bottom extends nearly 20 m below sea level. Having no connection to the sea, the level of the lake fluctuates as a result of the dynamic balance between inflow and outflow. Continuous recording of lake level has been carried out since 1969 and these records together with rainfall and evaporation data collected from 1969 were used by Pitman & Hutchison (1975) to compile a water budget and a computer simulation of the lake level. The simulation was then used together with earlier rainfall records to estimate lake levels between 1914 and 1969. Pitman & Hutchison's (1975) report together with Hill's (1969, 1975) data on bathymetry and unpublished information makes it possible to describe the bathymetry of the lake in both high and low level conditions.

Plate 1. A view of the lake looking northwards along the dune forest.

Bathymetry and Morphometry

Since lake levels fluctuate it is necessary to refer them to a fixed datum. This datum is 18.65 m above geodetic mean sea level and is the chart datum at a water level recorder (W7M02) operated in the south-east corner of the main basin of the lake. Depths are referred to as being below or above this datum.

The lake can be divided into five regions (Fig. 1). The Main Basin is the largest of these, comprising 56 to 59% of the area of the lake and containing the deepest water. In the south there are two small subsidiary regions, the South East and South West Basins accounting for about 9% of the lake area. The remaining two regions are the extremely dendritic Northern and Western Arms making up 12 to 20% of the lake area. Underwater, the two arms and the South Western Basin can be seen to be continuous with valleys which run across the floor of the Main Basin (Fig. 2a,b,c). The South East Basin shows no continuation in bathymetric profile with the Main Basin and

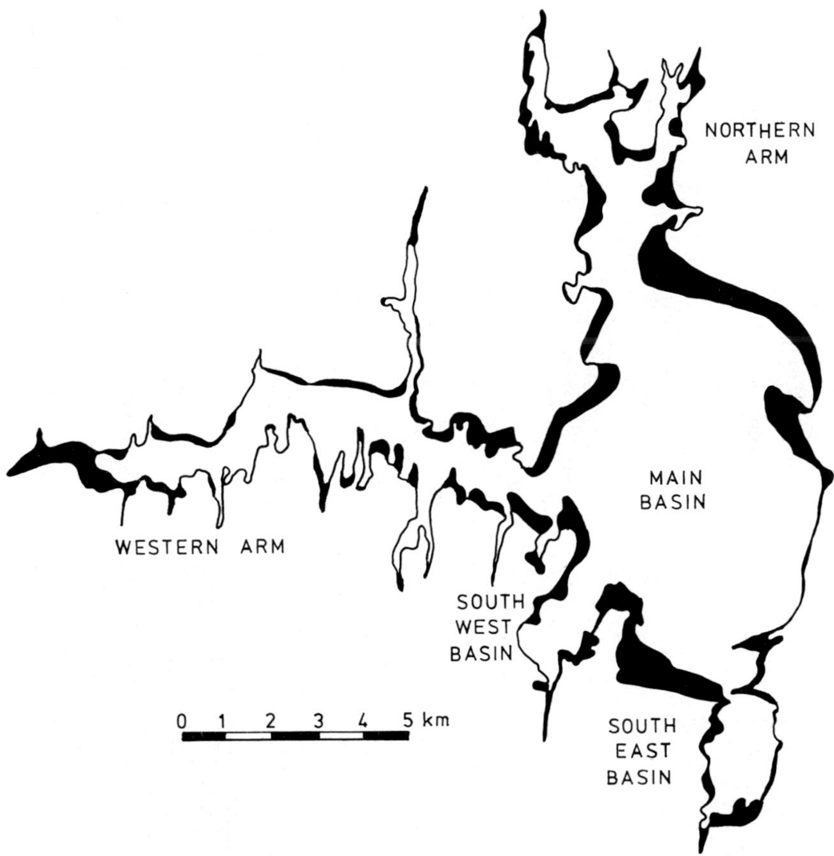

Fig. 1. Lake Sibaya. Shoreline of lake at high (outer) and low (inner) levels. Shaded area is submerged at high level. Low lake level as in July 1964 (1.7 m above datum) and high level as in March 1977 (4.6 m above datum).

Fig. 2a. Lake Sibaya: Bathymetry recorded in January 1970. Depth contours at 5 m intervals.

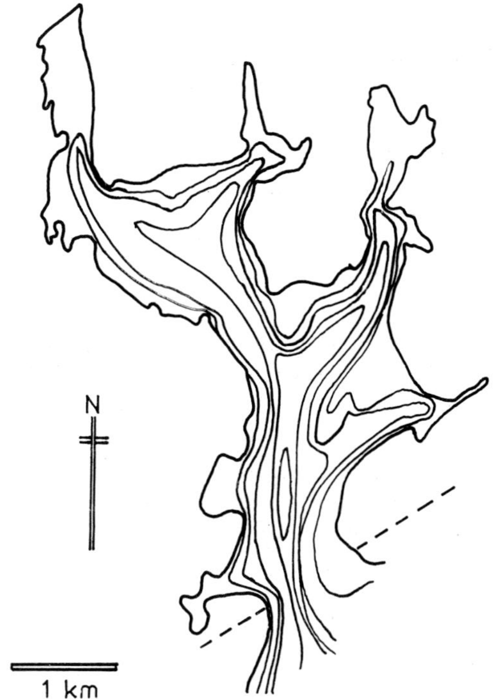

Fig. 2b. Lake Sibaya, Northern Arm. Bathymetry recorded in January 1965. Depth contours at 5 m intervals.

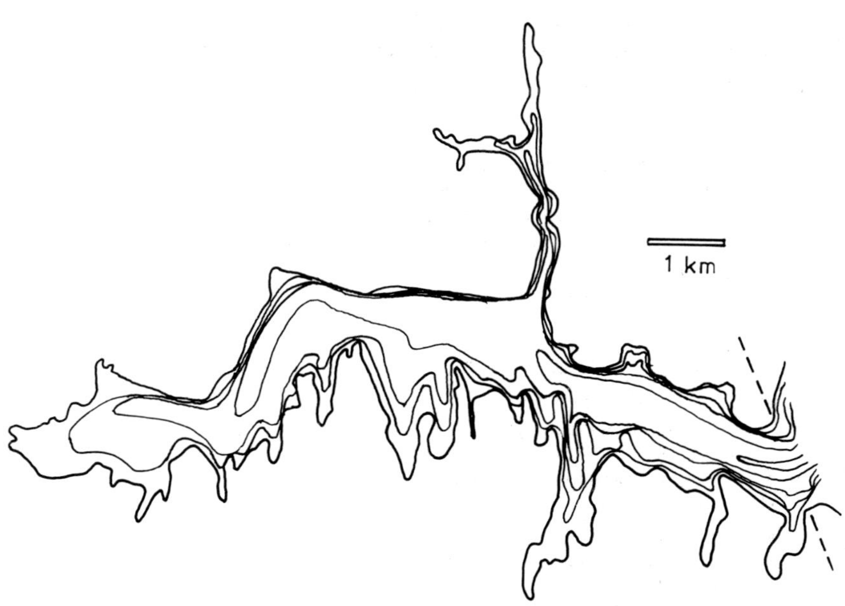

Fig. 2c. Lake Sibaya, Western arm: Bathymetry recorded in January 1965. Depth contours at 5 m intervals.

is nearly isolated from it. The connecting region between these two basins shows signs of considerable sand movement with the formation of two major sand spits. This sand movement could have obliterated any connection which might have formerly existed. The resulting segmentation in the south east corner of the lake was pointed out by Orme (1973) who stated that it was confined to the east because of the supply of sand from the longshore dune in this area. The large sandspit in the north east corner of the Main Basin known as Crocodile Point may have a similar origin. In the rest of the lake, deep water and a restricted input of sediments limit segmentation.

The narrow steep-sided valleys in the floor of the Main Basin (Fig. 2a) indicate that little sediment has accumulated. Similar conditions have been found in Lake Nhlange (Hill, 1969) and Lagoa Poelela (Hill, 1975) on the coastal plain to the north and contrast with the extremely heavy sedimentation found in most estuaries to the south (Orme, 1974). The absence of sediments in Sibaya is probably due to the small surface inflow and the fact that the small catchment (530 km^2) is sandy and low-lying.

Wave energy generated on the long open water of the Main Basin is dissipated by the gently sloping northern and southern shores. The fjord-like morphology of the Western Arm is not subject to much wave action because of the short fetch available to the predominantly northerly and southerly winds. Thus apart from fluctuations in water level and spit formation in the east the lake is morphologically in a fairly stable condition. Orme (1973) argues that the surface area of the lake has during the Holocene decreased by some 61%. Some extent of this infill is given in Fig. 6 in Chapter 1.

At high lake level water reaches the foot of the eastern sand dune; this accelerates erosion and increases the supply of sand to the eastern shore. Thus high levels encourage spit formation, and, as this progresses, the South East Basin might eventually become completely isolated from the rest of the lake.

Table 1. Lake Sibaya: Morphometry at low (1964) and high (1977) lake levels. Datum is 18.65 m above G.M.S.L. Bathymetric survey carried out in January 1970 when lake level was 1.2 m above datum; calculations take this into account.

	1964	1977
Surface level above datum (m)	1.7	4.6
Maximum length (km)	17.5	18.7
Maximum breadth (km)	16.3	18.3
Mean width (km)	3.3	4.1
Maximum depth (m)	40.0	43.0
Mean depth (m)	13.1	12.6
Area (km^2)	59.4	77.5
Shore length (km)	128.7	126.9
Shore development	4.7	4.1
Volume (10^6 m^3)	776	981
Volume development	0.98	0.88
Circularity ratio	0.26	0.27

Table 2. Lake Sibaya: Comparative areas and shoreline lengths of the five major regions of the lake at low (1964) and high (1977) levels.

Area (km^2)	Main Basin	N. Arm	W. Arm	S.W. Basin	S. E. Basin
1964	34.9	6.8	12.4	2.3	3.0
1977	43.2	9.7	17.7	3.1	3.9
% change	+24	+43	+43	+34	+29
Shoreline length (km)					
1964	27.4	29.7	55.8	7.8	8.3
1977	22.4	29.8	55.9	9.2	9.9
% change	−18	+0.1	+0.1	+18	+7

High and low lake level shorelines are shown in Fig. 1, areas exposed at low level and submerged at high level are shaded. A comparison of the morphology at the two levels is given in Table 1. At high level the lake area is about 30% greater than at low level due to flooding of low lying areas adjacent to the lake. This increase in area occurs in all five regions (Table 2). The total length of shoreline does not differ at the two levels (Table 1) but there are alterations in individual regions. The shoreline length of the Main Basin decreased at high level due to the drowning of spits, resulting in a more regular outline. In the Northern and Western arms, drowning of peninsulas also occurred, but water penetrated further up the numerous side arms and the net result was that shoreline length remained the same. In the southern basins the shoreline was lengthened due to flooding of narrow valleys. An increase in area with no change in total shoreline length results in a decrease in shoreline development from 4.7 at low level to 4.1 at high level. Although the length of shoreline remains the same, the area of shallow water (less than 3 m depth) is greater at high level – about 16 km^2, than at low level – about 10 km^2.

Hydrology

The following account is drawn from a report by Pitman and Hutchison (1975). Mean annual precipitation over the lake and catchment decreases in a westerly direction from 1200 mm in the south east corner to 700 mm in the west. The mean annual precipitation on the catchment (area 530 km^2) is 900 mm and on the lake (60 to 77 km^2) 1030 mm with an annual coefficient of variation of 23 to 30%. Although rain occurs throughout the year, most comes in summer with 43% falling in the 3 months January to March. Evaporation from the lake surface is about 1420 mm yr^{-1}. Thus annual evaporation exceeds annual precipitation directly on the lake surface.

The lake is situated on recent and Tertiary sands which are porous and groundwater can be expected to affect the water budget. Groundwater levels are closely correlated with lake levels and the lake can be regarded as an

Table 3. Lake Sibaya: Area-volume relationships.

Depth (m)	Area (km²)	Volume (10^6 m³)
4.6	77.5	205
1.7	59.4	261
−3.3	45.0	193
−8.3	32.0	136
−13.3	22.4	91
−18.3	13.9	54
−23.3	7.7	27
−28.3	3.1	11
−33.3	1.2	3
−38.3	0.09	

exposed portion of the water table. The lake surface is about 20 m above sea level so that there will be some seepage of water through the porous sand dunes to the sea one or two kilometres away. Seepage rates have not been measured at Sibaya, but are available for a similar substrate at Richards Bay to the south. Calculations based upon Richards Bay data gave an estimated annual value of 1 to 4×10^6 m³ for Sibaya. Groundwater inflow has also not been measured but can be estimated since the net result of seepage into and out of the lake can be calculated using data from years in which evaporation, precipitation and lake level were recorded. Total mean inflow to the lake was estimated to be in the region of 21×10^6 m³. In the absence of accurate measurements these figures must be regarded as approximation.

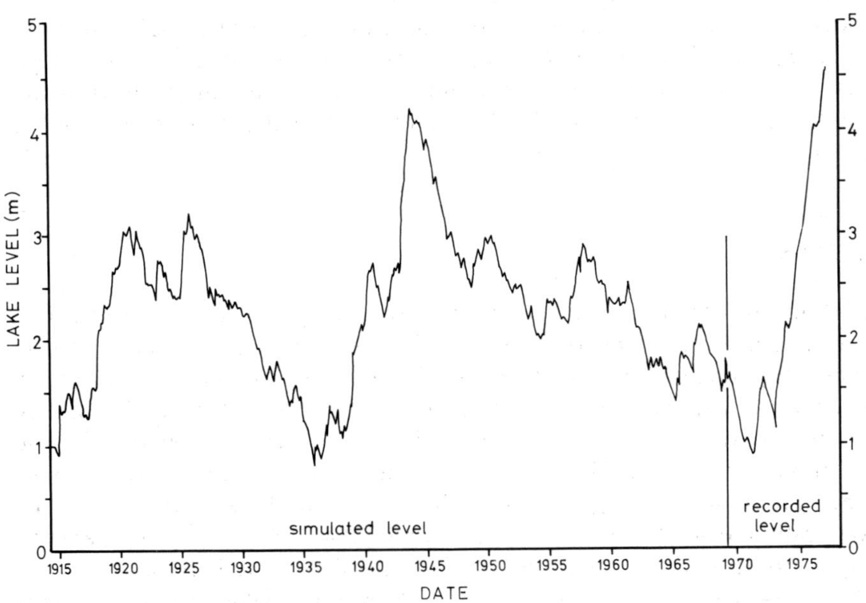

Fig. 3. Lake Sibaya: Level of the lake surface above datum (18.65 m above G.M.S.L.). Levels from 1914 to 1969 obtained by simulation (see text).

The water budget for the lake (inflow − outflow = change in storage) for any given period can be expressed by the following equation:–

$$(Q+PA)-(L+EA)=S_e-S_b$$

where Q = inflow to lake from catchment runoff and seepage (in 10^6 m^3) P = depth of rainfall on lake (in m), A = average area of lake (in km^2) L = seepage loss to sea (in 10^6 m^3); E = depth of evaporation loss from lake (in m); S_b = volume of water in lake at beginning of period (in 10^6 m^2); S_e = volume of lake at end of period (in 10^6 m^3).

Precipitation P and evaporation E have been measured for several years in the vicinity of the lake. The volume of the lake can be calculated from a depth/area relationship such as in Table 3.

Pitman & Hutchison (1975) assembled the available information on the water budget using continuous records of lake level, rainfall and evaporation from 1969 and rainfall alone from 1914. They then compiled a computer simulation of lake level from 1914 to 1969. The result is shown in Fig. 3 together with subsequent actual recorded levels. In the 63 years represented in this figure there have been three periods of sustained high level, *i.e.* when the level was greater than the median of 2.22 m above datum. The rate of change in water level varies widely and can be extremely rapid but in general changes of more than 1 m are spread over several years. Clearly lake level must be regarded as a dynamic factor with no particular state being 'normal'.

References

Hill, B. J. 1969. The bathymetry and possible origin of Lakes Sibaya, Nhlange and Sifungwe in Zululand (Natal). Trans. R. Soc. S. Afr. 38: 205–216.

Hill, B. J. 1975. The origin of southern African coastal lakes. Trans. R. Soc. S. Afr. 41: 225–240.

Orme, A. R. 1973. Barrier and lagoon systems along the Zululand Coast, South Africa. In: Coastal Geomorphology Ed. D. R. Coates. State University N.Y., N.Y. 1973, pp. 181–217.

Orme, A. R. 1974. Esturine sedimetation along the Natal coast, South Africa. Off. Naval Res. Tech. Dept. n.5, 62 pp.

Pitman, W. V. & I. P. G. Hutchison. 1975. A preliminary hydrological study of Lake Sibaya. Rep 4/75 Hydrological Res. Unit, Univ. Wits. 35 pp.

4 The physico-chemical limnology of Lake Sibaya

B. R. Allanson

Introduction

Since the establishment of the research station in 1965 it has been possible to record a number of simple meteorological conditions in the vicinity of the lake. Unfortunately measuring equipment was not available continuously throughout the 12 years the station was occupied. As a consequence data series are often truncated or fragmented. Nevertheless, enough information is available to describe in general terms the climatological ranges and the effect they had upon the lake.

Meteorological Conditions

The coastal peneplain of northern Natal (KwaZulu) is subtropical and windswept. Some indication of the daily variation in temperature during the cool and hot season is given in Fig. 1, while in Fig. 2a the average monthly temperature is recorded for 1972–1977. The high peaks of temperature in Fig. 1 usually occurred during periods of anticyclonic activity which result in warm north easterly winds. Such periods are invariably followed by cooler conditions brought about by southerly to south westerly winds associated with the easterly movement of low pressure fronts which arise in the south Atlantic. The cool season in KwaZulu is normally dry with daily temperatures at the lake varying in June 1963, for example, between 10 and 25°C. The seasonal trough in the average temperature as shown in Fig. 2a occurs in July and August and during the period of observation this seasonal minimum fell consistently from 1972 to 1976. Associated with this decrease is an increase in the number of cloudy days (Fig. 2d) and a marked decrease in the number of hours of sunshine per day during 1974, 1975 and 1976. Tyson (1978) has drawn attention to the meteorological conditions responsible for this inverse relationship, arguing that lower temperatures and cloud cover with high rainfall are due to greater cyclonic activity and increased convergence. Coincident with these environmental changes, the lake level rose steadily to stand at 5 m in March 1977 above the datum of the level recorder *W7MO2* at the research station which was sited at 18.65 m above GMSL. This increase in lake level reflects the influence of a quasi-20-year oscillation in rainfall which has been established by P. D. Tyson and his colleagues for South Africa and in particular for the summer rainfall region. Tyson (1978) reports that the decades in which cumulative rainfall has been below normal have centred upon 1911, 1930, 1949, 1967, while those decades in which cumulative rainfall was greater than normal have centred about 1921,

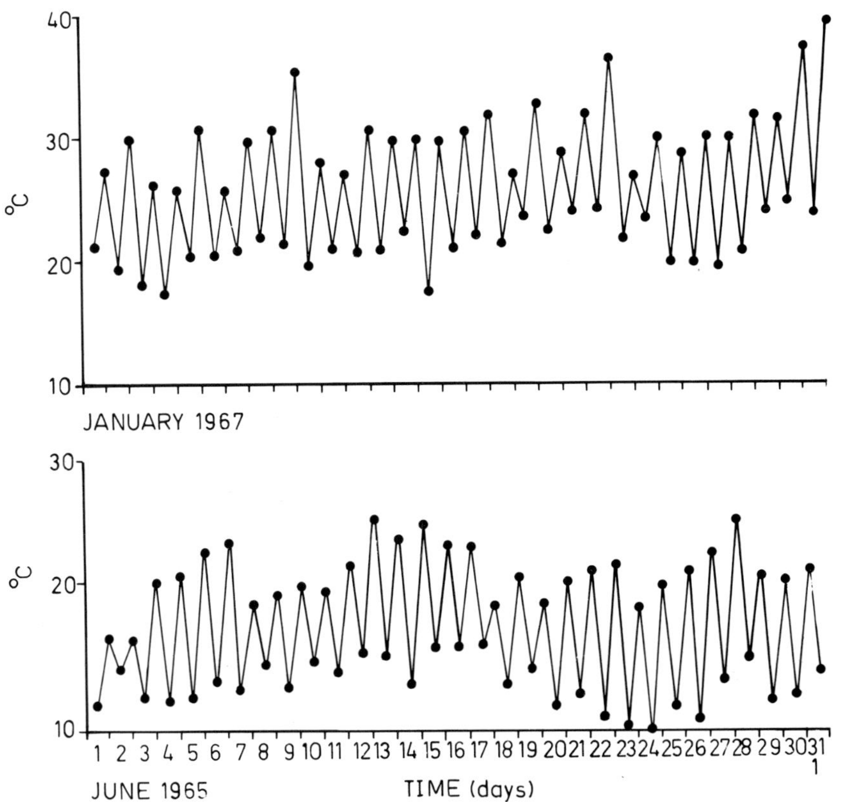

Fig. 1. The daily variation in temperature at 0800 and 1400 hrs at Lake Sibaya, June 1965 and January 1967.

1940, 1958 and 1976. If the level of the lake surface is in any way to mirror these changes it would be reasonable to expect an increase in elevation as from 1972, the observed end of a minimum decade. At the end of 1969 the lake level stood at 1.55 m above station datum, at which level the lake is ringed by extensive white sand beaches (Fig. 2b, Plate 1a). Compare this with conditions in 1977 (Plate 1b) just after the station was vacated, when lake level stood at 5 m above station datum. During this period Tyson (1978) draws our attention to the fact that the positive elevations in rainfall predicted for the summer rainfall region by Dyer (1976) have been realised. In Tyson's Fig. 10, p. 67, the maximum elevation is given for 1976/77 when the lake experienced its most rapid increase in level.

It is perhaps surprising that the lake and its rainfall did mirror quite reasonably the conclusions of Tyson & Dyer. In this regard Pitman & Hutchison (1975) have succeeded in providing a preliminary simulation of lake level from October 1914–September 1974, and once again one is struck by the likelihood of a quasi-20 year oscillation between peaks and troughs, although they appear not to be exactly in phase with what has been proposed for the summer rainfall region as a whole by Dyer (1976).

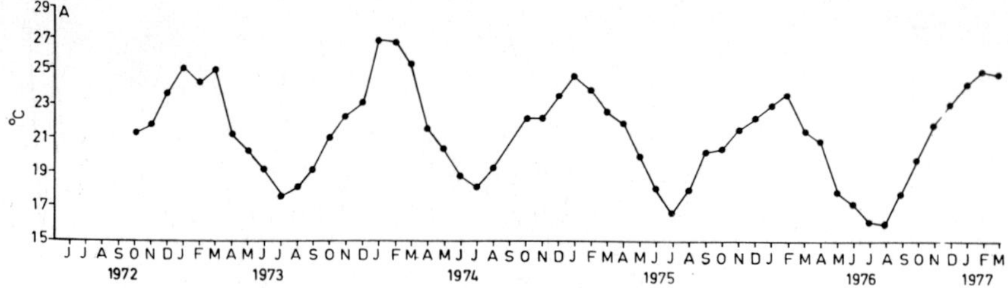

Fig. 2a. The mean monthly air temperature at Lake Sibaya.

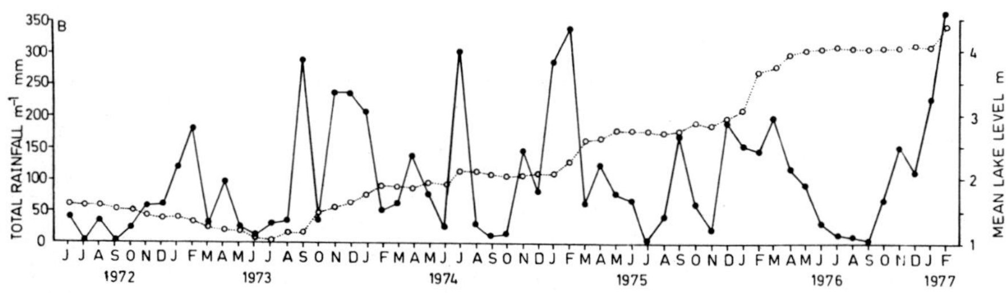

Fig. 2b. Variation in rainfall and mean lake level.

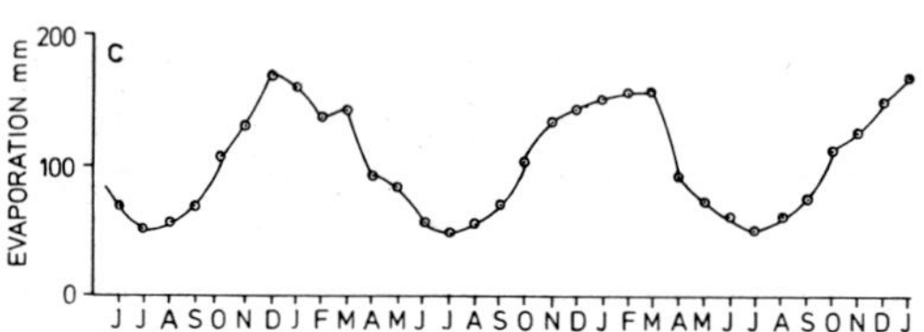

Fig. 2c. Evaporation 1972–1974.

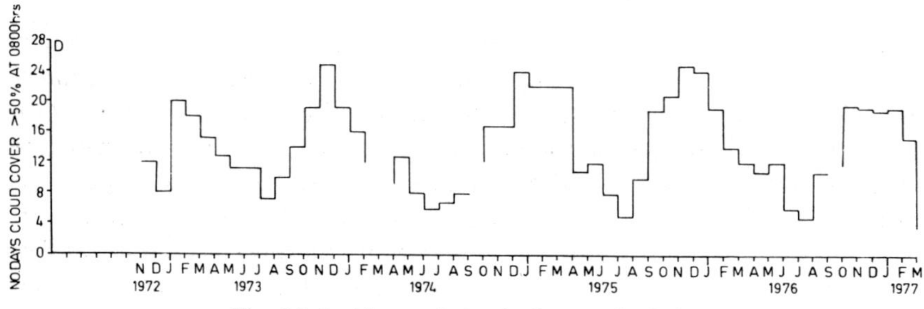

Fig. 2d. Incidence of cloudy days at the Lake.

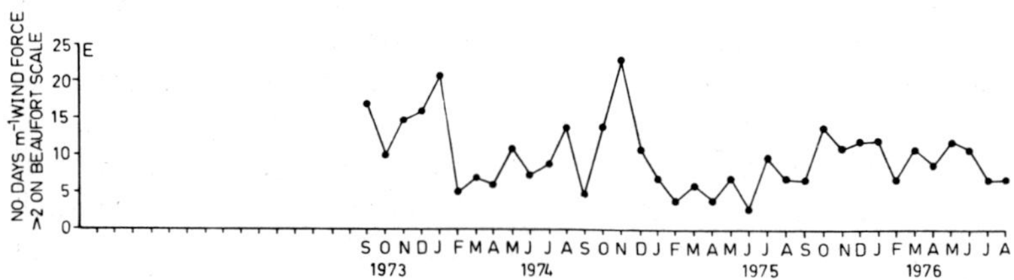

Fig. 2e. Variation in wind speed measured on the Beaufort scale.

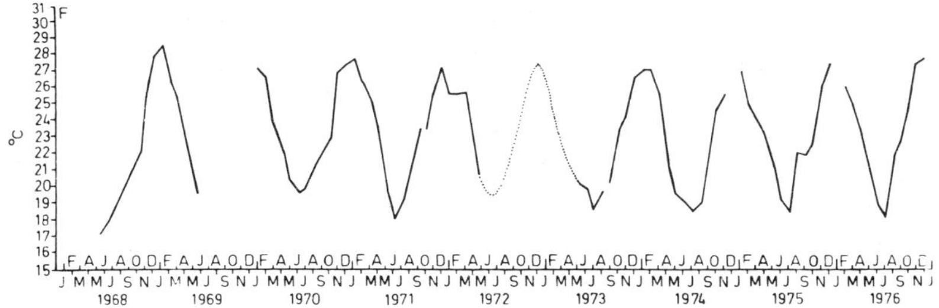

Fig. 2f. The mean monthly variation in lake temperature.

Plate 1a

Plate 1b

Plate 1a. The sand beaches of Lake Sibaya at low water level 1967–1969. b: Inundation of the lake research station and our eventually ineffectual means of stemming the flooding!

Nevertheless, the temporal similarity between the lowering and elevation of lake level in Lake Sibaya and the desiccation and inundation phases of Lake Chilwa, (Dyer 1979) points to both lakes acting as sensitive indicators of climatological change in the Eastern subcontinent.

The influence of direct precipitation on lake level is marked as Fig. 2b shows. As the lake lies in an endorheic basin, this is to be expected and although evaporation is high (Fig. 2c) the steady inexorable rise in lake level reflects, as B. J. Hill, Chapter 3 (this volume) has noted, the importance of flow from underground aquifers within its catchment, the state of charge of which is reflected by the overall rain pattern and cycling.

The next major meteorological factor is wind. Winds blow predominantly from the south and north in response to cyclonic or anticyclonic activity in the atmosphere above southern Africa and mainly over the eastern summer rainfall region. Wind speeds are generally strong, with the mode shifting between 4.5 and 8.5 m sec^{-1} as shown in Fig. 3. They also blow for long periods. In Fig. 2e the number of days per month the wind force is greater than 2 on the Beaufort Scale is recorded. This figure also indicates that the period November–January during 1973, 1974 and 1975, was the windiest time of the year. Some indication of the directional component of the wind structure is given in Fig. 4.

As South Africa lies within the arid latitudes of the southern hemisphere insolation is high and the light energy delivered to the lake surface is a prime factor controlling its thermal characteristics. Variation in surface water temperature is shown in Fig. 2f from 1968 to 1976.

The intensity of and variation in the energy expressed as K J m^{-2} day^{-1} delivered to the lake surface, is given in Fig. 5 for 38 separate daily measurements from February 1976 to March 1977. During the cool season of 1976 (April → August) total energy delivered to the surface of the lake on a clear day was 10730 K J m^{-2} d^{-1}, while during summer a mean value of 28330 K J m^{-2} d^{-1} was recorded. These data include measurements on cloudy and clear days. The effect of cloud upon energy input is

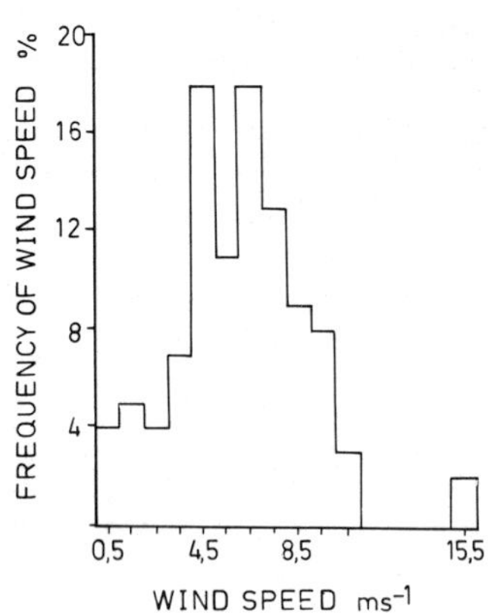

Fig. 3. Distribution of wind speed, m s^{-1} for 27 days during November–January 1968/69.

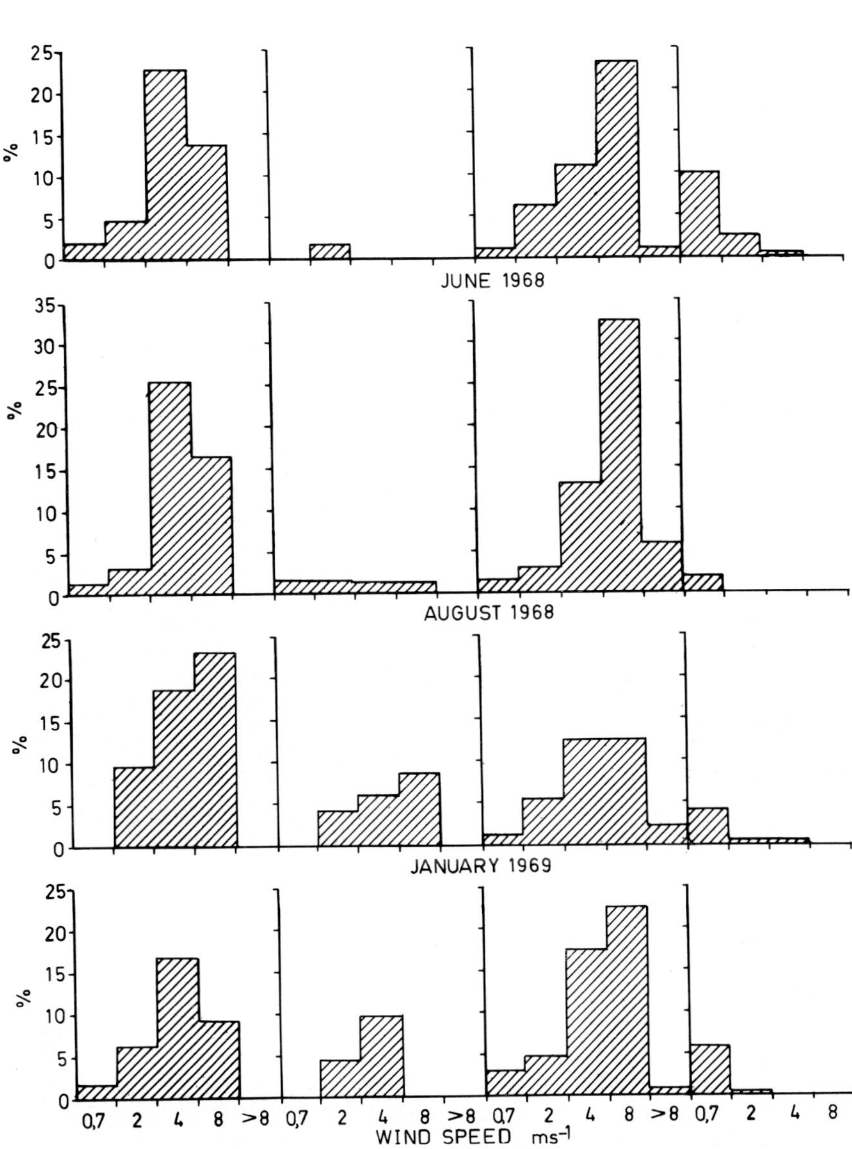

Fig. 4. The directional component of wind over Lake Sibaya during June and August 1968 and January and March 1969.

marked, particularly during the hot season when the incidence of cloudy days increases over that in the cool season (see Fig. 2d).

Not all of this energy is available as some fraction is reflected from the water surface. Table 1 records the values for reflected light on three occasions.

The per cent reflectances fall within the values used by Smith (1974) for

Loch Leven and by Hutchinson (1957) in discussions of Anderson's work who determined empirically for L. Hefner that as the angular height of the sun increased reflectivity as a per cent of total sun and sky radiation fell towards 5% on a clear day; lake surface disturbance had little effect upon the transmission of the radiation into the water column.

Wind and Waves

The most obvious effect of wind upon a fluid surface like water is the setting up of waves varying in amplitude and length depending upon the strength,

Plate 2a

Plate 2b

Plate 2c

Plate 2. Exposed sandy shore and wave cut platform towards the south east. b. Details of the exposed sandy shore. c. Sheltered well vegetated shoreline in South Basin.

duration and fetch of the wind. Table 2 records the height, length and frequency of waves under varying wind speeds.

The analysis given by Smith & Sinclair (1972) of properties of deep water waves in lakes was applied to our data from Lake Sibaya. The relation between the calculated significant wave height (H_s) using the wind speed data given in Table 2 and the mean wave height as observed in the lake (Fig. 6a) does suggest a nearly 1:1 relation notwithstanding the rather wide scatter. Furthermore the wave length and period observed for the lake match those given in Smith & Sinclair's Fig. 3 and 4 which relate on the one hand wave height, wind speed and fetch and on the other wave length, wind speed and fetch. This being the case, it provides a measure of confidence regarding the estimation of the wave mixed layer in the lake. The fetch of the lake was determined using Smith & Sinclair's (1972) method for three points, 1, 2 and 3, marked in Fig. 7. From our general experience of the direction of wave fronts on the lake, it appeared that waves which moved to either the western or eastern shore would be the result of winds either north or south blowing over a distance shorter than down the main axis of the lake, hence the position of point 1. The fetch for those winds blowing down the main axis of the lake was determined at point (2) or (3). A value of 3.7 km was obtained for point (1) and 6.1 m for (2) and (3).

At the time these measurements were made the depth of the east and west terraces stood at 2 and 3 m with marked rippling of the sand. This depth range matches the depth of the wave mixed layer under the stress of a 4.5 m s^{-1} wind given by Smith & Sinclair (1972) and reported in Fig. 6b. As the wave mixed depth is roughly equal to half the wave length, these data

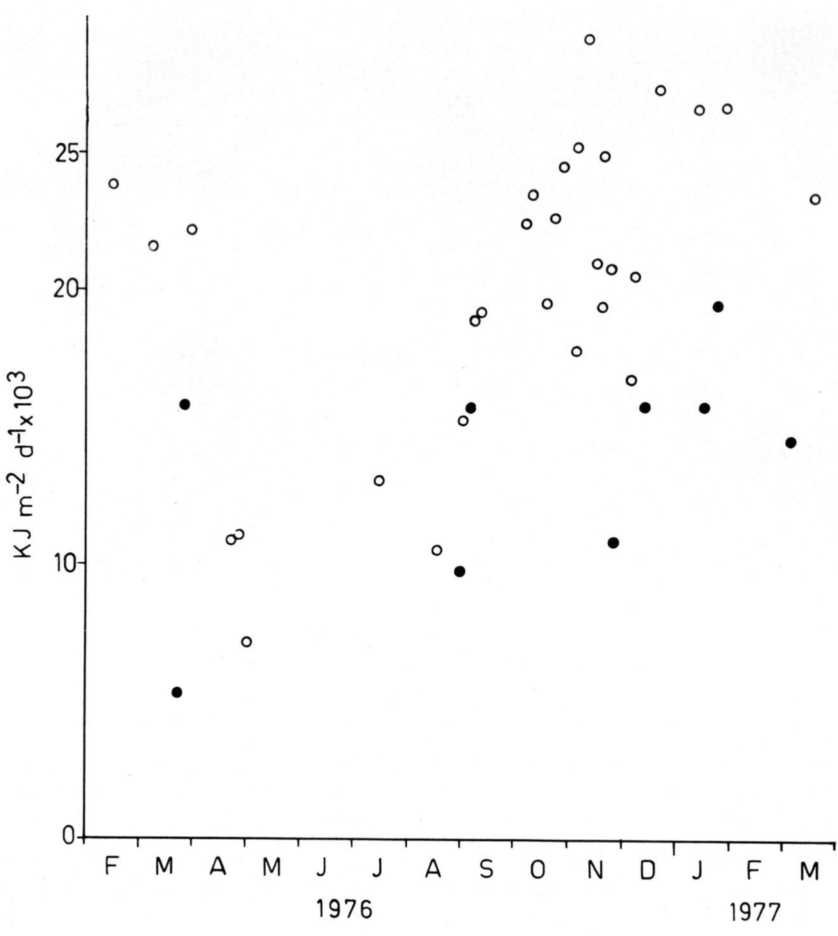

Fig. 5. The intensity of and variation in energy delivered to the surface of the lake in K J·m^{-2} day^{-1} on 38 days between February 1976 and March 1977. ○ clear days. ● cloudy days. The measurements obtained from a Lambda Radiometer sensor with integrator.

also provide an opportunity of comparing observed with predicted wave length. These data are given in Table 3.

The increase in the fetch, 6.1 m, of both the southerly and northerly winds as measured at points (2) and (3) in Fig. 7 implies a marked increase in the depth of the wave-mixed layer and it is significant in this regard that both the northern and southern shores of the lake are marked by extensive gentle

Table 1. Incident and reflected light in watts m^{-2}.

	Time	Incidence	Reflected	% reflectance	Lake conditions
18.11.76	1200	950	40	4.2	calm
10.12.76	1130	400	31	7.8	rippled
20.12.76	1345	1400	65	4.6	?

Table 2. The observed relationship between windspeed m s^{-1}, wave height (m) and period (s) in the southeasterly area of Lake Sibaya.

Date	Time	Wind speed	Wave Mean height	Period
15.10.76	1630	5.61	0.25	2
21.10.76	1100	4.82	0.25	2
23.10.76	0930	8.14	0.30	2–3
17.11.76	1000	2.37	0.15	2
24.11.76	1000	6.65	0.25	3
29.11.76	1030	6.44	0.07(?)	2
1.12.76	1200	4.29	0.20	3
10.12.76	1300	5.03	0.10	2
17.12.76	1330	4.93	0.05(?)	2
21.12.76	1230	6.84	0.12	1.5
19.1.77	1130	0.92	0.20	2–3
28.1.77	1300	5.29	0.15	2

slopes to at least 10 m depth, the sand of which is rippled to within this depth.

We can, therefore, recognize two primary wave-mixed areas in the lake. Firstly those associated with the east and western terraces with wave movement to 2–3 m and adequately defined by a mean fetch of 3.7 km Plate 2a and secondly the more extensive gentle slopes of the northern and southern shores subject to wave induced circulation to between 5–10 m defined by a fetch of 6.1 km.

The genesis of the slopes has been referred to by B. J. Hill, Chapter 3 and provides the opportunity to examine the resultant of the wind's stress upon the lake, namely its current regime.

Currents

The patterns and speed of wind induced currents in the lake have been determined by drift cards and underwater current direction sensors. Figs. 7 and 8 summarize the results of such studies on the circulation of the lake when northerly and southerly winds blow. Surface drift under southerly and easterly blows is indicated in the inset in Fig. 7. There appears to be no evidence of any marked horizontal circulation pattern as might be expected in shallower lakes, *e.g.* Loch Leven (Smith, 1974) and Lake George (Viner & Smith, 1973).

The influence of southerly winds is to produce a surface flow pattern down to 1 m towards the north west. At 5–10 m and below a return flow is generally experienced in the main basin, although this is complicated by basin morphometry, particularly near the opening of the south west basin, the entrance of the western arm and Etsheni Bay. The current pattern set up by a northerly wind as illustrated in Fig. 8 provides a somewhat clearer demonstration of the reverse current which confines the surface downwind

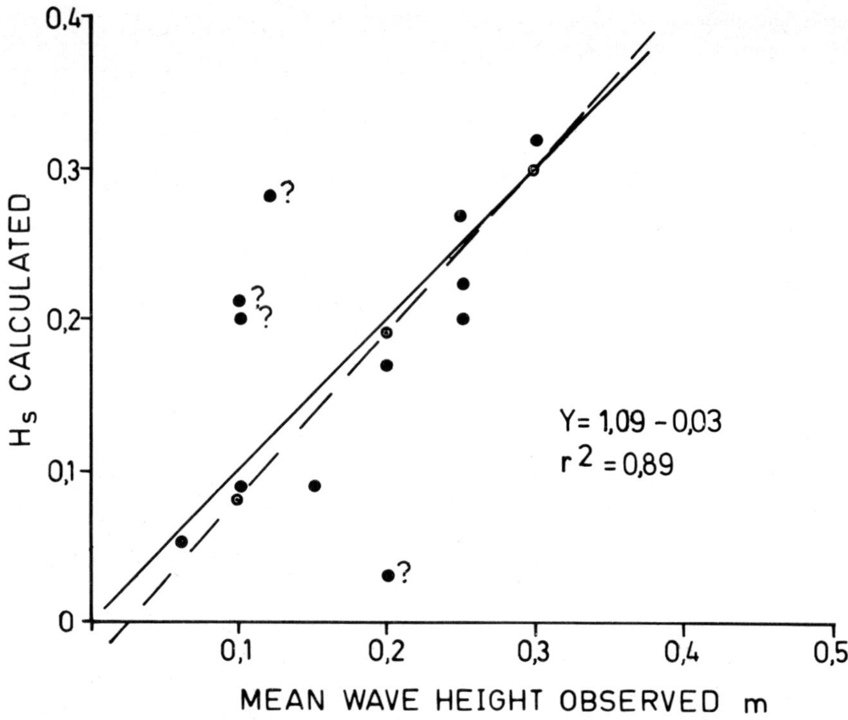

Fig. 6a. Relation between calculated significant wave height (H_s) and mean wave height in Lake Sibaya. The continuous line is drawn by eye through all co-ordinates. The regression applies to those at which ? is not shown.

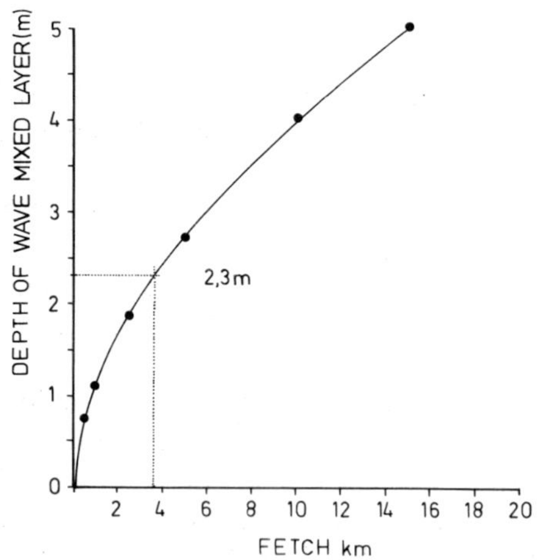

Fig. 6b. The relation between fetch distance and depth of wave mixed layer under the influence of a 4.5 m s^{-1} wind.

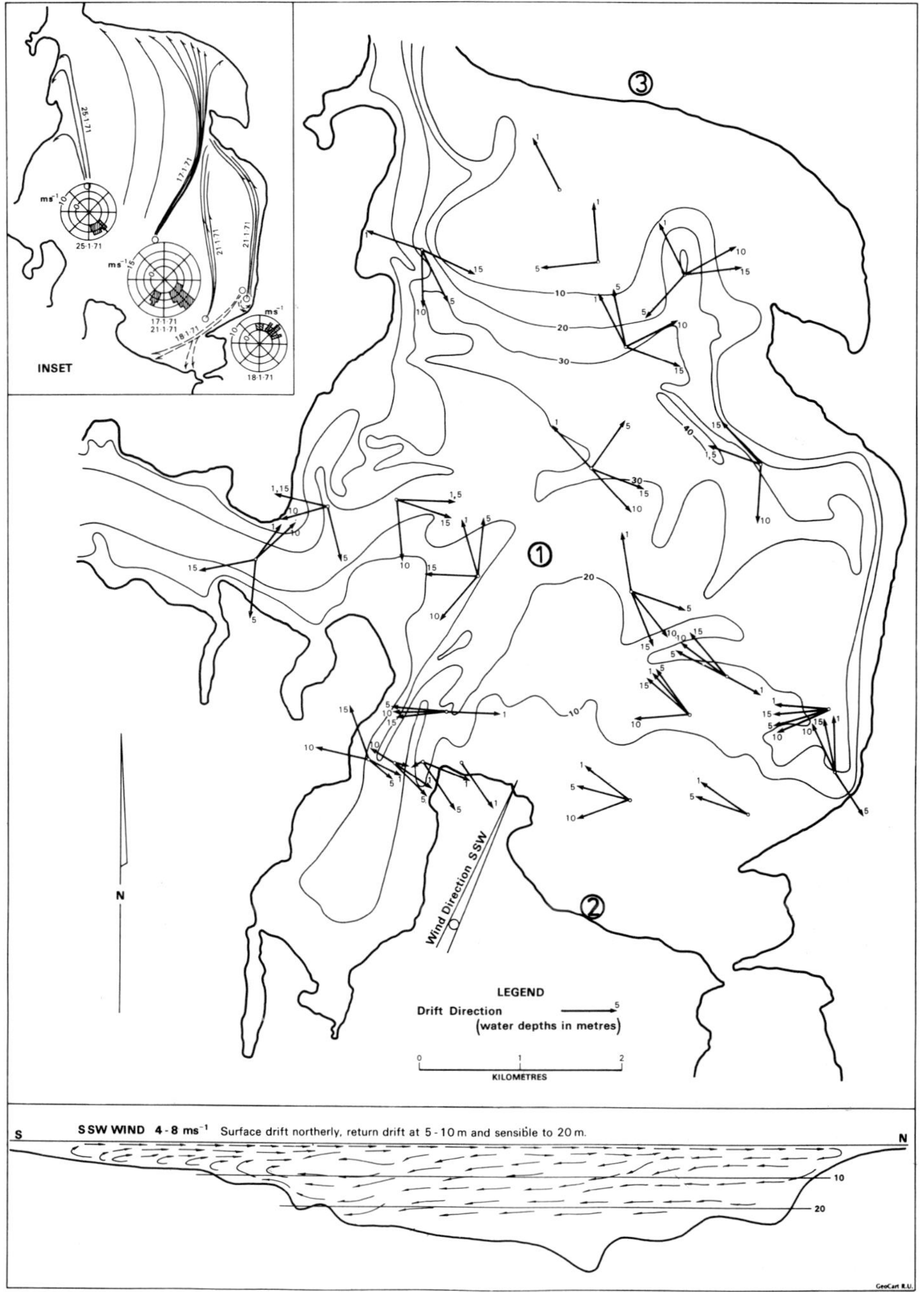

Fig. 7. The pattern of current movement in Lake Sibaya under the stress of southerly winds. Surface drift is given in the inset. The wind roses indicate wind strength and direction during the drift card release period.

Table 3. Calculation of T_s using the equation of $T_{s/w} = 0.46[g\,F/W^2]^{0.28}$ and comparing the wave length obtained from $\lambda = 1.56T_s^2$ with that from the ripple depth on the terrace.

Wind speed m s^{-1}	T_s	$\lambda = 1.56T_s^2$	λ from ripple depth, $D = 0.5\lambda$
4.3	1.68	4.4	4-6 m
8.1	2.54	10.1	>10

drift to within a metre of the surface. Thereafter the reverse current occupies the depth of the basin. The steep slopes along with the eastern edge of the basin are clearly the result of the sculpturing effects of currents set up by both southerly and northerly winds.

It would appear from our evidence that the inflection point (Fig. 9) lies at between 10–15 m. The return flow occurs within the depth of frictional resistance as defined by $D = 3.67\sqrt{W^3}/\sqrt{\phi}$ where W = wind speed in m s^{-1}; ϕ = latitude (Hutchinson, 1957). The depth of frictional resistance, 52 m, is in fact greater than maximum depth of the lake, $z_m = 40$ m. The reason for this is to be found in the comparative shallowness of the lake $\bar{z} = 13$ m and the windiness of the region in which it exists.

The velocity of the current set up by wind stress of this magnitude falls off with depth. Sometimes our data suggest a logarithmic decay as in Fig. 10, but this is by no means an easily reproducible result.

In January 1970 it was possible to remain on station over 20 m opposite the research station for a period of some 7 hrs, during which the behaviour and strength of the current was monitored under the stress of force 3–4 NNE wind. The composite results are given in Fig. 11 along with the temperature profiles for each observation time. The current direction has been presented in a somewhat different manner than in Fig. 9 but it does serve to demonstrate the variation found, and confirms the current reversal depth spanning vertical distance of some 10 m beginning in this series at 8 m. This is followed by a deep and narrow layer moving over the bottom in the direction of the surface current. The variation in current speed is once again marked and it is certainly not easy to assign a logarithmic pattern of decay. The temperature profile is remarkably uniform. The small variations observed were not linked to the striking current reversal referred to earlier.

The effect of this complex of moving water strata upon the turbulence within the water column is obviously considerable. As we have seen, mixing by wave action is limited largely to the steep east and west terraces and the more gently sloping north and south shores. This leaves the work of the wind to be largely translated into mixing the water column throughout its depth down to the maximum of 40–45 m. Superimposed upon this is another important energy input via wind stress, namely the setting up of a surface seiche. I am indebted to Mr and Mrs J. Minshull for this analysis of the seiche phenomenon in the lake.

Such short term effects of wind on the water level of the lake must also be considered in conjunction with the variation in barometric pressure as the

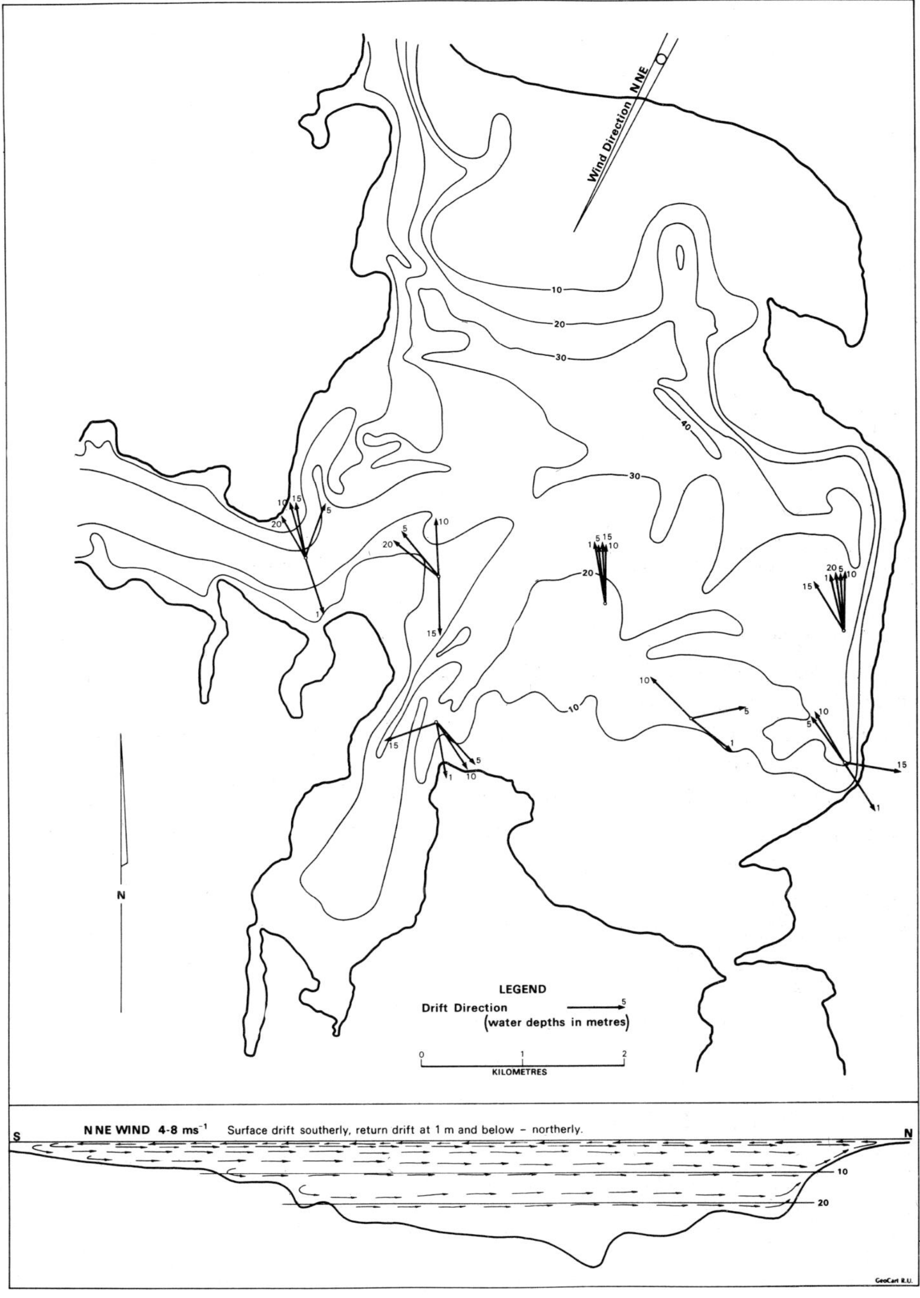

Fig. 8. The pattern of current movement in Lake Sibaya under stress of northerly winds.

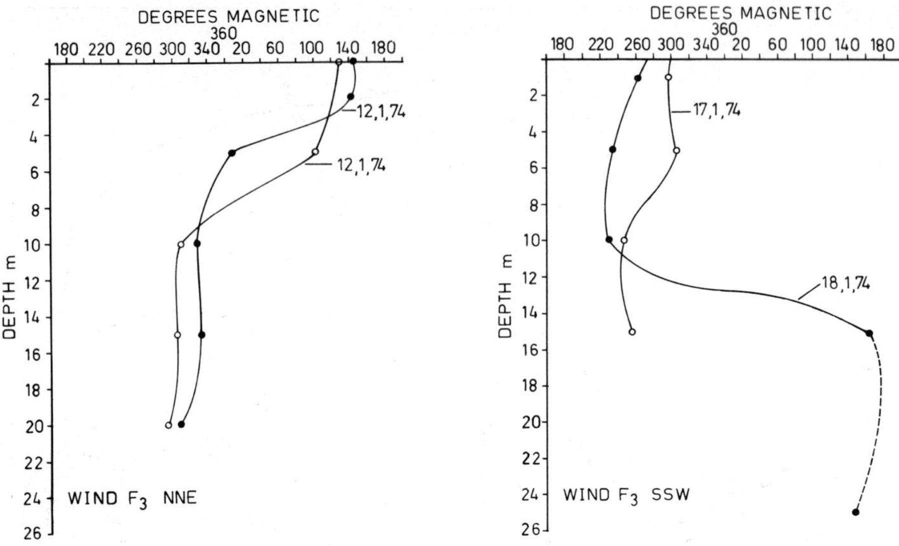

Fig. 9. Current direction profiles in Lake Sibaya under stress of southerly and northerly winds 4–5 m s^{-1} in the vicinity of maximum depth.

two are closely linked. In Fig. 12a the oscillation in lake level is recorded along with variation in barometric pressure and wind speed during the period 1–5 October 1978. On the morning of 1 October the north wind, starting from an initially calm period with high barometric pressure, increased at a rate of 1 m s^{-1} hr^{-1} to a speed of 8–9 m s^{-1} by 0700 hrs. The lake remained undisturbed until wind speeds exceeded 5 m s^{-1} when the lake level began to oscillate. The lowered wind speed on 2 October was

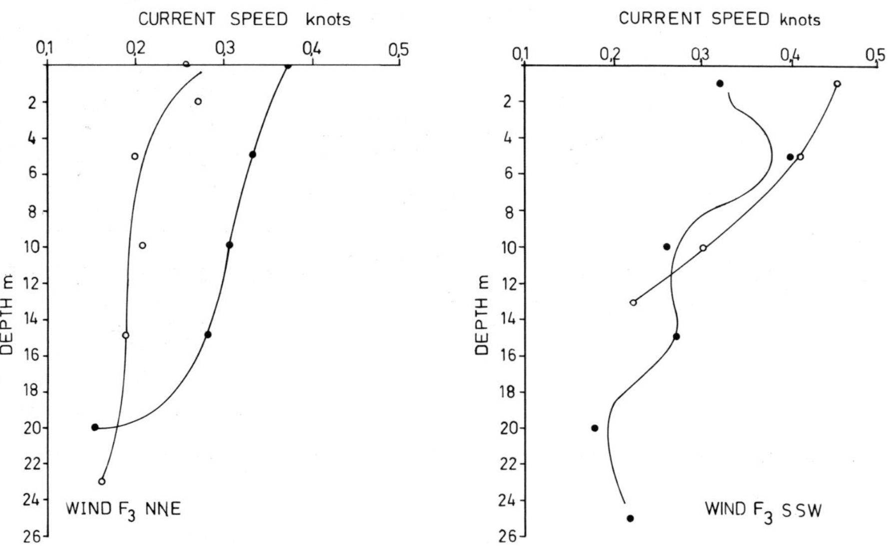

Fig. 10. Decrease in current speed (knots) in Lake Sibaya at the same dates as given in Fig. 9. Southerly and northerly winds 4-5 m s^{-1}. A 'Hydroproducts' current meter was used.

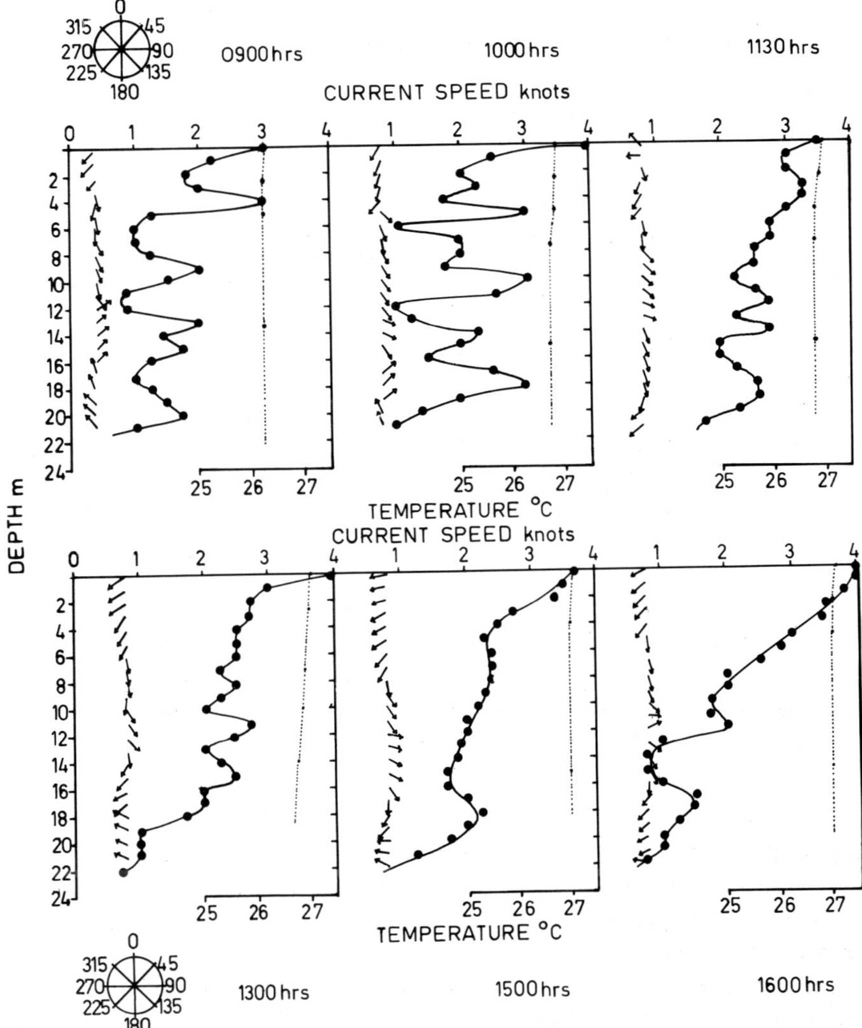

Fig. 11. Variation in current direction and speed during a 7 hour period of observation over 20 m opposite the research station at Lake Sibaya. Temperature profiles co-incident with the times of current measurement are also given. The temperature profiles are drawn from bathythermograph traces. Wind was NNE, F 3–4.

accompanied by a sharp fall in barometric pressure heralding a reversal in wind direction from north to south at 1400 hrs. By 1600 hrs the wind had attained a speed of 15 m s^{-1} and this high wind caused a depression of approximately 6 mm in lake level at the southern end of the lake. With reduction in wind speed the water resumed its normal level. On 3 October southerly winds of 15–20 m s^{-1} again lowered the water level by 6 mm. The subsequent drop in wind speed was accompanied by an increase in barometric pressure and a decrease in the amplitude and frequency of lake level oscillation. As Fig. 12a shows, as soon as the north wind, which started to

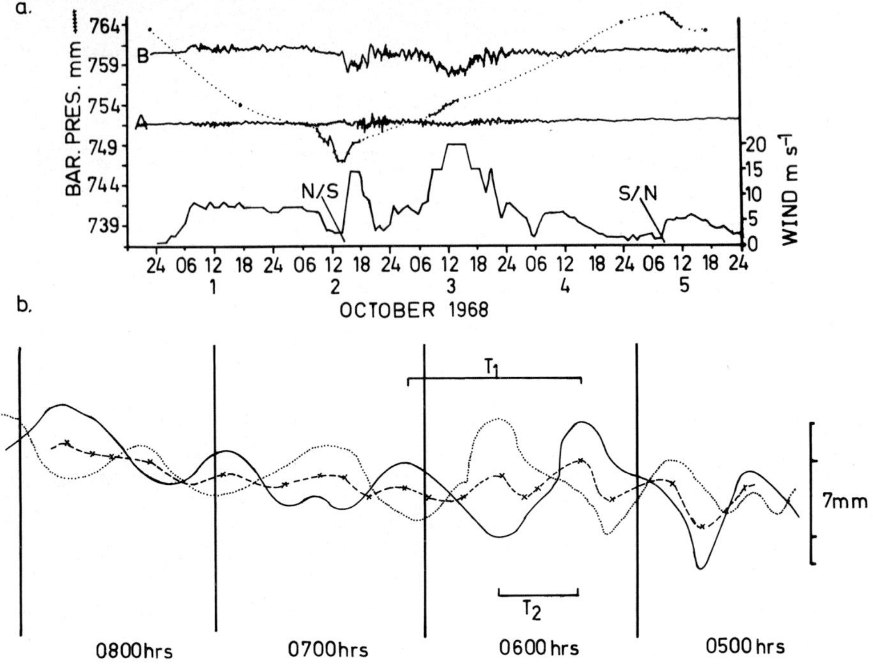

Fig. 12. a) Lake level oscillation as recorded at the research station (A) and Banda Banda (B) under the influence of north and south winds and barometric pressure fluctuations for the period 1–5 October 1968. b) Residuation of part of curve (A) in Fig 12a.

blow from 0800 hrs on 5 October reached 5 m s^{-1}, lake level oscillation began. Trace B in Fig. 12a illustrates level fluctuations as recorded at the level recorded in Banda Banda bay in the south basin. The amplitude of oscillation is exaggerated because this recorder is situated at the apex of the narrow bay, Plate 2c.

These oscillations seem to be surface seiches. From an inspection of them, the periodicity of the oscillation appears to be of the order of 40 minutes. Calculations of the periodicity of a uninodal seiche using Hutchinson's (1957, p. 301) equation and the lake dimensions given by Hill (1969) gives a value of 48 minutes for a complete oscillation. The amplitude of the oscillation is variable but it appears that oscillations of large amplitude are associated with a rapid decrease in wind speed and a rate of change in barometric pressure.

Inspection of the lake traces shows a complex seiche pattern. It has, however, been possible using the method of residuation to identify in addition to a uninodal period T_1 of 48 min, a binodal seiche T_2 of 21 min as represented in Fig. 12b. These should be considered as very approximate estimates as neither level recorder A nor B was ideally sited for such a study.

It has not been possible in this simple study to separate the relative effects of wind and barometric pressure. Clearly they are both important in setting up seiche activity in Lake Sibaya and as might be expected the surface pile up of water at the axial limits of the lake results in a return flow following cessation of wind stress or steadying of barometric pressure. It would be expected that these oscillations would contribute in some small measure to turbulent increment in the upper layers of the lake but possibly of greater importance is the maintenance of existing channels, as for example between the main and south basin (B. J. Hill, this volume) against the slow but inexorable march of the basin sands.

The Thermal Regime

We have, so far, examined the factors and their magnitude which are responsible for the thermal structure of the lake. The thermal structure of the lake has been described by Allanson & van Wyk (1969) and in terms of mean monthly surface temperatures in Fig. 2f. Some indication of the temperature changes in the water column are given in Fig. 13. No obvious thermocline is present throughout the summer and the stratification which is set up is shortlived and the homothermal conditions are usually re-established under the stress of wind, assisted particularly during the onset of the hot season by an increase in sensible and evaporative heat loss (Fig. 14). This uniformity is emphasized by examination of the temperature-time diagram of Fig. 15. Rapid heating of the surface does from time to time bring about fairly marked temperature differentials, but they are rapidly eroded under the influence of the wind.

In Fig. 16 evidence is provided to show something of the rate at which water heating occurs and the immediate source of the calorific increase. In January 1969 temperature variation at the station opposite the research station and over 18 m of water was measured over 24 hours. A rapid depression of thermal discontinuity occurred between 1300 and 1700 hrs.

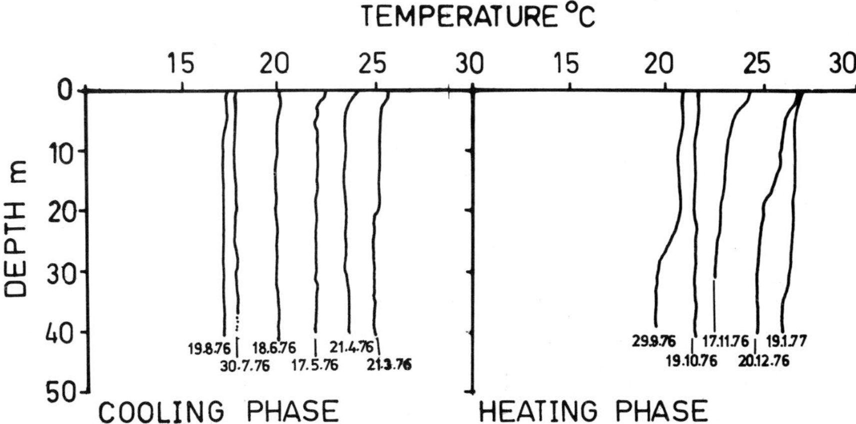

Fig. 13. Temperature changes in the water column at station 4, Lake Sibaya, showing vertical variation and seasonal change.

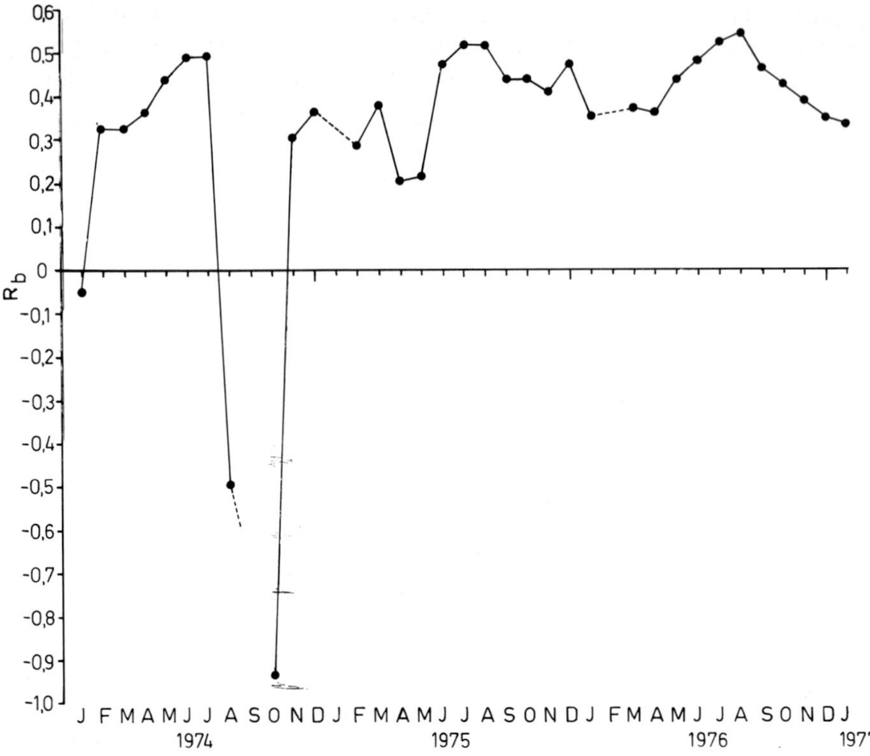

Fig. 14. Variation in Bowen's ratio R_b in Lake Sibaya 1974–76, based on mean monthly temperature. The high negative value between August and October 1974 implies that heat was transferred from the atmosphere to the lake.

This correlated with an increase in wind speed and a change in wind direction from south to north between 0800 and 1400 hrs. While the depression of the 26.9, 27.0, 27.1 and 27.5°C isotherms to 17 m is in part related to the turbulence set up in the water column by increasing wind activity, estimation of the heat gain by the water column in the vicinity of

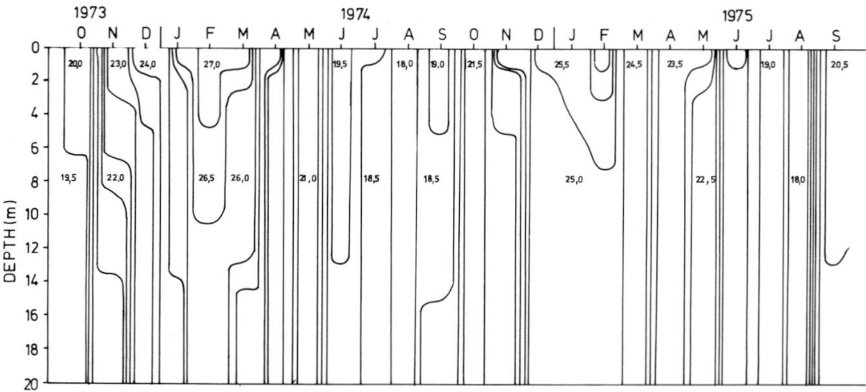

Fig. 15. Temperature isopleths for 1974–75, Lake Sibaya.

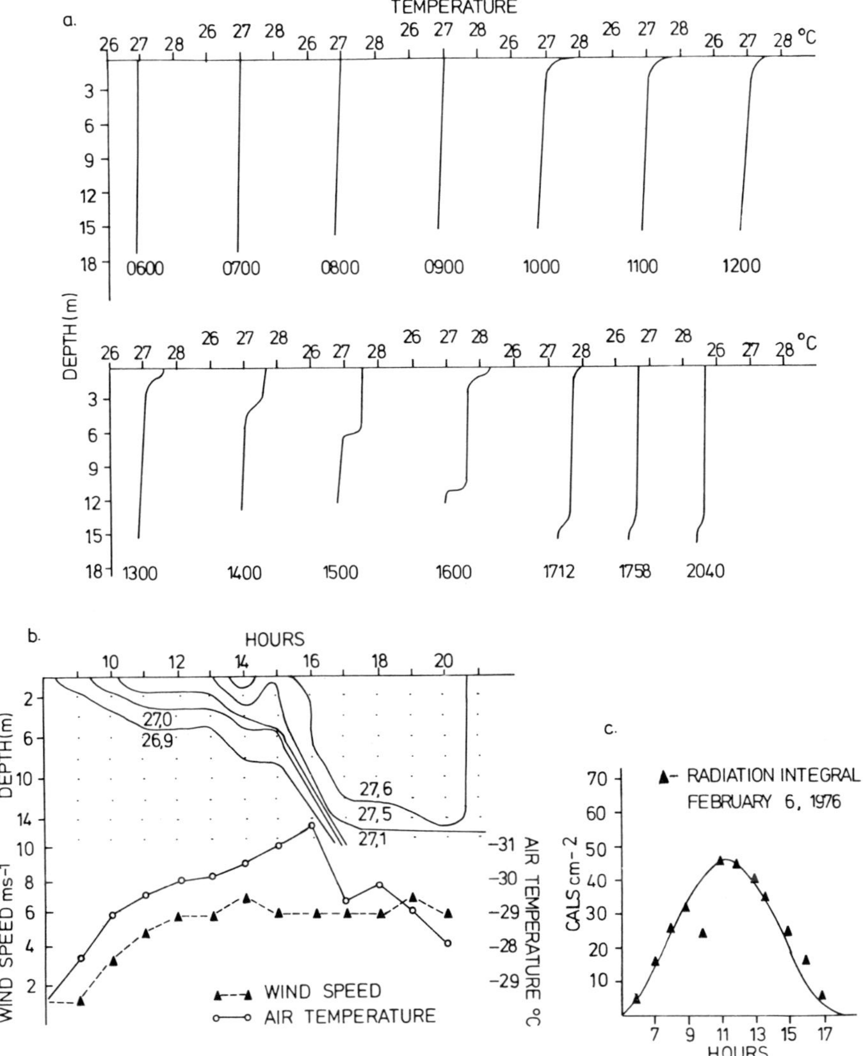

Fig. 16. a) Hourly temperature profiles at stations between 0600 and 2040 hrs–25 January 1969. b) Vertical depression of isotherms in water column between 0800 and 2040 hrs. (○) air temperature °C and (▲) wind speed between 0600 and 1900 hrs – 6 February 1976, Lake Sibaya, compared with c) radiation integral at Durban (January means: 1951–1962).

the observation station, particularly between 0800 hrs and 1700 hrs indicated that solar radiation incident upon the lake during this period was too small to account for the increase observed. The determination of the heat increase was made using Shonting's (1964) expression $Q_{is} = C_p \sum_{n-1}^{n} = 1\, L_n (T_{n-1} + 0.05° – 26.9°)$ where Q_{is} is the amount of heat absorbed, C_p is the specific heat of water, L_n is the thickness of the layer between two adjacent isotherms of cross sectional area 1 cm^2. T_n is temperature of the 'n'

isotherm, $T_{n-1}+0.05$ is the average temperature between $n-1$ and n isotherms as they are 0.1°C apart.

The total increase in heat at the station between 0800 hrs and 1700 hrs, the time of maximum isotherm penetration, was 903 cal cm^{-2}. At the time of these studies the field station at the lake was not equipped for incident solar energy measurements. The nearest measure of the integral of incident radiation, 294 cal cm^{-2}, was obtained for the latitude of Durban during January, from Archer (1964). These data are shown in Fig. 16c together with actual incident energy measurements of 288 cal cm^{-2} determined at Lake Sibaya on February 6 1976, using a Lambda pyranometer sensor LI-200S. The closeness of fit of these somewhat disparate data is encouraging and provides a measure of confidence in the application of Archer's mean January integral of incident solar energy, albeit at the latitude of Durban some 400 km to the south. This being the case the heat gain of 609 cals cm^{-2} (803-294 cal cm^{-2}) indicates that the rapid heating of the water down to a depth of 17 m must be due to translocation of warm water from other parts of the lake under the stress of a 6 m s^{-1} north easterly wind. The observation buoy is sited in 18 m of water close to a submerged slope which rises sharply at an angle of 25–31° to the horizontal (Boltt *et al.* 1969) to a shallow shoreline terrace. The isotherm pattern given in Fig. 16b is consistent with a pattern of warm surface water moving towards a shelving shore and descending to the lake floor. While this is not surprising, what is significant is the magnitude of the heat transfer, 68 cals cm^{-2} hr^{-1}. It supports the observations made earlier that in Lake Sibaya we experience very rapid distribution of heat energy throughout the depth of the lake.

Moving away from the special conditions of the lake shore to the open water provides us with the opportunity of determining a number of important components of the heat budget of the lake.

In Table 4 the summer heat income Θ, work of the wind B and total stability S are given. The depth distribution of B and S are shown in Fig. 17. Their sum equals G (the work required to maintain homothermal conditions throughout the heating season) which in Lake Sibaya is small relative to that for large stratified lakes where values of $G = 11{,}340$ to 25,241 are recorded (Hutchinson, 1957, p. 511, Table 59). This provides a clue as to why the lake does not exhibit summer stratification. The relation between the work of the wind which promotes turbulence $\partial u/\partial z$ and the density differential $\partial\rho/\partial z$ which suppresses it is given by Richardson's number, $R_i = g(\partial\rho/\partial z)/\rho(\partial u/\partial z)^2$ where g is the gravity constant, ρ is density, u is current velocity and z is depth. Because of the difficulty of measuring

Table 4. Summer heat income, work of the wind and total stability for Lake Sibaya during 1972.

A_0	z_m	$\bar{z}$	Θs	B	S	$G(B+S)$
km^2	m	m	cal.cm^2	gm cm.cm^2	gm cm.cm^2	gm cm.cm^2
71.4	40	13	11088	2280	2020	4300

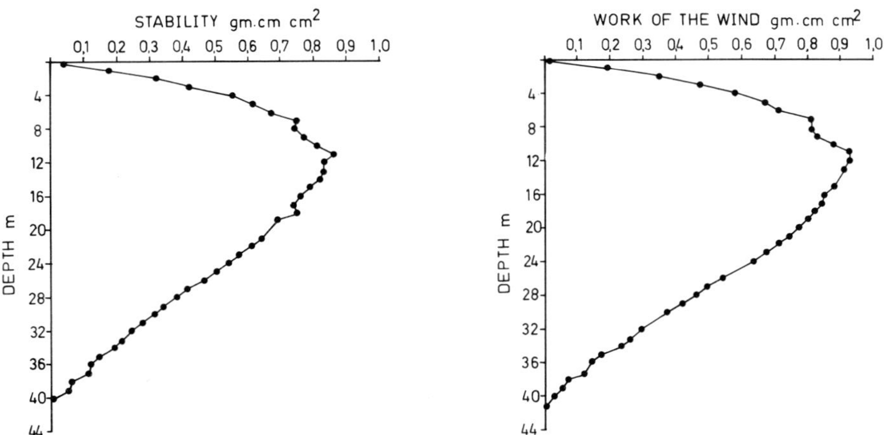

Fig. 17. a) Stability in Lake Sibaya and b) The depth distribution of the work of the wind.

synoptically the variation in current throughout the water column, no absolute value can be given to this ratio. However, it is known that when $R_i < \frac{1}{4}$, turbulent motion dominates, and because the current differential is squared, increases in velocity will have a marked effect upon the magnitude of the ratio and our estimates of the instability of the system.

Howard-Williams & Allanson (1978) in an analysis of ectogenic meromixis in a southern African coastal lake related the stress in gm.cm.cm^2 imparted to the surface of the lake to the force required to deepen the mixolimnion. Using the equation provided by Hutchinson (1957), namely that wind stress $T_a = 3.2 \times 10^{-6}\ W^2$ where W is the wind speed expressed in cm s^{-1}, they demonstrated that under the wind regime of the coastal area in which the work was done, erosion of the monimolimnion did occur with winds of >5 m s^{-1} blowing for al little as 1–2 hrs per day. By adopting a similar approach, but with a wind speed of 6 m s^{-1} blowing for at least 12 hrs per day during summer to represent typical conditions at Lake Sibaya, values of $T_a = 49.8$ gm cm.cm^2.d^{-1} were obtained. From the values given earlier for the work of the wind and stability, the minimum work required per day for the heating period of 150 days is $B = 15.2$ gm cm.cm^2.d^{-1} and $S = 13.5$ gm cm.cm^2.d^{-1}. Clearly in terms of the work required to distribute the summer heat income, the energy imparted by the wind is greatly in excess of what is required.

Two important components of an analytical heat budget of the lake have been determined from the temperature data available. The change in heat storage is reported in Figure 18 for three years. Heat storage gain occurs between August and December with the most rapid period between August and September. The period December through to May is a period of rapid heat loss, although the record is by no means consistent and reflects variation in the balance of the incoming radiation (Fig. 5) and evaporation (Fig. 2). Compared with a high latitude lake such as Loch Leven (Smith, 1974) the heat storage change is small, certainly not greater than $\pm 1 \times 10^4$ KJ.m^{-2}.d^{-1}.

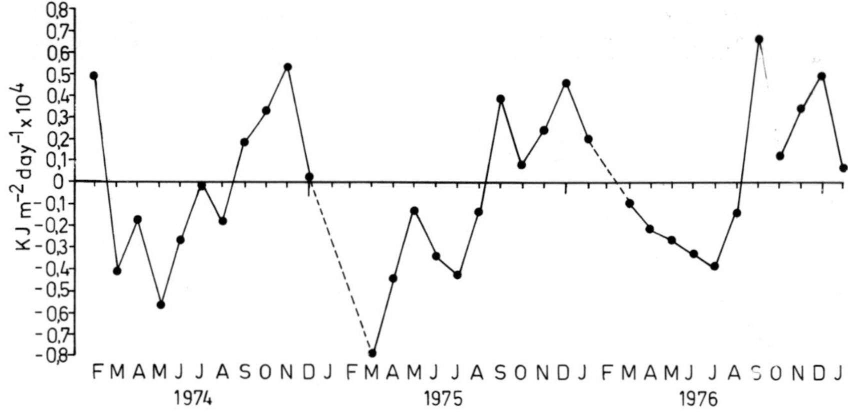

Fig. 18. Change in heat storage during 1974, 1975 and 1976 in Lake Sibaya.

Another important thermal property which it has been possible to calculate is the ratio $R_b = Q_s/Q_E$ or Bowen's ratio where Q_s = sensible heat transfer and Q_E = heat transfer due to evaporation. R_b can take either a negative or positive value. A negative value indicates that heat is conducted from the atmosphere to the lake. The average value for oceans (Sverdrup *et al.*, 1942) is 0.1, indicating that about 10% of the heat surplus of the ocean is lost as sensible heat. The remainder, 90%, is evaporative heat loss.

In Lake Sibaya our data (Fig. 14) suggest that based on mean monthly temperatures for 1974–76 Bowen's ratio varies predominantly between 0.3 and slightly in excess of 0.5, indicating that sensible heat loss, unlike that in oceans, is an important component of the radiation from lake surfaces. As might be expected, loss of stored heat is correlated with positive values of R_b.

In Fig. 19 mean water temperature is plotted against mean air temperature. During the period of cooling the temperature differential will result in considerable dissipation of sensible heat. The value of this has been obtained from the regression of Q_s against the temperature differential in Fig.

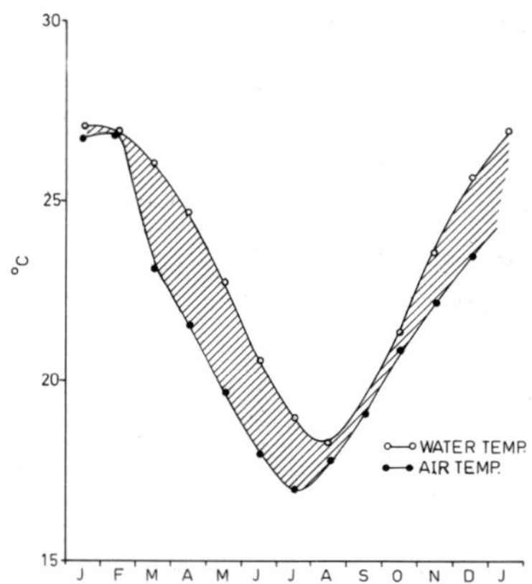

Fig. 19. Mean air and water temperatures over 3 years at Lake Sibaya.

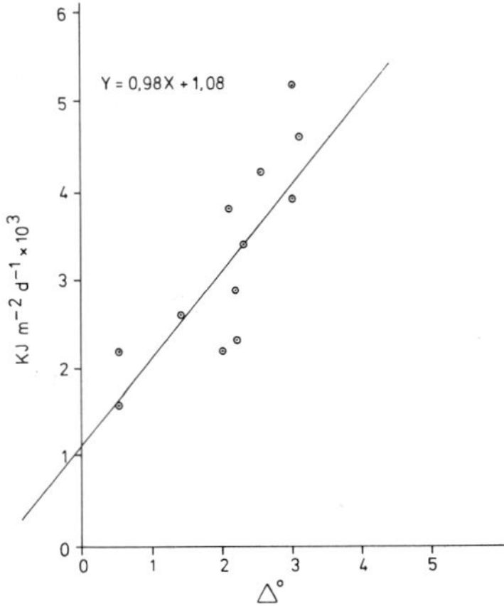

Fig. 20. Regression of sensible heat loss (Q_s) as determined from Bowen's ratio for Lake Sibaya against the water/air temperature differential.

20. From these data a 1°C differential is equivalent to 2060 KJ.m^{-2}.d^{-1} loss from the lake by conductance.

Some of the elements of the radiation balance during the months of August and December which are representative of the cool and hot periods of the year respectively are given in Table 5. The long wave length heat loss has been determined by difference.

It appears from this particular analysis that heat loss by conductance is of greater importance in Lake Sibaya than Stanhill & Neumann's (1978) data for Lake Kinneret and other mid-latitude lakes indicate. In how far this difference is real is difficult to assess as the data for Lake Sibaya are derived from a single lake-side evaporation tank. The energy gain ($Q_S + Q_H$), because of the measuring equipment used, represents both global and sky radiation in the sense of Stanhill & Neumann (1978). This value is, however, a rather poor mean so that an estimate of daily energy input is likely to be an underestimate. It is for this reason that the other important loss via longwave radiation is underestimated for this subtropical lake.

Table 5. Radiation balance for the cool and hot season in Lake Sibaya: all values are given in KJ.m^{-2}.d^{-1} × 10^3.

	Hot (Dec. 1974)	Cool (Aug. 1975)
Energy gains: ($Q_S + Q_H$)	20.2†	10.2†
Energy loss:		
Reflection Q_R	1.0	0.5
Evaporation Q_E	13.1	5.0
Sensible heat Q_S	5.2	2.6
Longwave radiation Q_W	0.9*	2.1*

* By difference
† Means of data given for the relevant months in Fig. 5.

The Light Regime

The Secchi disc transparency of the lake is 3.2 m ($n = 36$). The most penetrating component of the visible energy spectrum belongs to the green wave lengths followed by red and blue. The extinction curves for the spectral bands produced by Schott filters BG 12, RG 1 and VG 9 are given in Fig. 21. The variation shown in Fig. 22 in the extinction of both photosynthetic active radiation (PAR) and white light (LUX) is due to changes in transmission brought about by as yet unspecified changes in

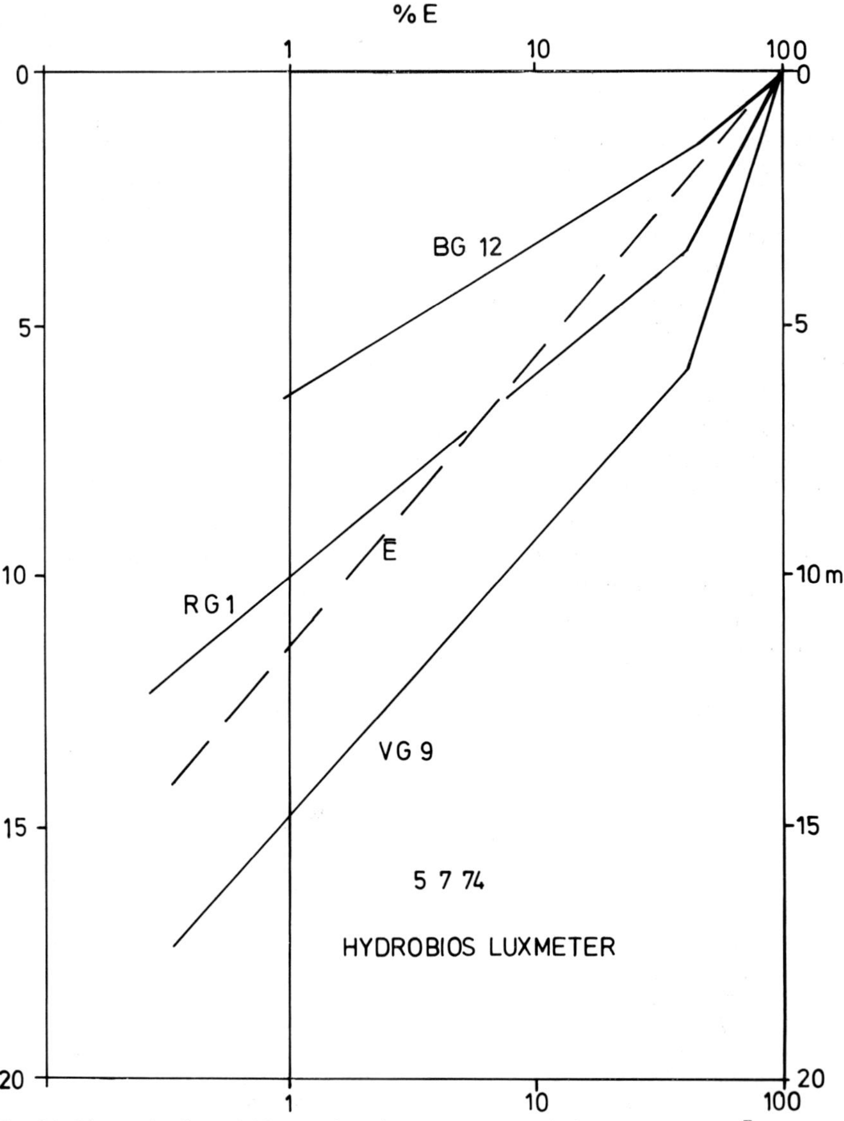

Fig. 21. The extinction of blue, red and green wavelength in Lake Sibaya. $\bar{E}$ represents the mean for white light.

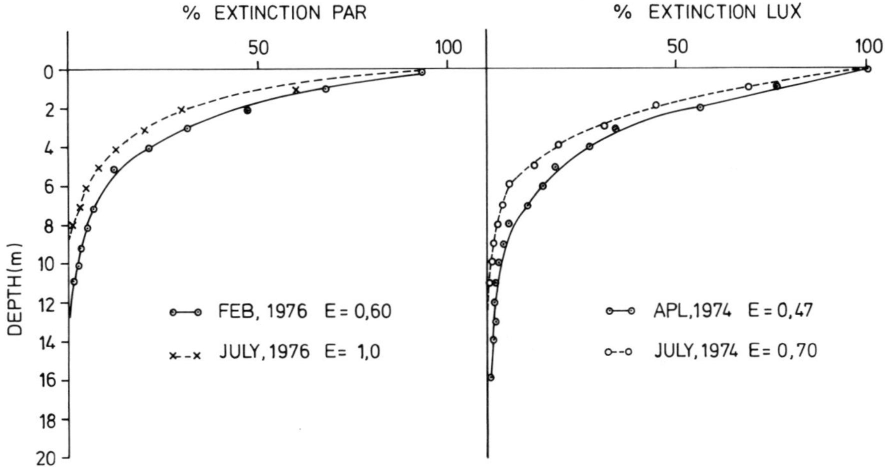

Fig. 22. The variation in extinction of A) PAR and B) white light during the hot and cool season. The coefficient of extinction E is given for each curve.

phytoplankton and suspended matter in the water column. Due to the turbulent nature of the lake as a result of wind stress, suspension of surficial material is to be expected. The mass of this material (tripton) was measured during February 1976 by collection in a sampling device similar to that used by Pennington (1974) in her study of tripton in the lakes of the Lake District. The sampling tubes were set at 4 levels in the water column. At each depth 3 collecting tubes with a total collecting area of 143.4 cm^{-2} were set and after one week the sample was raised and the contents filtered onto dry preweighed Whatman GF/C filters. The results of this study are given in Table 6.

It would appear from an examination of Pennington's data that the dry weight accumulation shown above is very much larger than that reported for the Lake District lakes, even Blelham Tarn. These data stress the importance of the resuspension of surficial material upon the transparency and the spectral of quality of the lake water.

The Dissolved Gases

Measurements by unmodified Winkler in Lake Sibaya by Allanson & van Wyk (1969) established that no serious departure from saturation occurred

Table 6. The mass of tripton collected over an area of 143 cm^2 during the 6–14 February 1976 in Lake Sibaya.

depth (m)	dry mass (mg) tripton
4	889
14	1210
24	584
34	860

in the water column during the cool season months. In summer, however, quite marked decreases in dissolved oxygen concentration could occur in deeper waters. Some indication of the magnitude of these changes is given in Table 7 for 1967, 1968 and near the maximum of lake level, January 1977.

These data show the presence of photosynthetic activity down to 15 m, below which oxygen concentration decreased quite markedly during the hot season, suggesting an increase in oxygen utilization in the deeper waters of the lake. This is to be expected on two counts, firstly from oxygen demand of the sediments which Boltt (1969) has shown to possess an increasing mass of fine subsieve (>0.04 mm) generally with depth, and secondly the vertical distribution of surficial material (Table 6) and its associated microflora along with the deep circulating phytoplankton component which, because it is below its compensation depth, is presumably actively respiring.

In Fig. 23a and b data are presented which show the variation in dissolved oxygen at a station representative of the pelagic zone and a shallow weeded bay known as Etsheni Bay during April 1974. At both stations there was a marked increase in dissolved oxygen during the day followed by a decrease at night. During this particular diurnal sampling period (11–12 April 1974) weather conditions were ideal with a gentle easterly breeze and variable cloud cover. With stronger winds, which are more characteristic of the lake, it is unlikely that the oxygen pulse would have been demonstrated as our failure to determine such a pulse in the submerged macrophyte beds along the eastern terrace during southerly blows has shown.

The importance of the total biological community, both benthic and pelagic in bringing about variation in the concentration of oxygen and carbon dioxide is demonstrated by these data. Due to the high buffering capacity of the lake water, marked changes in molecular CO_2 are not possible to show except at the surficial interfaces in deep water. R. E. Boltt (unpublished data) has demonstrated that free CO_2 is generated at a sufficient rate to influence the locomotory behaviour of tubiculous amphipods, particularly *Grandidierella lignorum* and consequently their distribution (R. C. Hart, Chapter 7, this volume).

The Dissolved Nutrients

The first determinations of NO_3(N) and soluble reactive phosphorus (SRP) were reported by Allanson & van Wyk (1969). Later work by Hart (unpublished) and Hart & Hart (1977) gave concentration values somewhat different from the previous authors. It is unlikely that these differences represent real changes, rather they are as a result of the differences in the application of the same analytical technique. In any event the levels of SRP and nitrate are low and account for these difficulties, table 8.

Neither Allanson & van Wyk (1969) nor Hart & Hart (1977) found any evidence of significant variation in the concentrations of these nutrients downwards in the water column, emphasizing again the overall chemical uniformity of the lake. These low concentrations of essential nutrients are

Table 7. The variation in dissolved oxygen (d.o.) concentration in mg.l^{-1} and percent saturation in Lake Sibaya. After Allanson & van Wyk (1969) and unpublished data of R. C. Hart.

	January 1967			July 1967			January 1968			March 1970			January 1977		
d(m)	*t*°C	d.o.	%	*t*°C	d.o.	%	*t*°C	d.o.	%	*t*°C	d.o.	%	*t*°C	d.o.	%
0	25.7	7.6	95	19.2	8.9	100	28.4	7.3	95	26.5	7.9	99	27.4	8.0	100
5	26.0	7.8	98	19.2	8.9	100	27.6	7.5	96				27.3	—	
10	26.0	8.0	100	19.0	8.7	97	27.6	6.8	87				27.2	7.5	94
15	26.2	8.1	102	19.0	8.5	95	27.6	6.4	82	26.1	7.7	97	27.2	—	
20	26.3	8.1	102	19.0	8.4	94	—	—	—				27.2	7.4	93
25	—	—	—	19.0	8.4	94	—	—	—				27.2	—	
30	25.6	6.9	85	18.9	8.4	94	—	—	—	26.1	7.0	88	27.2	6.9	87
35	27.7	6.6	82	18.9	8.4	94	—	—	—				27.2	4.8	60
38	—	—	—	—	—	—	27.0	5.4	68				—	—	
39													27.2	4.5	57

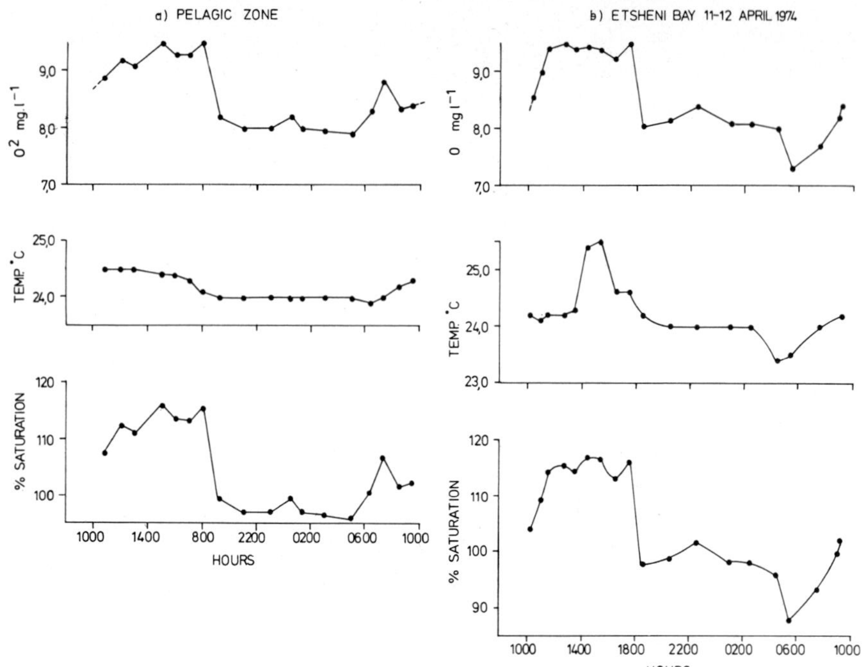

Fig. 23. Diurnal changes in dissolved oxygen in the pelagic zone (a) and in a shallow bay (b) during 11–12 April, 1974.

reflected in the level of algal standing crop and primary productivity. Hart & Hart (1977) record for 1970/71 that α chlorophyll varied between 1 and 4 $\mu g.l^{-1}$ at both the surface (0–5 m) and in the bottom waters (30 m).

A useful series of Kjeldahl nitrogen analyses of water and lake sediments are available for the period January to March 1971. These data are recorded in Table 9.

As might be expected, the deep sediments of the lake are high in Kjeldahl nitrogen due to a high subsieve fraction which is largely organic in origin. The littoral sediments are wave washed and while surface organic debris may be rapidly transferred to the deeper sediments Bowen (1976) has demonstrated a highly significant loss in protein content of detritus with

Table 8. Concentrations of SRP and NO_3(N) in Lake Sibaya from data given in Allanson & van Wyk (1969), Hart & Hart (1977) and previously unpublished.

	SRP $\mu g\, l^{-1}$	NO_3(N) $\mu g\, l^{-1}$
1967 (L)	25–55 ($n = 10$)	17–32 ($n = 36$)
1968 (L)	10–20 ($n = ?$)	
1970 (L)	14–25 ($n = 150$)	
1979 (H)	9 ($n = 9$)	50 ($n = 5$)

L refers to low lake level, H to high lake level

Table 9. Kjeldahl nitrogen in lake: water ($\mu g\, l^{-1}$) and sediments ($mg.Kg^{-1}$).

Lake water	Deep sediments	Littoral sediments	Pool sediments
11.2–28.0 ($n = 3$)	427 ($n = 3$)	4.2–62.5 ($n = 16$)	37–126 ($n = 2$)

depth. His studies showed clearly that the elevation of protein in shallow water detritus was due not only to diatoms but to dense concentrations of bacterial cells. Indeed he proposes that the diatom cells are primarily responsible for the production of DOM upon which the bacterial flora depend. The coincidence between decrease in diatom density and bacterial flora with depth is due to the reduction in DOM as depth increases.

These data are presented primarily as a yardstick against which to judge future changes in the concentration of these nutrients in the lake as the peneplain of Maputaland undergoes development. While at the moment I have argued that the lake from the viewpoint of algal standing crop and primary production (see Chapter 5), lies at the oligo-mesotrophic boundary, it will presumably quite quickly respond to allochthonous inputs of nitrogen and phosphorus due to its relatively high transparency and pattern of circulation.

The Inorganic Ions

Talling & Talling (1965) have, from a detailed analysis of 65 lakes in Africa, divided these lakes into three classes using among other criteria, conductivity in μmho at 20°C. Lake Sibaya with a conductivity of ~630 μmhos falls into Class II (600–6000 μmho). In this respect it is essentially similar to Lake Mohasi in Ruanda in which there is an elevation of both Cl^- and Ca^+ ions. The origin of the elevated chloride ion of these two lakes is different. In Lake Mohasi local salt sources in the vicinity of the lake are important, whereas in Lake Sibaya cyclical salt from the sea and from fossil sources within the tertiary sands of its catchment are largely responsible. Even the increase in volume during this decade has not lowered the chloride concentration excessively, supporting the view that seepage into the lake maintains the concentration of this ion in the lake. In Table 10 the chloride ion concentration is given for a number of years through the decade.

Table 10. Chloride ion concentrations and lake volume.

	μmhos 25°C	Cl^- mg l^{-1}	$M^3\ 10^6$
1967/68	598	136–138 (n=34)	552
1970/71	630	132–149 (n=36)	590
1975	645	125–127 (n=5)	680
1979	700	142 ($n = 3$)	776

Table 11. Cation and anion concentration in mg/l in Lake Sibaya in 1967 (low lake level) and 1979 (high lake level).

	Cations 1967	1979		Anions 1967	1979
Na^+	86	?	HCO_3^-	152	178
K^+	7	6	Cl^-	135	142
Ca^{++}	28	7			
Mg^{++}	9	13	SO_4^{--}	24	—
pH	8.3	8.3	SiO_2	16	—

These data, and particularly the range for 1970/71 and the single sample taken in February 1979 suggest that the change in concentration of the chloride ion in the lake is not a simple function of dilution as lake volume increases. Surface inflows are small relative to that contributed by ground water seepage. If, as has been suggested, fossil sources of salt water within the tertiary sands are responsible, at least in part, for the elevated chloride ion concentration in the lake, it follows that the varying rates of lake level elevation during the decade will be reflected in the delivery of salts into the system. Presumably the elevation of Cl^- around lake level maximum in 1979 reflects the contribution from sand horizons previously not involved in underground flow. Pitman & Hutchison (1975) have delineated the underground catchment of the lake but its integrity, especially during periods of such high water table elevation is questionable. It has been our experience that many previously disconnected surface swamp systems during the early part of the new decade are now contiguous and that even the lake which has been described as endorheic does overflow into lowlying swamps near the base of the dune forest. Such surface changes may well reflect a highly complicated network of underground aquifers responsible for the transport of water of varying ionic composition into the lake basin.

The pH of the lake has varied within a very narrow range during the decade. During both the cool and hot season it has been found to vary between 8.2 and 8.3 units. Likewise the concentration of the major cations and anions. In Table 11, the concentrations of these ion groups are given for the lake in 1967 and 1979.

A more detailed account of the chemical composition of the lake is given by Allanson & van Wyk (1969).

Discussion

The importance of the raised Cl^- levels in the lake, whether at low or high lake levels, is reflected in the surprising number of estuarine faunal relict species which appear in the lake. The fact that none of these taxa is represented in the typically freshwater of southern Africa implies on the one hand a boundary limit beyond which they are incapable of crossing and consequently invading the truly freshwaters and on the other a remarkable degree of ionic tolerance. Dr R. C. Hart has dealt with this aspect in more detail in Chapter 7.

Dr S. H. Bowen in his researches (Bowen, 1976, 1978) has pointed to yet another limiting chemical feature – the quality of the food resources available in the lake, particularly for the fish community. There seems no doubt that the overall low level of protein in the pelagic areas of the lake seriously inhibits the growth of fish, in particular *Sarotherodon mossambicus* after their first year. The difficulty we have experienced in determining $NO_3(N)$ in the free waters of the lake, together with the low values recorded for sediment, Kjeldahl nitrogen suggests that the lake and its community is nitrate limited. Silica is not a limiting resource in the lake. Support for this view comes from the appearance of the chlorophycean alga *Botryococcus braunii* which has repeatedly shown evidence of yellowing of the cells collected in surface-plankton tows, and the interpretation given by Fogg (1965) of the red stationary phase when the alga is brought into culture as nitrate limited.

The current patterns and thermal regime of the lake gives a picture of a lake in complete circulation with substantial current movements, set up by wind stress, sufficient to maintain a complex bathymetry in the absence of silt loads from surface inflow. We have argued earlier (Allanson & Hart, 1975) that the lack of a hypolimnion during the hot season may be one of the contributory factors to the low production in the pelagic zone. Sequestration of nutrients in deep water is prevented by the wind and thermal regime so that at no time during the year is the phytoplankton assemblage provided with a pulse of nutrient rich water, and we see a community geared to the rapid cycling of nutrients between its compartments within which biomass levels and production rates are set by the concentration of immediate nutrients.

It follows from this that increased production in the lake will only occur if the allochthonous inputs of nutrients are increased. Natural events such as the inundation of the *Acacia karroo* fringe of itself appears to make very little difference to the level of SRP or $NO_3(N)$ concentration in the lake water. This would, however, be surprising if, in a system already operating at near minimum levels, these nutrient ions were to remain in the water column. The increase in lake volume would effectively compensate for any increase in SRP and $NO_3(N)$ from the decay of such plant material so that effectively the available concentrations may be expected to remain the same or even decrease.

The increase in lake level as we have experienced over the past 10–12 years, now that we understand its dependence upon larger climatological events, provides an excellent opportunity of examining a variety of fundamental limnological events matching in some respects, though not to the same degree, the remarkable events which occur in the cycle of Lake Chilwa. It is my hope, therefore, that the research station will be rebuilt at another site on the lake shore to provide the facilities for the study of the lake during the next decade of *decreasing* lake level.

References

Allanson, B. R. & J. D. van Wyk. 1969. An introduction to the physics and chemistry of some lakes in Northern Zululand. Trans. R. Soc. S. Afr. 38: 217–239.

Allanson, B. R. & R. C. Hart. 1975. The primary production of Lake Sibaya, Kwazalu, South Africa. Verh. Internat. verein. Limnol. 19: 1426–1433.

Archer, C. B. 1964. The relationship between radiation and solar altitude in Southern Africa. Notos 13: 21–24.

Boltt, R. E. 1969. The benthos of some southern African lakes. Part II. The epifauna and infauna of the benthos of Lake Sibaya. Trans. R. Soc. S. Afr. 38: 249–269.

Boltt, R. E., B. J. Hill & A. T. Forbes. 1969. The benthos of some southern African lakes. Part I: Distribution of aquatic macrophytes and fish in Lake Sibaya. Trans. R. Soc. S. Afr. 38: 241–248.

Bowen, S. 1976. Feeding ecology of the cichlid fish *Sarotherodon mossambicus* in Lake Sibaya, KwaZulu. Ph.D. thesis, Rhodes University, Grahamstown.

Bowen, Stephen H. 1978. Benthic diatom distribution and grazing by *Sarotherodon mossambicus* in Lake Sibaya, South Africa. Freshwater Biol. 8: 449–453.

Dyer, T. G. J. 1976. Expected future rainfall over selected parts of South Africa. S. Afr. J. Sci. 72: 237–249.

Dyer, T. G. J. 1979. Pseudo-periodicities in lake level changes. In 'Lake Chilwa.' Eds. Margaret Kalk, C. Howard-Williams and A. J. McLachlan. pp. 47–49. Dr W. Junk. The Hague.

Fogg, G. E. 1965. Algal cultures and phytoplankton ecology. The University of Wisconsin Press, Madison. pp. i–xiii 1–126.

Hart, R. C. & R. Hart. 1977. The seasonal cycles of phytoplankton in subtropical Lake Sibaya: A preliminary investigation. Arch. Hydrobiol. 80(1): 85–107

Howard-Williams, C. & B. R. Allanson, 1978. Swartvlei Project Report: Part II. The limnology of Swartvlei with special reference to production and nutrient dynamics in the littoral zone. Institute for Freshwater Studies Special Report No. 78/3, Rhodes University.

Hutchinson, G. E. 1957. A treatise on limnology. Vol. 1. Geography, physics and chemistry. xiv, 1015 pp. John Wiley & Sons, Inc. New York.

Pennington, W. 1974. Seston and sediment formation in five lake district lakes. J. Ecol. 62: 215–251.

Pitman, W. V. & I. P. G. Hutchison. 1975. A preliminary hydrological study of Lake Sibaya. Report No. 4/75 Hydrological Research Unit, Department of Civil Engineering, University Witwatersrand, Johannesburg. pp. i–iii, 1–35.

Shonting, D. H. 1964. Some observations of short term heat transfer through the surface layers of the ocean. Limnol. Oceanog. 9: 576–588.

Smith, I. R. 1974. The structure and physical environment of Loch Leven, Scotland. Proc. R. Soc. Edinb. (B) 74: 81–100.

Smith, I. R. & I. J. Sinclair. 1972. Deep water waves in lakes. Freshwater Biol. 2: 387–399.

Stanhill, G. & J. Neumann. 1978. Energy balance and evaporation. In: Lake Kinneret. Ed. C. Serruya. pp. 173–181. Dr W. Junk bv. Publishers. The Hague.

Talling, J. F. & Talling, I. B. 1965. The chemical composition of African lake waters. Int. Revue ges. Hydrobiol. 50: 421–463.

Tyson, P. D. 1978. Rainfall changes over South Africa during the period of meteorological record. In: Biogeography & Ecology of Southern Africa. Ed. M. S. A. Werger. pp. 53–69. Dr W. Junk bv. Publishers. The Hague.

Viner, A. B. & J. R. Smith. 1973. Geographical, historical and physical aspects of Lake George. Proc. R. Soc. B. 184: 253–270.

5 The phytoplankton and primary productivity of the lake

B. R. Allanson

There have been few long term studies of pelagic primary productivity in the lakes of southern Africa. The studies of Hart & Hart (1977) and Allanson & Hart (1975) and of Howard-Williams & Allanson (1978) and Allanson (in press) have provided much of the data for this chapter, in which I have attempted to relate changes in the physico-chemical climate of the lake with those of primary productivity.

In Chapter 4 I have shown that the wind induced current regime and circulation in the lake do not allow the development of extended periods of thermal stratification so that there is no sequestration and concentration of nutrients in a hypolimnion which, from time to time would be the source of nutrient pulses in the upper photosynthetic zone of the lake. This must have a significant influence upon the pattern of photosynthetic carbon fixation in the illuminated waters of the lake.

The chemical environment of the lake has been described in Chapter 4. Neither soluble reactive phosphorus nor nitrate $NO_3(N)$ stand at levels which would result in eutrophic conditions developing. Given that these nutrient stocks are in equilibrium with the lake environment (which seems to be a reasonable assumption) the standing crop of algae in the water column as measured by α-chlorophyll must represent the magnitude of this equilibrium. Thus the primary fixation of carbon and its essential corollary, the fixation of carbon in excess of respiration (net production) is controlled by three principal components, (Talling, 1971): (1) the time course of surface irradiance, (2) the relative penetration of radiant energy with depth, (3) the proportions of light and 'dark' water in the water column.

The Light Regime

In the lake the most penetrating component of the visible energy belongs in the green wave lengths as measured by a 'Hydrobios' luxmeter in association with Schott Filter VG9. The level of 1% PAR (photosynthetically active radiation) as measured by a Lambda submersible quantum meter lies at a maximum depth of 11–12 m with ε varying 0.43–0.60 m^{-1} during summer. In winter 1% lies at 7–8 m and $\varepsilon = 0.996\ m^{-1}$. The relative penetration of this energy is influenced by factors such as Secchi disc transparency ($\bar{x}$ 3.2 m$\pm$0.3, n = 36) and colouration. For example, the reduction in the depth of 1% surface irradiance in the lake as compared with that of Lake Poelela to the north (Allanson, in press) is due to suspended particles. In addition, while the latitude of the lake prevents there being a marked

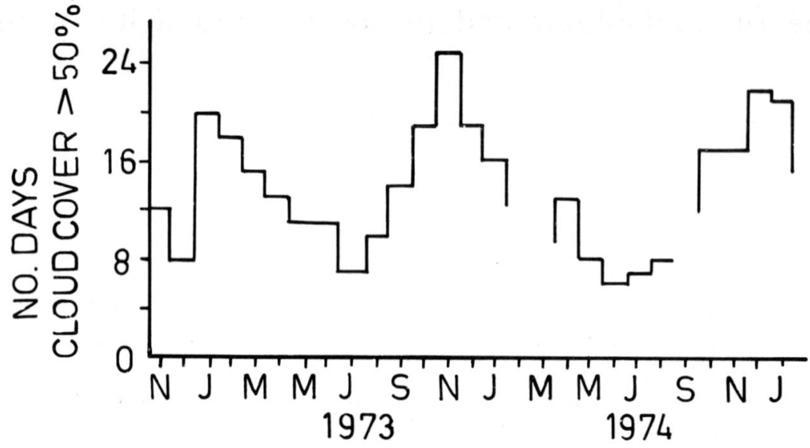

Fig. 1. Seasonal change in the number of cloudy days per month (cloud cover >50%) November 1972 to January 1975 at Lake Sibaya.

difference between summer and winter in the amount of energy delivered to the lake surface, changes are effected by seasonal variation in cloud cover. The number of cloudy days (>50% cloud cover) is shown in Fig. 1 and is higher during summer than winter. This is largely due to the development of coastal depressions which move eastwards, cyclonic activity in the Mozambique channel and the formation of convection cloud. Some indication of the magnitude of this variation is given in Fig. 2.

The Algal Community

The structure of the net phytoplankton community is described by Hart & Hart (1977). From a cell volume and therefore total biomass viewpoint the community is dominated by *Closterium* spp., *Synedra acus*, *Anabaenopsis* sp., *Melosira granulata* and *Anabaena* sp. in order of importance. An indication of the net phytoplankton structure is given in Fig. 3 while in Fig. 4 the numerical importance of the nannoplankton in a vertical series of bottle samples taken throughout the depth of the lake in July 1973 is demonstrated.

Hart & Hart (1977) have drawn attention to the likelihood that in view of high numbers, the volumetric contribution of nannoplankton is of major importance to the operation of the community. They calculate on the assumption of a spherical configuration of 20 μm diameter than their nannoplankton counts represent 4190–16760 $mm^3 \cdot m^{-3}$, greatly in excess of the total volume of the principal net phytoplankton.

The variation in cell volume within this algal community, as Fig. 3b shows, has little if any influence upon the total α-chlorophyll content of the lake. As such this chlorophyll component provides a useful measure of algal standing stock which is responsible for the photosynthetic fixation of carbon.

Pelagic Primary Productivity

Photosynthetic production was determined by the carbon-14 technique. We found that the difference in carbon fixation by the total algal community and that fixed by phytoplankton passing 35 μm nitex was almost negligible. This result emphasises the importance of the nannoplankton component in the production of the lake.

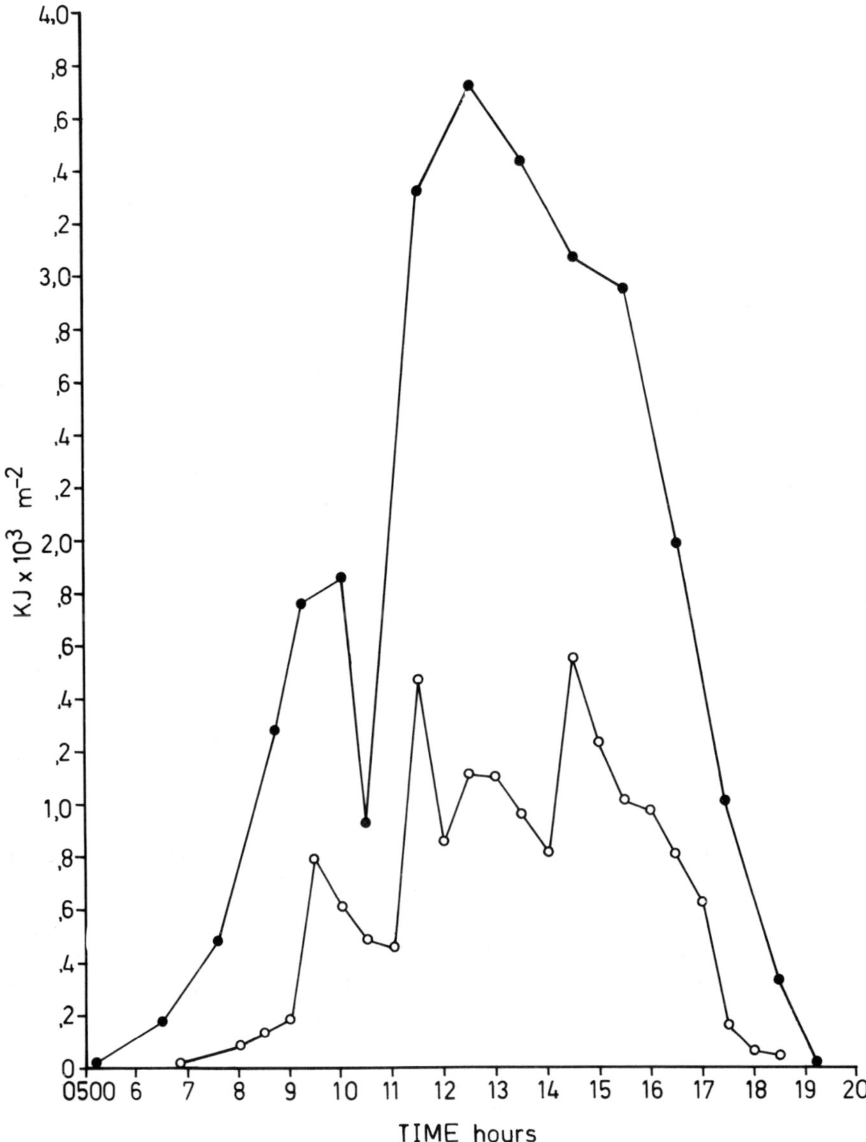

Fig. 2. Variation in diel energy delivered to the surface of Lake Sibaya on a clear (●) and cloudy day (○) in the summer of 1976/77.

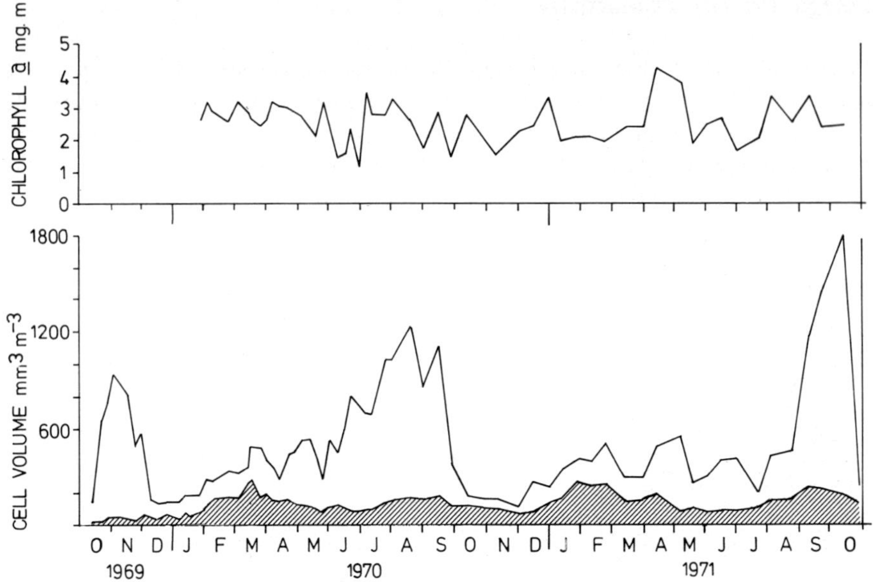

Fig. 3. Seasonal variation in α chlorophyll and the dominant net phytoplankton *Closterium* spp. in relation to the remaining net plankton volume. After Hart & Hart (1976).

Carbon assimilation, while showing a cool season depression, is largely light limited in the lake. With rare exceptions there is strong surface inhibition with the maximum light saturation photosynthetic rate occurring at between 3–4.5 m (Allanson & Hart, 1975) overlapping the value of Secchi disc transparency. The compensation depth varies with season, being maximal during the hot season at 14–15 m, which is somewhat deeper than the depth of 1% PAR.

The influence of variation in the energy reaching the lake surface upon carbon assimilation rate and the photosynthetic integral is shown in a qualitative manner in Fig. 5. Unfortunately we were unable to measure the incident PAR at the time these carbon assimilation experiments were performed but the effect upon assimilation of cloudiness appears to be marked. Note in particular 6 December 1973, 8 February, 14 September and 21 November 1974. The total integral is markedly reduced relative to the measurements of the previous month and as compared with the cool season months of June and July. That of 8 February 1974 seems to provide an example of the effect of continuous cloudiness for long periods. Dr M. N. Bruton reported (pers. comm.) overcast conditions during the early part of February which resulted in the lowest photosynthetic integral of the whole period of measurement. In support of this conclusion, the results from Swartvlei, a coastal lake to the south may be cited. Thus on April 13 1977 it was clear and sunny with a surface irradiance of 1352 $J.m^{-2}\,day^{-1}$, while on April 14 there was 100% cloud cover with rain and irradiance fell to 816 $J.m^{-2}\,day^{-1}$. The decrease in photosynthetic integral was 6.2%.

Some indication of the change in photosynthetic rate with depth and time is given in Fig. 6. This time-depth diagram is a coarse one in which with the exception of January 1975 a single incubation series is deemed representative of each month. This is clearly insensitive of the real changes which occurred. For example, the variation in photosynthetic rate on three consecutive days in January 1975. Nevertheless, it suggests that the greater proportion of carbon assimilation in the water column occurs in general above 6–7 mm at which depth the 5 mg C m^{-3} hr^{-1} isopleths lie and that pulses of carbon assimilation occur at varying periods of the year. According to the analysis of Allanson & Hart (1975) this is roughly representative of the depth at which attenuation of the irradiance (ε) at maximum light saturation photosynthetic rate has reached 50%. By far the greater proportion of photosynthetic fixation occurs above this irradiance level in the lake.

The vertical circulation pattern of the lake referred to in Chapter 4 will certainly bring a proportion of the algal cells below their compensation depth for some period of time. Indeed, the current regime of the lake is such that a fixed proportion of the algal biomass will be maintained constantly below this depth as Hart & Hart's (1977) data on α-chlorophyll indicates.

The ratio of mixed depth z_m to euphotic depth (z_{eu}) is dependent upon the dimension assigned to z_m. In Chapter 4 the depth of frictional resistance, *D*, is far greater than the maximum depth so that by taking $z_m = 40$ m and $z_{eu} = 14$ m in summer the ratio $z_m/z_{eu} = 2.9$ increasing to 3.3 in the cool season.

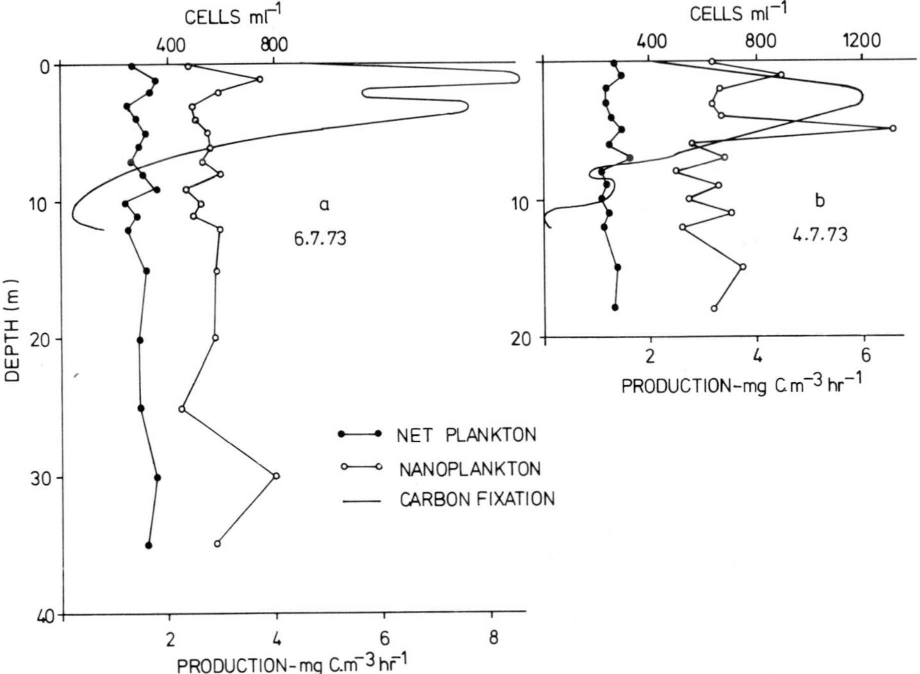

Fig. 4. Vertical changes in net and nannophytoplankton at Station 1 in Lake Sibaya, July 1973. The photosynthesis-depth profiles of this community are also given. After Allanson & Hart (1975).

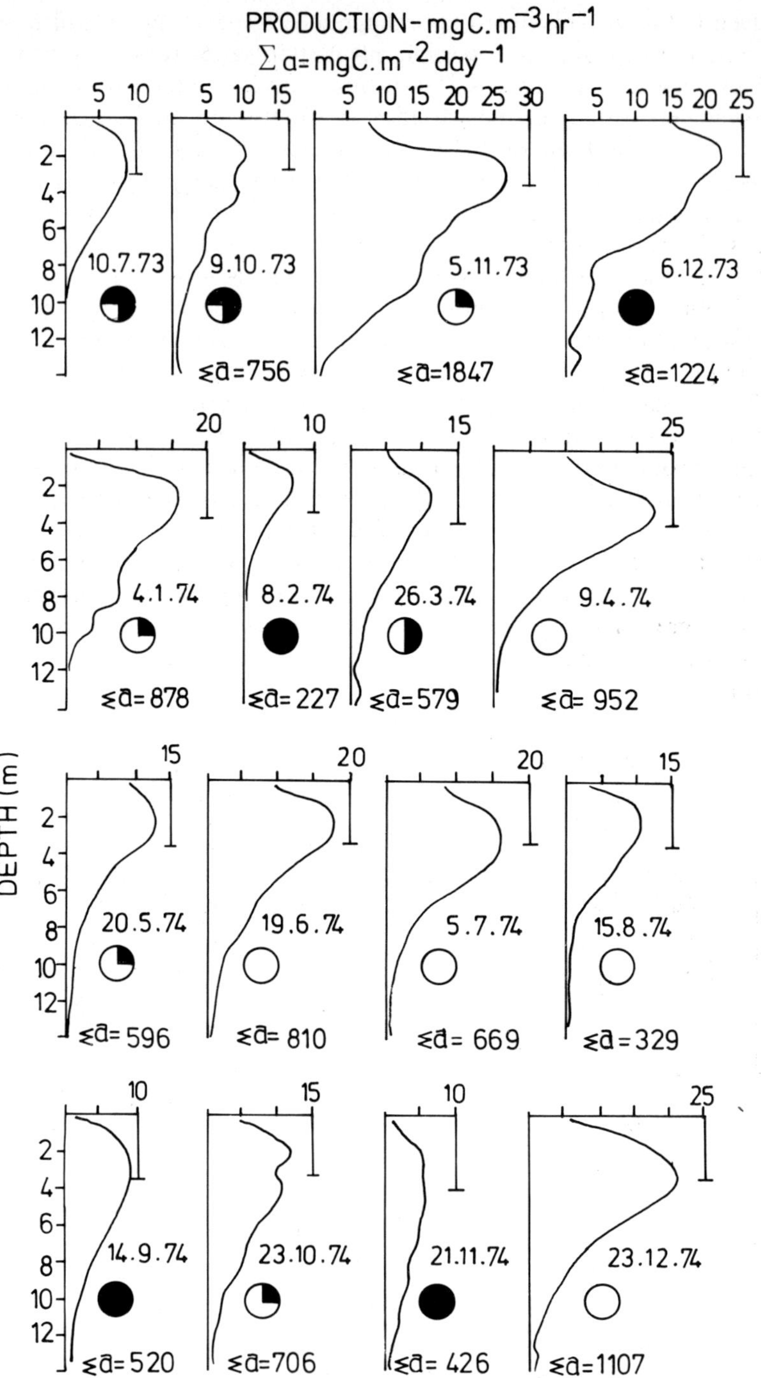

Fig. 5. The photosynthesis-depth profiles for Lake Sibaya during the year 1973/74 correlated in a qualitative manner with the degree of cloudiness (>50% cloud cover).

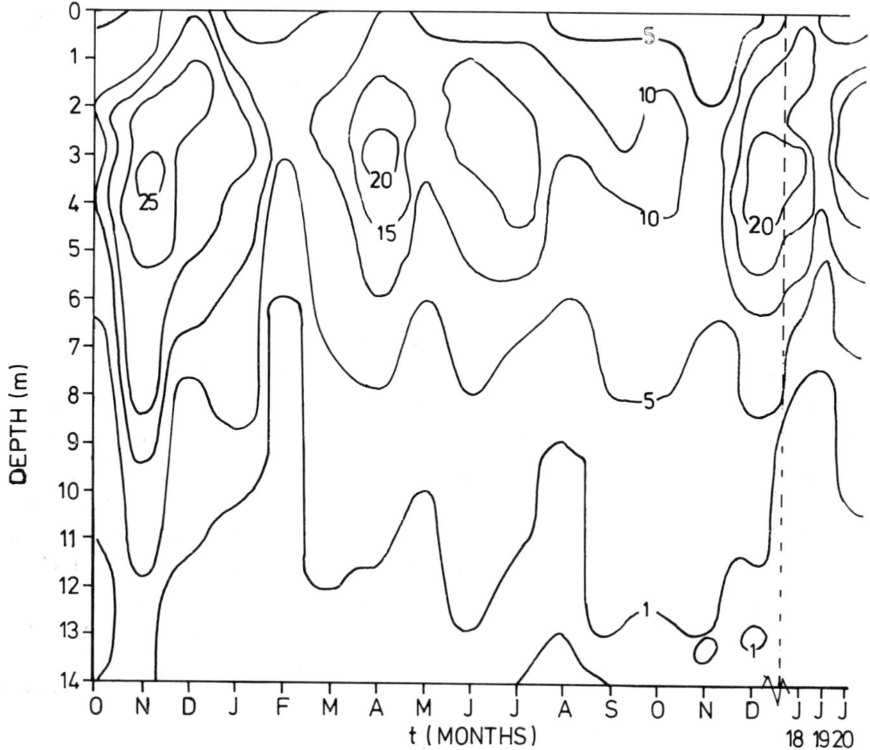

Fig. 6. A time-depth diagram of photosynthetic rate in mg C · m^3 · hr^{-1} for Lake Sibaya during 1973/74/75.

It will be appreciated that in lakes possessing distinct thermal stratification the mixed layer is often less than the depth of the euphotic zone, *i.e.* $z_m/z_{eu} < 1$. Under these circumstances the surface integral of photosynthesis ($\sum a$) will reach its maximum with no 'dilution' (Talling, 1971) due to deeper and presumably non-photosynthesizing algal cells – a situation which never pertains in this lake. In Lake Sibaya, however, in which wind induced circulation is never limited by a thermocline, the possiblility that dilution by non-photosynthesizing cells could influence the relation between total carbon fixation and respiration needs examination. Talling (1957b and 1971) has developed a photosynthetic model in which the underwater light climate is defined by three groups of factors. They are surface solar irradiance, vertical attenuation and depth of mixing. Talling (1971) proposes a dimensionless quantity 'q' which expresses, for a homogeneously vertically distributed phytoplanktonic community, the combined action of these three groups of factors. 'q' is basically the ratio between the logarithmic function of daily subsurface radiation expressed as light division hours per day [L.D.H.]^{day}, to the optical depth, $\varepsilon_{min} \cdot z_{\acute{m}}/1n\,2$.
For Lake Sibaya

$$q_c = \frac{[\text{L.D.H.}]^{day}}{\varepsilon_{min}\, z_{\acute{m}}/1n\,2} = 3.18,$$

where the minimum extinction coefficient of P.A.R., $\varepsilon_{min} = 0.37$, the effective mixed depth, $z_{m} = 18.5$ m, and $[\text{L.D.H.}]^{day} = 31.4$.

The relative respiration rate, $R/GP = r$, for Lake Sibaya was determined on 9 February 1976 as

$$\frac{1.8 \text{ mg m}^{-3} \text{ hr}^{-1}}{32.7 \text{ mg m}^{-3} \text{ hr}^{-1}} = 0.055.$$

If this value is substituted in the expression $q = 32r$ (Talling, 1971) which defines a critical value of q, (q_c) at which production equals respiration: q_c for the lake = 1.76. This value being less than the calculated value above, namely 3.18, implies that on this date the phytoplanktonic community was capable of autotrophic growth. If, however, r increases, which is likely if the value of ε_{min} increases during the cool season or with a decrease in transparency, then q_c becomes larger and column respiration approaches that of production and may even surpass it in Lake Sibaya. While I have no evidence that this could happen in the lake, it does point clearly to the importance of these environmental factors upon photosynthetic control even in subtropical and tropical lakes. The application of this type of analysis in Lake George by Ganf & Viner (1973) has been particularly successful in demonstrating how delicately balanced is ecological stability where 'the magnitude of net production is not simply a function of the daily fluctuation of solar radiation but is closely related to the mixing regime of the water column'.

While it may seem strange to assert that the water column in a lake at the latitudes and with the morphometric characteristics of Lake Sibaya may exhibit a net negative production, I believe it points towards the need to reassess our approach to a) the significance of current estimates in photosynthetic carbon fixation, and b) the physiological status of algal phytoplankton in lakes where nutrient loads of N & P are particularly low, and in which a significant proportion of the algal biomass is maintained below the photosynthetic compensation point by wind induced current regimes.

Epipsammic Primary Production

With the completion of these initial studies on pelagic productivity I turned my attention to other significant sites of primary energy fixation within the lake. While Howard-Williams (Chapter 6, this volume) concluded that the importance of the macrophytes was small in this regard, Bowen (1978) had established the distribution and density of diatoms on the terrace sands of the lake littoral and his data hinted strongly that the terrace littoral could be an important zone of energy fixation.

In view of the highly developed shoreline of the lake, this production, calculated on an areal basis, would make a significant contribution to the overall energy fixation in the lake ecosystem.

Accordingly field experiments were designed to determine the primary productivity of the littoral sands using light and dark bottle techniques.

Three sites perpendicular to the shore were chosen at 1 m, 2 m and 3 m depth while a site at 10 m was worked on one occasion. Sand samples to a depth of 1 cm were collected by divers using a 'Hookah' air supply (Plate 4, Chapter 12) and transferred at the collection site to 100 ml 'Pyrex' light and dark bottles. An initial dissolved oxygen concentration was determined in the usual way for each series. At least 3 replicates of the light bottle were set up and two for the dark bottles. Photosynthetically active radiation measured in μ Einsteins $m^{-2}.sec^{-1}$ was measured on the sand surface.

The results are reported in Table 1, which demonstrates clearly the magnitude of the energy fixation by the epipsammic plant community. The values for gross production shown in Fig. 7 match satisfactorily the diatom cell density as reported by Bowen (1978). Some indication of the diatom structure of this sand community is given in Plate 1. The relatively low production at 1 m and 10 m is likely to be due to light inhibition and attenuation respectively. Maximum values occurred at 2–3 m and matched quite closely the depth of a maximum production for the phytoplankton

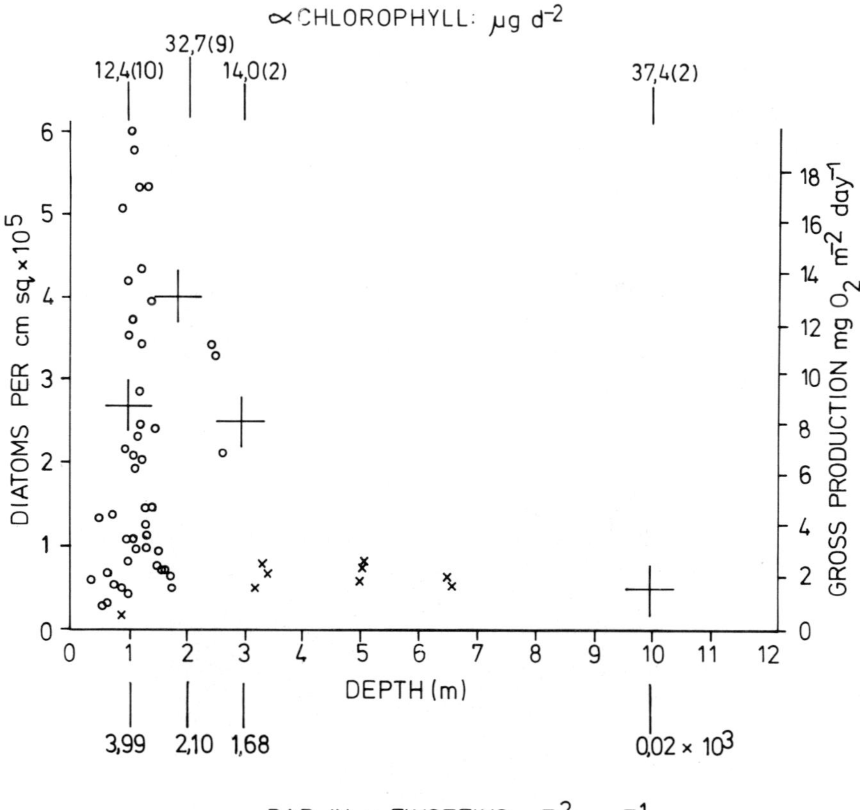

Fig. 7. The distribution of diatom cells (O) over the terrace of Lake Sibaya and the gross production (+) of the community in mg $O_2.m^{-2}.day^{-1}$ as measured in situ using Winkler bottles.

Table 1. Epipsammic production in the terrace sands in Lake Sibaya January 1977 expressed in mg $O_2 \cdot m^{-2} \cdot day^{-1}$ at 28°C.

	1 m	2 m	3 m	10 m
	$n=8$	$n=10$	$n=4$	$n=2$
$\bar{X}_{GP}$	8800±1200	13,066±4414	8048±5759	1655
$\bar{X}_R$	6348±4103	8388±4813	6978±1239	1728
$GP:R$	1:0.69	1:0.62	1:0.87	1:0.96
$\bar{X}_{NP}$	2752	5244	1062	—

Assuming ^{14}C photosynthetic fixation approximates net production pelagic production as compared with epipsammic production is –

Pelagic	Epipsammic
1847 mgC $\cdot m^{-2} \cdot day^{-1}$	1630 mgC $\cdot m^{-2} \cdot day^{-1}$

community reported by Allanson & Hart (1975). The maximum summer photosynthetic integral as reported in this paper was 1847 mg $C.m^{-2}.day^{-1}$: the maximum recorded for the period in January 1977 for epipsammic production was 1630 mg $C.m^{-2}.day^{-1}$.

The phytoplankton pelagic integral is a summation of carbon fixation over a depth of 14 m, while that for the terrace sands is available within a sand layer of 1 cm thick and emphasises the importance of the terrace littoral in the feeding of *Sarotherodon mossambicus*. The utilization of this trophic niche by *S. mossambicus* is described by M. N. Bruton in Chapter 8.

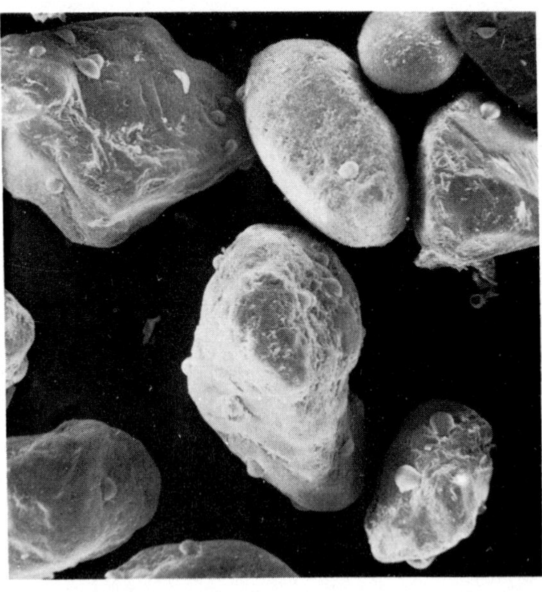

Plate 1a. The distribution of diatom cells on sand grains × 100 from the terrace of Lake Sibaya.

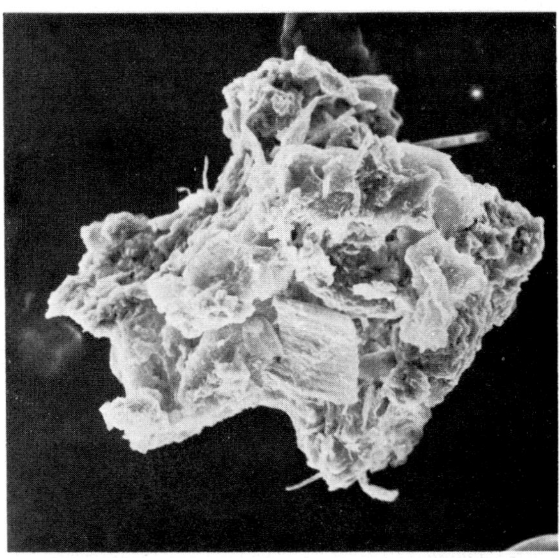

Plate 1b. The distribution of diatom cells on detrital floc × 400 from the terrace of Lake Sibaya.

Enough has been said throughout this volume to emphasise the comparatively low photosynthetic rate in Lake Sibaya. The overall integral of pelagic productivity places the lake at the oligo-mesotrophic boundary. The utilization of the primary producer resources in the lake has been described in particular in Chapters 5 and 8 but for a synthesis of a number of ideas which emerge from this and the other contributors' work the reader is referred to Chapter 11.

Addendum

A preliminary checklist of the benthic and planktonic algae of Lake Sibaya.

	Benthic	Planktonic
1. Chlorophyta		
Botryococcaceae		
Botryococcus braunii		×
Scenedesmaceae		
Scenedesmus sp.		×
Desmidiaceae		
Closterium ? aciculare		×
Closterium sp.		×
2. Chrysophyta		
Bacillariophyceae		
Achnanthes minutissima	×	
Anomoeneis exilis	×	
Caloneis silicula v. truncatula		×

Cocconeis thumensis	×	×
Cymbella kapii	×	
C. microcephala	×	
C. muelleri	×	
C. pussila	×	
C. turgida	×	
Diploneis subovalis	×	
Epithemia sorex	×	
Gomphonema lanceolatum	×	
Mastogloia elliptica v. dansei	×	
M. smithii v. amphicephala	×	
M. smithii v. lacustris	×	
Melosira granulata		×
M. granulata v. angustissima	×	×
Navicula cryptocephala	×	
N. pupula	×	
N. pseudony		×
N. seminoloides		×
N. tenelloides	×	
N. vividula	×	
N. zanonii	×	
Nitzschia amphibia	×	×
N. confinis		×
N. epiphytica	×	
N. fontifuga	×	
N. kützingiana	×	×
N. obsoleta	×	
N. palea		×
N. perminuta		×
N. perspicua	×	
N. vanoyei	×	
Rhopalodia gibba	×	
R. gibberula	×	
Synedra acus v. angustissima		×
S. ulna	×	×
3. Cyanophyta		
Chroocaccaceae		
Merismopedia ? glauca		×
Nostocaceae		
Anabaena sp.		×
Anabaenopsis sp.		×

References

Allanson, B. R. (in press). The coastal lakes of Southern Africa. In: Estuaries. ed. by J. H. Day.

Allanson, B. R. & R. C. Hart. 1975. The primary production of Lake Sibaya, KwaZulu, South Africa, Verh. Internat. Verein. Limnol. 19: 1426–1433.

Bowen, S. H. 1978. Benthic diatom distribution and grazing by *Sarotherodon mossambicus* in Lake Sibaya, South Africa. Freshwater Biol. 8: 449–453.

Ganf, G. G. & A. B. Viner. 1973. Ecological stability in a shallow equatorial lake (Lake George, Uganda). Proc. R. Soc. B., 184: 321–346.

Hart, R. C. & R. Hart. 1977. The seasonal cycles of phytoplankton in subtropical Lake Sibaya: A preliminary investigation. Arch. Hydrobiol. 80 (1): 85–107.

Howard-Williams, C. & B. R. Allanson. 1978. Swartvlei Project Report: Part II. The limnology of Swartvlei with special reference to production and nutrient dynamics in the littoral zone. Institute for Freshwater Studies Special Report No. 78/3, Rhodes University.

Talling, J. F. 1957b. The phytoplankton population as a compound photosynthetic system. New Phytol. 56: 133–149.

Talling, J. F. 1971. The underwater climate as a controlling factor in the production ecology of freshwater phytoplankton. Mitt. Internat. Verein. Limnol., 19: 214–243.

6 Distribution, biomass and role of aquatic macrophytes in Lake Sibaya.

C. Howard-Williams

The majority of the Southern African coastal lakes are shallow with a large proportion of the area available for colonization by aquatic macrophytes. The shallow side arms of Lake Sibaya and the extensive littoral terrace of the main basin provide an area of about 20 km^2 or 26% of the lake area of less than 5 m in water depth. Yet, surprisingly one does not feel on an initial trip to Lake Sibaya that macrophytes are particularly abundant, especially when compared with some of the other coastal lakes (Howard-Williams, 1977). Perhaps it is for this reason that in comparison with the other major groups of aquatic organisms (Chapters 7 and 8) the aquatic macrophytes of Lake Sibaya have not been well studied.

It is convenient when looking at any group of organisms in an ecosystem to consider their ecology under the following headings:

1) The species present
2) Their abundance
3) The factors controlling abundance
4) Stability or change
5) The role macrophytes play in the system.

This essay is an attempt to provide an introduction to the aquatic macrophytes of Lake Sibaya by considering their ecology under each of these headings in turn, except for (4) which is discussed throughout.

The Species Present*

Before considering the presence of species it is necessary to define clearly the area under consideration because as one progresses from a permanently flooded to a seasonally flooded or moist habitat the number of species often tends to increase (Howard-Williams, 1975). Therefore, for the purposes of this study, only plants rooted underwater during the sampling period are considered. It must be remembered however that much of the area under consideration has only been flooded for, at the most, two years due to the recent rise in lake level (Chapter 3). This has resulted in a number of anomalies in the vegetation, and, as most of the work on the other groups of

* The taxonomic authority is given in Appendix 1.

organisms in Lake Sibaya was done previous to this high water period, much of this macrophyte study is in fact out of phase with the other work on the lake.

From the point of view of macrophyte biology two periods in the lake's recent history can be distinguished, the *low water period* from 1964 – 1973 and the *high water period* from 1974 – 1977 (See also Fig. 1, Chapter 3).

Low Water Period

A qualitative account of the aquatic macrophytes during this period is given in Boltt, Hill and Forbes (1969), Breen and Jones (1971), Allanson *et al.* (1974), and more recently in connection with catfish (*Clarias gariepinus*) habitats by Bruton (1979), and in connexion with Mollusc habitats by Appleton (1977).

The dominant emergent plants of the exposed shores of the lake were *Scirpus littoralis* and *Phragmites mauritianus.* In sheltered areas *Typha latifolia* subsp. *capensis, Cyperus papyrus* and the grasses *Sporobolus virginicus* and *Dactyloctenium geminatum* were particularly abundant.

Submerged macrophytes ringed the entire lake from a depth of 1 to 7 m, with *Potamogeton* spp. predominating in exposed areas and *Myriophyllum spicatum* and *Ceratophyllum demersum* in the sheltered shallows. Fig. 1a and d reflect the general structure of the macrophyte community in exposed and sheltered areas respectively, during the low water period. During this time the 1% level for white light in Lake Sibaya was 12 m (Allanson and Hart, 1975).

High Water Period

The most significant change in the vegetation resulting from the rise in water level can be seen in the main basin of the lake, particularly on the exposed shores. The dominant emergents (*Typha, Phragmites*) of the low water period had gone or were very sparse growing in 3 m of water (*Scirpus*). The edges of the entire lake, except for river inflow areas were defined by a distinct zone of *Cyperus natalensis.* Due to increased turbidity at the high lake level the macrophytes no longer grew down to a depth of 7 m as they did during the low lake period but were restricted to the areas above 5.5 m. At high water level the depth of penetration of photosynthetically active radiation was 7–8 m (Chapter 5). As the lake level has risen 3 m it can be appreciated how the entire zone of submerged macrophytes has moved upwards some 5 m vertically from the low distribution. The northern and southern shallows of the main basin which, during low water had abundant macrophyte populations (Boltt, Hill and Forbes, 1968), were almost devoid of macrophytes in 1977. Therefore, although new areas have been made available for colonization by macrophytes due to the recent water level rise, the increase in turbidity has meant that the actual colonisable area has not altered much from the low to the high water periods.

Fig. 1b is a profile diagram of a shoreline of the main basin in 1977. The

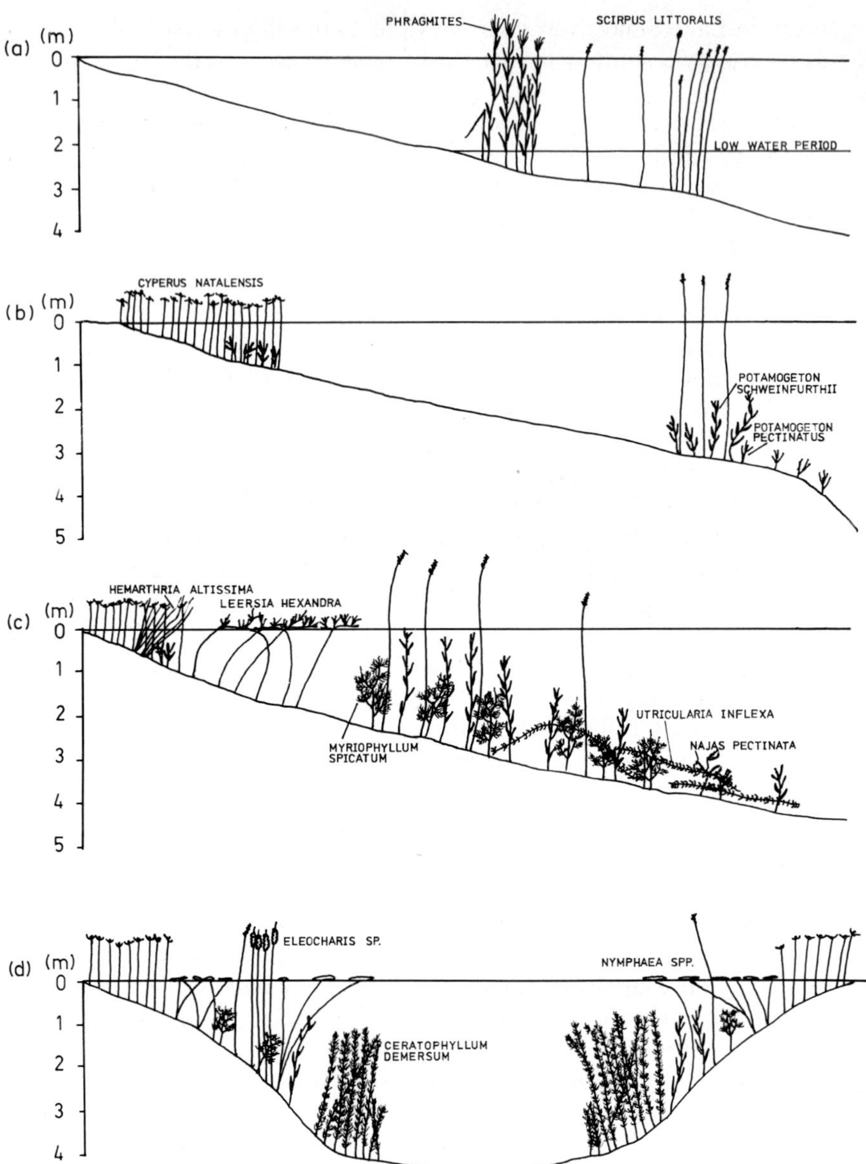

Fig. 1. Diagrams showing the distribution of aquatic macrophytes on various shorelines in Lake Sibaya. a) Distribution of emergent macrophytes on exposed shores in the low water period. b) The same shore during the high water period. c) Gently sloping shores in the bays of the southwest and southeast basins–high water period. d) Sheltered side arms of the northern and western arms–low and high water periods.

Cyperus natalensis fringe on the shore has some *Potamogeton* spp. associated with it in shallow water although it was generally associated with the grasses *Hemarthria altissima & Ischaemum arcuatum.* A wave washed zone follows down to about 2 m and then a narrow band of *Potamogeton schweinfurthii* with some associated *Myriophyllum* and *P. pectinatus* occurs. *Juncus krausii* and *Panicum meyerianum* were found on wave washed terraces in very shallow water. On gently sloping shores (Fig. 1c) in the bays of the southwest and southeast basins there is a continuous and very dense growth of submerged plants from the *Cyperus* zone down to 5 m. *Leersia hexandra* floats in the shallow water and very dense *P. schweinfurthii* and *Myriophyllum spicatum* beds occur deeper. Long strands of *Utricularia inflexa* occur on the bottom as well. In the deepest waters *Najas pectinata* occurs in clumps. In the sheltered side arms of the North and West arms of the lake a typical zonation as shown in Fig. 1d occurs. Very dense beds of *Ceratophyllum demersum* occur in the deepest water from 3 to 5 m. Further up the slope *Potamogeton schweinfurthii* and *Myriophyllum* occur merging into dense water lily beds (*Nymphaea capensis* and *N. caerulea*). Emergents here are *Eleocharis* sp. and *Scirpus littoralis.* Once more the waters edge is characterised by *Cyperus natalensis.*

The Abundance of Each Species

The abundance of the aquatic macrophytes of Lake Sibaya was measured in terms of biomass. Biomass sampling was carried out at stands located all around the lake as shown in Fig. 2.

Selection of stands was done by choosing sampling areas before-hand on a 1 : 50,000 map so as to give a fairly regular spread around the lake. Within each sampling area the sampling stand was located by throwing the anchor overboard and sampling where the boat came to rest. Sampling was done by placing twenty 25 × 25 cm quadrats parallel to the shoreline at 1 m intervals and hand cutting the above ground material. In deepwater this process was carried out using SCUBA or Hookah diving gear, or where there was a very real danger from crocodiles or *Hippopotamus*, submerged biomass was obtained using the rotary sampler of Howard-Williams and Longman (1976). This was recalibrated on Lake Sibaya by hand cutting parallel samples with the aid of SCUBA, and the hooks on the sampler moved until no significant difference ($P = 0.05$ $n = 20$) was found between hand cut samples and those obtained with the rotary sampler. Generally the standard error of total biomass at the sampling stands varied between 5 and 20% of the mean. After cutting, the material was dried for five minutes on a wire mesh frame and wet weighed. Subsamples were taken for dry-wet conversions. These were done by briefly soaking the samples in dilute (5%) HCl to remove marl encrustations and then oven drying the material at 100°C.

Table 1 shows the mean biomass values for the 22 most common species encountered in the sampling stands in different areas of the lake. The species are grouped into submerged, floating leaved, emergent and flooded grassland species.

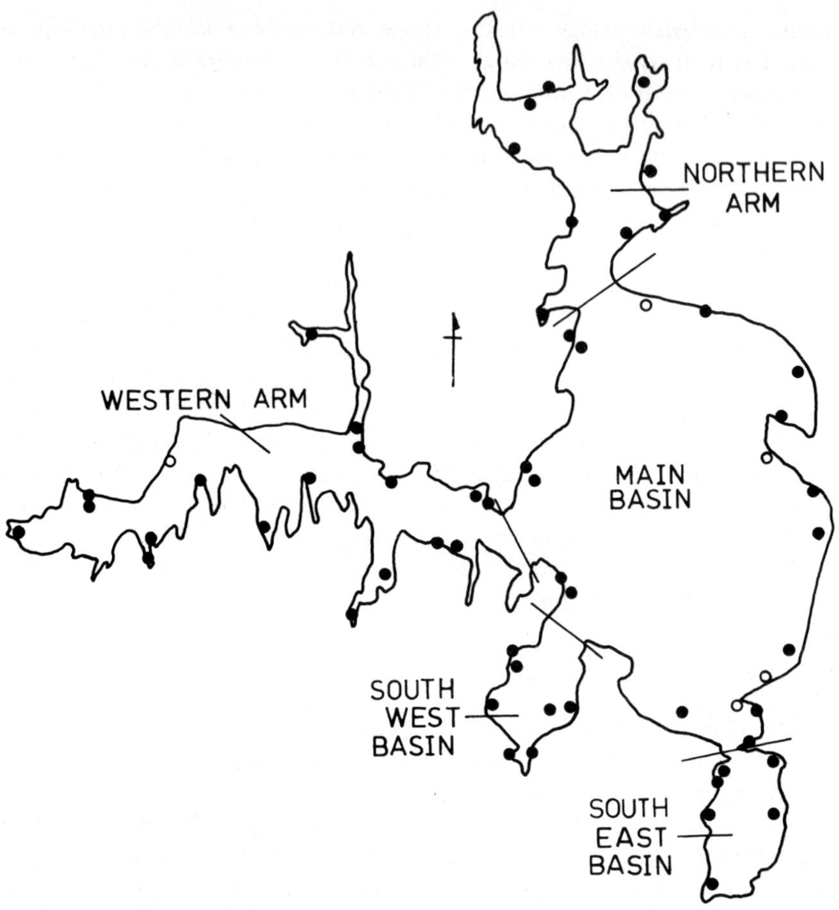

Fig. 2. The distribution of sampling stands around Lake Sibaya. Open circles indicate positions where no macrophytes were found.

Of the submerged species, the highest biomass was shown by *Ceratophyllum*. This species occurred most frequently in the sheltered bays of the western and northern arms of the lake. In the main basin *Ceratophyllum* occurred infrequently and with a very low biomass. In contrast, the highest mean biomass of *Myriophyllum* was found in the main basin, although it is not particularly dense even here (48.3 g m^{-2}). The mean biomass of the most common macrophyte species in Lake Sibaya, *Potamogeton schweinfurthii* varied from 22 g m^{-2} in the main basin to 59 g m^{-2} in the southeast basin. Again these values are fairly low. The values for the other submerged species are extremely low. For instance *Potamogeton pectinatus* in Swartvlei, another coastal lake, reaches values of 2000 g m^{-2} dry mass (Howard-Williams, 1978), whereas in Lake Sibaya the values are around 5 g m^{-2}. In terms of biomass (and ultimately production) the only submerged macrophyte species of Lake Sibaya which contribute significantly to the littoral zone are *Ceratophyllum*, *Myriophyllum*, and *Potamogeton schweinfurthii*.

Table 1. Mean biomass values ($g\,m^{-2}$) ± Standard error in stands where the 22 most common species were recorded. Values are presented for different areas of the lake. Species occurring in less than 3 stands or with a mean biomass in all the stands of less than 5 $g\,m^{-2}$ have been omitted to simplify the data. Location of stands in the different areas is shown in Fig. 2.

Type	Species	Main Basin $g\,m^{-2}$ ± S.E.		South East Basin $g\,m^{-2}$ ± S.E.		South West Basin $g\,m^{-2}$ ± S.E.		Northern Arm $g\,m^{-2}$ ± S.E.		Western Arm $g\,m^{-2}$ ± S.E.	
SUBMERGED	Ceratophyllum demersum	6	2.50	228	220.0	8	—	196	190.0	829	449.0
	Myriophyllum spicatum	48	29.6	47	20.5	30	17.7	13	18.9	11	4.6
	Najas pectinata	3	—	—	—	10	6.5	—	—	20	—
	Potamogeton pectinatus	5	1.5	2	1.0	6	2.5	3	—	6	2.8
	Potamogeton schweinfurthii	22	12.4	59	25.3	44	23.6	22	11.1	26	10.0
	Utricularia inflexa	1	—	11	3.3	12	2.3	16	7.0	8	3.2
FLOATING LEAVED	Nymphaea caerulea	—	—	1	—	2	1.0	9	7.5	3	0.5
	Nymphaea capensis	—	—	128	—	9	7.0	21	5.3	50	24.4
EMERGENT AQUATICS	Cyperus natalensis	193	54.5	144	—	197	124.4	252	71.1	217	30.1
	Eleocharis sp.	—	—	3	—	16	—	10	6.5	9	7.5
	Fuirena sp.	—	—	—	—	2	1.0	43	21.1	25	27.4
	Juncus krausii	102	52.2	—	—	1	—	22	13.9	22	13.9
	Leersia hexandra	1	—	16	—	11	10.3	151	85.4	15	8.0
	Phragmites mauritianus	28	20.0	32	—	21	9.3	—	—	22	12.0
	Pycreus polystachyus	—	—	—	—	—	—	14	2.5	3	1.0
	Scirpus littoralis	9	5.9	21	7.0	23	8.7	5	1.5	88	72.2
FLOODPLAIN GRASSES	Eragrostis chapelieri	—	—	—	—	—	—	16	9.0	—	—
	Hemarthria altissima	—	—	—	—	—	—	3	2.0	16	5.7
	Imperata cylindrica	—	—	—	—	3	—	4	1.0	3	2.0
	Ischaemum arcuatum	258	25.4	—	—	25	10.7	112	1.0	121	45.8
	Panicum meyerianum	96	16.0	—	—	—	—	91	85.8	—	—
	Panicum repens	—	—	—	—	—	—	528	—	38	22.5

Plate 1a. A section of the marginal *Scirpus littoralis* zone which has recently become inundated by rising water levels. *Phragmites mauritianus* dies out in these conditions. A few dying specimens can be seen in the left and centre foreground.

Plate 1b. A sandy beach below the dune forest showing the dark line of organic debris of terrestrial origin along the wave wash zone. This allochthonous matter is thought to contribute significantly to organic matter turnover in Lake Sibaya.

The floating leaved species were never found in the main basin, and biomass values in the other areas remained fairly low. In the Western arm a mean value of 50 g m^{-2} for *Nymphaea capensis* was recorded.

Of the emergent species, by far the most abundant was *Cyperus natalensis*. Biomass values for this species, were remarkably consistent, irrespective of area in the lake. Values varied from 144 g m^{-2} dry mass in the southeast basin to 252 g m^{-2} in the northern arm. Although *Scirpus littoralis* is a much bigger plant, it was very thinly distributed with densities of less than 5 shoots m^{-2}. Hence biomass values for this species were always less than 100 g m^{-2} (as compared with 1675 g m^{-2} in other South African coastal lakes, Howard-Williams unpublished). During the low water period on Sibaya the density (and hence biomass) of *Scirpus littoralis* was probably much greater than at high water.

Phragmites, once the dominant emergent in the lake was only found in a few areas during the high water period, and these stands consisted usually of very thin almost leafless shoots (Plate 1a) growing in 2 m of water, hence the low biomass values in Table 1. *Juncus krausii*, especially common on exposed sandy shores of the main basin, had biomass values of over 100 g m^{-2} where it occurred.

Ischaemum arcuatum was the most abundant of the flooded grassland species. It was a common constituent of the *Cyperus natalensis* zone and was especially prevalent in the main basin where the mean biomass value was 258 g m^{-2}. *Panicum repens* was very dense in parts of the Northern Arm. *Imperata cylindrica* which was a very prominent feature of parts of the shoreline at low water was scarcely recorded during the high water study.

To sum up, therefore the most important species in terms of biomass in Lake Sibaya during the high water period were *Ceratophyllum*, *Myriophyllum*, *Potamogeton schweinfurthii*, *Leersia hexandra*, *Cyperus natalensis*, *Ischaemum arcuatum* and *Panicum repens*.

The Control of Abundance

It is generally agreed that the primary factor controlling not only macrophyte species distribution but also abundance in a given water body is water depth (see references in Sculthorpe, 1967) with its associated changes in light regime. Exposure to wind and wave action has also been shown to be very important (Sculthorpe, 1967), as have changes in substrate texture (Pearsall 1920, Curtis 1959, Spence 1964). Seddon (1972) found that in a given water body physical factors (depth, light, exposure *etc.*) were the most important in influencing macrophyte growth, but that chemical factors accounted for differences in macrophyte growth between different water bodies. Howard-Williams and Walker (1974) found that in Lake Chilwa changes in water chemistry particularly conductivity and pH were important in defining the types of macrophyte vegetation. In Lake Sibaya, the substratum texture throughout the lake basin is a uniform sand, so it is unlikely that soil texture would influence macrophyte growth. The following environmental factors were therefore measured at each sampling stand to see which accounted for most of the variance in species biomass; water depth, exposure to wind and wave action, (calculated as in Chapter 7), soil organic matter, soil pH, water pH, soil conductivity, water conductivity. The methods of analysis are given in Appendix 2.

There was a very large number of zero values in the original species – stand matrix. The presence of zero values in phytosociology has been recognised as a problem for some time (Lambert and Dale 1964) because to allocate zero values to a species in a number of clearly different environmental areas implies that the species has an equal probability of occurring in all those areas which is obviously not true. However, one cannot just ignore the stands where the species was not recorded. In other words, the zero values need to be adjusted so as to present a degree of absence for the species in each stand where a zero value was given. Swan (1970) proposed a method whereby the degree of absence of a species from a stand could be calculated from available data in the matrix. In the following analysis of species – environment interactions the biomass values were adjusted to account for zero values by Swans' (1970) method. In this section of the essay I have therefore used the term 'relative biomass'.

In order to evaluate how well relative biomass could be related to the measured environmental variables, correlation coefficients between relative biomass and these variables were computed, and a stepwise multiple regression analysis (see *e.g.* Lee 1971) between relative biomass and all the environmental variables was performed on each species in turn.

In this analysis the regression between relative biomass and one of the environmental factors is obtained, then, each environmental factor is added in turn to the regression. After the addition of each factor, the standard error of the regression line is examined, and variables which result in an increase in the standard error can be considered as unimportant and are excluded. The relative importance of the remaining variables can be expressed by calculating what percentage of the total variation in the relative biomass can be accounted for by each of these variables. It must be stressed, however, that as with all regression techniques, the changes in the dependent variable (in this case relative biomass) are not necessarily *caused* by variations in the independent variables, even though they may be related. Thus no cause-effect relationship is proved by a multiple regression analysis. This must be borne in mind when interpreting the analysis.

Correlation coefficients between relative biomass and environmental variables are shown in Table 2. *Potamogeton pectinatus* and *Leersia hexandra* are the only species showing no correlation with water depth. *Leersia* is known to be tolerant to a wide range of water depths (Junk 1970, Howard-Williams and Walker 1974). However, the lack of correlation between biomass of *P. pectinatus* and water depth is puzzling as in other coastal lakes the biomass of this species changes considerably with water depth (Howard-Williams and Allanson 1978). The biomass values of *P. pectinatus* in Lake Sibaya are extremely low anyway throughout the depth range and it does not appear tolerant of a wide range of depths. The remaining species in Lake Sibaya can be divided into two groups according to their responses to

Table 2. Correlation coefficients between the 22 most common species in Lake Sibaya and the seven measured environmental variables. Only values significant at the P = 0.05 level of probability are shown. (P = 0.01, $r = 0.342$; P = 0.001, $r = 0.429$) — = Not significant $N = 55$.

	Species	Water Depth	Exposure	Water pH	Water Conductivity	Soil pH	Soil Conductivity	Organic Matter
Submerged	Ceratophyllum demersum	0.410	—	—	—	−0.333	0.470	0.410
	Myriophyllum spicatum	0.591	−0.266	0.313	0.269	—	0.474	—
	Najas pectinata	0.679	—	0.386	0.386	—	0.370	—
	Potamogeton pectinatus	—	—	0.379	0.267	0.569	—	—
	Potamogeton schweinfurthii	0.265	—	0.340	0.295	—	—	—
	Utricularia inflexa	0.574	−0.357	0.271	—	—	0.299	—
Floating Leaved	Nymphaea caerulea	0.643	—	—	0.291	—	0.460	0.420
	Nymphaea capensis	0.407	−0.474	—	—	−0.466	0.599	0.499
Emergent Aquatics	Cyperus natalensis	−0.771	—	−0.272	—	—	−0.346	—
	Eleocharis sp.	0.730	−0.343	—	—	—	0.474	0.452
	Fuirena sp.	−0.724	—	−0.412	—	—	−0.347	—
	Juncus krausii	−0.663	0.332	−0.346	−0.310	—	−0.293	—
	Leersia hexandra	—	—	—	—	—	—	—
	Phragmites mauritianus	0.324	—	0.331	0.271	0.280	—	—
	Pycreus polystachyus	−0.836	0.333	−0.310	—	—	−0.453	−0.342
	Scirpus littoralis	0.697	—	0.305	—	—	0.283	—
Floodplain Grasses	Eragrostis chapelieri	−0.550	—	—	—	—	−0.343	—
	Hemarthria altissima	−0.704	—	−0.473	−0.355	—	−0.354	−0.278
	Imperata cylindrica	−0.786	—	−0.489	−0.417	−0.301	−0.356	−0.268
	Ischaemum arcuatum	−0.487	—	−0.385	—	—	—	—
	Panicum meyerianum	−0.514	0.317	−0.461	−0.582	—	—	—
	Panicum repens	−0.426	—	−0.407	—	—	—	—

water depth. All the submerged and floating leaved species (Table 2) increase in relative biomass with water depth to the maximum water depth measured which was 400 cm, and in almost all the remaining species, biomass decreases with depth.

Table 2 also shows that species least able to tolerate exposed conditions are *Nymphaea, Utricularia, Eleocharis* and *Myriophyllum*. The biomass of *Juncus, Pycreus polystachyus* and *Panicum meyerianum* increased in exposed shores. *Najas, Myriophyllum* and *Potamogeton* spp. increased with increasing water pH, whilst most of the grasses and sedges showed higher values in lower pH conditions. As there was no relationship between water depth and pH ($r = 0.058$) the apparent influence of pH on these species could be real. Howard-Williams and Walker (1974) showed that pH was an important factor in explaining the variation in the vegetation of Lake Chilwa.

There was very little variation in water conductivity around Lake Sibaya (coefficient variation 12%, $n = 55$) but soil conductivity was generally lower in the newly flooded areas than in deeper water. Hence a number of flooded grassland species were negatively correlated with soil conductivity, whilst positive correlations were observed with submerged species. These results are consistent with the findings of Musil *et al.* (1973) who found that in the nearby Pongola pans submerged species favoured more alkaline conditions than rooted emergent macrophyte species.

There were few correlations between relative biomass and soil organic matter although *Ceratophyllum, Nymphaea,* and *Eleocharis* appeared to grow best in organic soils, whilst the relative biomass of *Imperata* and *Hemarthria,* in contrast, increased as soil organic matter decreased.

Fig. 3 shows what proportion of the variation in relative biomass of each species can be accounted for by the measured environmental factors. The first point to note from Fig. 3 is the relatively small proportion of the variance in biomass which is accounted for at all. In some species, such as *Leersia hexandra, Potamogeton schweinfurthii* and *Phragmites mauritianus,* less than 30% of the variation in relative biomass can be explained by the measured environmental variables. Howard-Williams and Walker (1974) found that only 47% of the variance in the vegetation of a shallow tropical lake could plausibly be accounted for by the measured environmental factors and suggested that a recent dry period in the lake as well as chance variation accounted for much of the remaining variance. Chance variation as discussed in Howard-Williams and Walker (1974) is always important in wetland vegetation, but clearly in Lake Sibaya the recent rapid rise in water level, and the present shifting of the vegetation into new habitats must be responsible for much of the unexplained variation in biomass. It is, however, obvious from Fig. 3 that water depth accounts for most of the explained variation in biomass in most species. However, with respect to *Potamogeton,* most of the explainable variation in the performance of this genus in Lake Sibaya can be accounted for by pH differences, whilst soil and/or water conductivity can be used to most accurately predict the performance of *Ceratophyllum, Nymphaea capensis* and *Panicum meyerianum.*

In general, the environmental factors which accounted for most of the

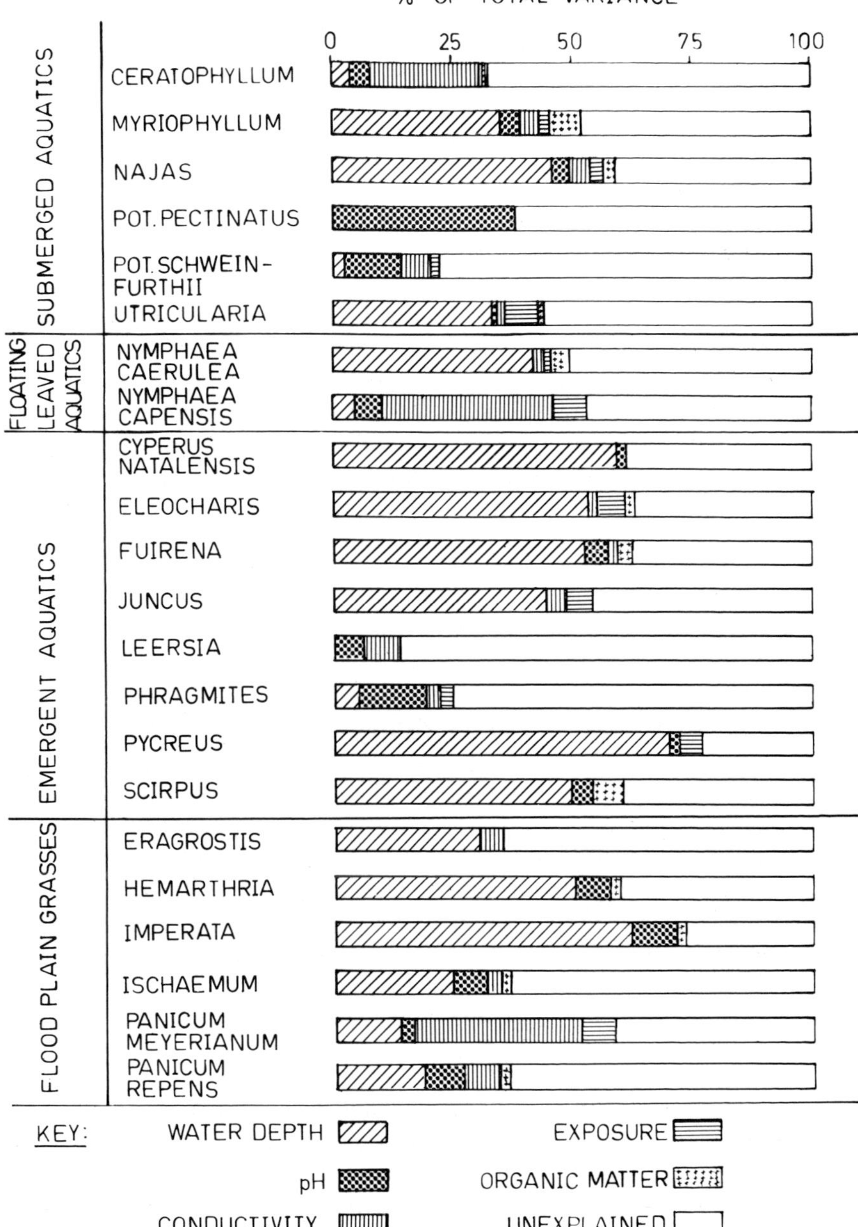

Fig. 3. Percent of the total variance in biomass of the common macrophytes of Lake Sibaya which was accounted for by various environmental parameters.

explainable variance in macrophyte biomass at high water levels in Lake Sibaya were, in decreasing order of importance: water depth, conductivity, pH, exposure, soil organic matter. This was at first somewhat surprising as Boltt, Hill and Forbes (1969) and Allanson *et al.* (1974) emphasised the apparent importance of exposure to macrophyte growth, yet this analysis showed exposure to be relatively unimportant. Three facts may account for this. Firstly, the visually striking species in the sheltered areas, namely *Cyperus papyrus*, *Typha capensis* and *Polygonum tomentosum* were not sampled in my analysis. They occupy very small areas at the ends of bays only. Secondly although the presence of a species may be subjectively correlated with sheltered conditions, the biomass values may not always correspond with such conditions. For instance although *Ceratophyllum* reaches its maximum biomass in the very sheltered areas of the western arm, very low biomasses of this species also occur here. Thirdly, and probably most important, the measurements of exposure (based only on fetch) are complicated by the recent rise in water level, and the resulting dense band of *Cyperus natalensis* along the shores. This plant acts as a wave barrier so although a shore may have a high exposure rating, plants such as *Potamogeton* and *Myriophyllum* can grow in the shelter of the *Cyperus* zone. During low lake levels when exposed shores were characterised by sandy beaches this shelter was not available.

The Role of the Macrophytes in the Lake

The role of aquatic macrophytes in lakes has come under increasing attention in recent years, Wetzel and Hough (1973) and Howard-Williams and Lenton (1975) summarised some of the recent literature on this. Apart from providing a diverse habitat for aquatic animals, in which to shelter, breed and feed, macrophytes in the littoral zone can influence the chemical budget of the open lake and contribute (often a major portion) to the autotrophic production of a lake. Because of the peripheral position in lakes macrophytes also play a role in sieving and trapping allochthonous and autochthonous matter.

Let us consider first the influence of macrophytes in Lake Sibaya in the sieving and trapping of allochthonous and autochthonous matter. Because of the soft sandy nature of the catchment area there is very little runoff so there is a very small load of allochthonous particulate matter entering the lake which could be trapped anyway. However, heavy rain over the seaward dune forest will bring down locally concentrated inputs of allochthonous material which becomes deposited on the terrace floor. Plate 1b shows such debris on an exposed part of the shore during low lake level. Wave action during storms will wash this material into the lake. In the main basin and all exposed shores in other areas of the lake, the macrophyte zone is subjected to strong and continuous current action. Attempts to obtain diurnal oxygen production curves in dense beds of *Potamogeton schweinfurthii* in the main basin have proved futile because of these currents. During routine diving observations I have noticed the plants even down to 5 m being bent by water currents. Hence, there is very little, if any, accumulated allochthonous or autochthonous material in the macrophyte zone of the main basin, or in exposed shores in general. Schiemer and Prosser (1976) suggested that the importance of the macrophyte zone (in their case *Potamogeton pectinatus* and *Myriophyllum spicatum*) in the open lake of the Neusiedlersee in Austria was in the reduction of water turbulence and the consequent sedimentation of organic and inorganic material. In Lake Sibaya, this particular function of macrophytes is virtually non-existent.

The influence of macrophytes in the chemical budget of a lake can be of two types – macrophytes can alter the nutrient status of inflowing waters, and they can act as reservoirs of nutrients in the lake itself. The influence of macrophytes, especially emergent species on nutrients in flowing waters has been discussed by Gaudet (1976) and Howard-Williams and Howard-Williams (1978). Gaudet (1976) noted that *Cyperus papyrus* swamps played an important role in reducing dissolved nutrient loads in tropical rivers. It is significant that all the inlets to Lake Sibaya are blocked by *Cyperus papyrus*. At present, nutrient inputs from the catchment area are low and the papyrus is likely to reduce these even more before they enter Lake Sibaya. I would suggest, in the light of Gaudet's (1976) findings that even if dissolved nutrient (N & P compounds) levels in the inflowing streams were to increase due to increasing human activity in the catchment, the papyrus 'sieves' in Lake Sibaya would be able to maintain the low nutrient levels of the lake for some time, (assuming of course that these nutrients do not enter in the ground water).

The influence of macrophytes as reservoirs of nutrients in lake basins depends on the maximum biomass of macrophytes, and their distribution in a lake. For instance, in Lake Warniak (Poland) Bervatowicz (1969) suggests that macrophytes play a major role as nutrient reservoirs. Howard-Williams and Junk (1977) in discussing the role of macrophytes in tropical South American floodplain lakes, showed that there was 8 times more potassium, 4 times more phosphorus and 3 times more nitrogen in an area covered by macrophytes than in the same area of water. Howard-Williams (1977) in another South African coastal lake showed that the reserve (as $g\,m^{-2}$) of both N and P in the macrophytes was greater than that in the water. In all these studies, macrophyte biomass values in excess of 1 kg and often $2\,kg\,m^{-2}$ (dry mass) were encountered, and all these lakes were shallow. However, in the shallow Neuseidlersee Schiemer and Prosser (1976) showed that because of the low macrophyte biomass, the N and P reservoirs in the macrophytes were negligible. In Lake Sibaya the low biomass values of macrophytes in general (Table 1) and the large volume of water in the lake means also that the nutrient reservoir provided by macrophytes here can be considered negligible.

Of the major fish species in Lake Sibaya all utilized shallow areas for breeding purposes (Chapter 8). Bruton (1977) found that *Clarias* never spawned in submerged macrophytes, but preferred the marginal emergent vegetation for this purpose. *Sarotherodon mossambicus* nests were usually

associated with submerged macrophytes but were never seen in dense macrophyte beds. They were also found in shallow areas without macrophytes. However, Boltt, Hill and Forbes (1969) and Bruton (1977) suggest that this species prefers open sandy areas. *Tilapia sparrmanii* and *Pseudocrenilabrus philander* show a preference for well vegetated habitats but the latter species probably occur in deep water also (Boltt, Hill and Forbes 1969). In contrast to Lake Chilwa in Malawi where Bourn (1972) found that macrophyte remains made up a major proportion of the diet of the three common fish species, in Lake Sibaya Bruton (1977) found that vegetation and detritus was unimportant in *Clarias* diet, and Bowen's 1976 work showed that the major cichlid species of Lake Sibaya, *Sarotherodon mossambicus* fed almost exclusively off benthic diatoms, and not macrophytes. The work of Hart shows how the most abundant benthic invertebrate of the lake has its densest populations amongst emergent flooded grassland and *Cyperus* zone, but the population density is apparently influenced mostly by periphyton abundance rather than macrophytes. The importance of macrophytes to the major consumer organisms of Lake Sibaya may be only physical in that they act as a substratum for periphyton which is utilised by benthic invertebrates and fish, and as a shelter for young fish. Their biological role in this respect does not appear to be very significant in the littoral so it is worth considering the importance of macrophytes as autotrophs in the whole lake, and comparing the contribution by macrophytes to that of the algal components.

As no measurements of macrophyte production in Lake Sibaya have been made I have provided an estimate of production based on the biomass data in Table 1. From Table 1, and the discussion earlier in this chapter I concluded that on the basis of biomass values the following species were likely to be the major macrophyte producers in Lake Sibaya: *Ceratophyllum demersum, Myriophyllum spicatum, Potamogeton schweinfurthii, Nymphaea capensis,* at all times, *Phragmites mauritianus* and *Scirpus littoralis* at low water periods and *Cyperus natalensis* and *Ischaemum arcuatum* at high water periods.

In most lakes where production measurements of *Potamogeton* species have been made, the production to biomass (P/B) ratio is seldom much greater than 1 (Jupp *et al.* 1974, Bervatowicz 1969, Howard-Williams 1978), whilst for *Nuphar,* a floating leaved plant, there are 5 above ground biomass turnovers per year (Blanton 1976). Wetzel and Van Thuy (1971) suggested a P/B ratio of 3 for submerged macrophytes in a marl lake. Of the emergents, *Phragmites* in Europe generally has a P/B ratio of close to 1.0 (Kvet 1971) but in the tropics Thompson (1975) found that *Cyperus papyrus* has a P/B ratio greater than 2.5. Westlake (1975) in his summary of production of freshwater macrophytes found that in general tropical emergent and submerged vascular plants had a P/B ratio of 3. I feel that it is reasonable to assume that in Lake Sibaya macrophyte production will lie between 1 and 3 times the mean summer biomass values (Table 1). Production values for various macrophytes in Lake Sibaya can therefore be calculated using this conversion. Total biomass values for various regions of the lake have been

Table 3. Mean total macrophyte biomass values in different areas of Lake Sibaya, areas occupied by macrophytes in 1977, and total biomass of macrophytes for each area. N = Number of stands. Biomass = above ground dry mass

		Emergents			Submerged		
	N	Mean Biomass $g\,m^{-2}$	Area km^2	Total Biomass kg	Mean Biomass $g\,m^{-2}$	Area km^2	Total Biomass kg
Main Basin	15	369	2.76	1018×10^3	72	5.00	360×10^3
Northern Arm	9	659	0.97	639×10^3	275	3.12	858×10^3
South East Basin	6	156	0.30	47×10^3	271	0.89	241×10^3
South West Basin	7	411	0.27	110×10^3	163	0.92	150×10^3
Main Channel	(12)				117	3.00	351×10^3
Western Arm:	18	405	1.77	717×10^3			
Side Arms	(6)				1473	3.26	4801×10^3
TOTAL			6.06	2531×10^3		16.21	6761×10^3

estimated by calculating the areas covered by aquatic macrophytes in Lake Sibaya in 1977, and multiplying this by the mean macrophyte biomass encountered in the stands in each area. The results are shown in Table 3. The main basin has a large emergent macrophyte stock, due mostly to the recently flooded flat areas to the north and south, but due to the very low mean biomass of submerged macrophytes here the total contribution by submerged macrophytes to the main basin is relatively small. The high total biomass value in the side arms of the Western Arm of Lake Sibaya shows clearly that these sheltered areas produce more than half of the macrophyte biomass of the whole lake. Taking into account the P/B ratio of 3 discussed earlier, the maximum contribution by macrophytes to the autochthonous organic matter inputs to Lake Sibaya would amount to a total of 27,876 tons dry matter yr^{-1} for the whole lake of which 4134 tons (15%) comes from the main basin. Allanson and Hart (1975) give an annual mean photosynthetic integral for phytoplankton in Lake Sibaya of 11997 $KJ\,m^{-2}$ which is equivalent to 674 $g\,m^{-2}$. Phytoplankton production in the main basin alone of Lake Sibaya would therefore amount to something in the region of 248,000 tons (dry matter) per year – about five times more than macrophyte production. If we add to this the algal epipsammic production (Chapter 5) and periphyton production, this figure would be increased considerably. In spite of localised high macrophyte biomass values, the data shows that in general algae, and even just the phytoplankton component are a much more important source of organic matter to Lake Sibaya than aquatic macrophytes.

Conclusions

The aquatic communities of Lake Sibaya have changed considerably in recent years due to the recent rise in water level. Of particular importance is the newly established *Cyperus natalensis* emergent zone around the lake. Emergents now make up 25% of the total macrophyte biomass of Lake Sibaya. The most important individual species in terms of biomass are *Ceratophyllum demersum, Myriophyllum spicatum, Potamogeton schweinfurthii, Cyperus natalensis, Leersia hexandra, Ischaemum arcuatum* and *Panicum repens.* The latter two species are floodplain grasses. The main centres of macrophyte production are the Northern and Western Arms of the lake. Macrophyte biomass in Lake Sibaya corresponds largely with water depth which is assumed to be the main factor controlling biomass. Exposure to wind and wave action is shown to be important in only a few species, but difficulties in measuring this factor may be to some extent responsible for this result.

Because of strong currents in Lake Sibaya the role of macrophytes in the reduction of water turbulence and the resulting trapping of allochthonous and autochthonous matter is negligible. Macrophytes can also be considered negligible as a nutrient reservoir in Lake Sibaya – but the small *Cyperus papyrus* swamps at the inflows may play a role in nutrient removal from surface inflows. Again, due to the appreciable groundwater flows (Chapter 3) this role may not be significant to the lake as a whole. Macrophytes or macrophyte detritus do not appear to form a food source for the most abundant fish species (*Clarias gariepinus* and *Sarotherodon mossambicus*), but they may be important as fish nesting areas and for shelter for young fish. Algae in Lake Sibaya are far more important than macrophytes as primary producers to the lake system.

If we consider all these points together I would suggest that in Lake Sibaya, Shelford's highly controversial statement on the role of macrophytes (Shelford 1918) is valid, and that if macrophytes in Lake Sibaya were replaced with inert structures of the same surface area the effect on the ecology of the lake as a whole would hardly be noticed.

Acknowledgements

I would like to thank Professor Allanson for considerable material and mental assistance. Robin Allanson helped tremendously throughout the long and sometimes tedious field sampling. I would like to thank also my colleagues in the Institute for Freshwater Studies, particularly Mike and Tony Bruton, Sandra Rudd and Rob Hart, for helpful discussions and field assistance. Peter Adamson wrote the multiple regression analysis programme. Nola Bruton typed the manuscript and I would like to thank Professor Brian Walker for criticism of it.

References

Allanson, B. R., M. N. Bruton and R. C. Hart. 1974. The plants and animals of Lake Sibaya, KwaZulu, South Africa: A checklist. Rev. Zool. afr. 88: 507–532.

Allanson, B. R. and R. C. Hart. 1975. The primary production of Lake Sibaya, KwaZulu, South Africa. Verh. Internat. Verein. Limnol. 19: 1426–1433.

Appleton, C. C. 1977. Freshwater mollusca of Tongaland, with a note on molluscan distribution in Lake Sibaya. Ann. Natal Mus. 23: 129–144.

Bernatowicz, S. 1969. Macrophytes in the Lake Warniak and their chemical composition. Ekol. Pol. Ser. A. 17: 447–467.

Blanton, L. R. 1976. Primary productivity and biomass distribution. In: Brinson, M. and Davis, J. Primary productivity and mineral cycling in aquatic macrophyte communities of the Chowan River, North Carolina. Report No. 120 of the Water Resources Institute, University of North Carolina.

Boltt, R. E., B. J. Hill and A. T. Forbes. 1969. The benthos of some southern African lakes. Part 1: Distribution of aquatic macrophytes and fish in Lake Sibayi. Trans. R. Soc. S. Afr. 38: 241–248.

Bourn, D. 1972. The feeding, diet and ecological relations of the three species of fish of economic importance in Southern Malawi. M.Sc. Thesis, University of Edinburgh.

Bowen, S. H. 1976. Feeding ecology of the cichlid fish *Sarotherodon mossambicus* in Lake Sibaya, KwaZulu. Ph.D. Thesis, Rhodes University, Grahamstown. South Africa.

Breen, C. M. and I. O. Jones. 1971. A preliminary list of Angiosperms collected in the vicinity of Lake Sibayi. Trans. R. Soc. S. Afr. 39: 237–245.

Bruton, M. N. 1977. The biology of *Clarias gariepinus* (Burchell, 1822) in Lake Sibaya, KwaZulu, with emphasis on its role as a predator. Ph.D. Thesis, Rhodes University, Grahamstown, South Africa.

Bruton, M. N. 1978. The habitats and habitat preferences of *Clarias gariepinus* (Pisces: Clariidae) in a clear coastal lake (Lake Sibaya; South Africa). J. Limnol. Soc. sth. Afr. 4: 81–88.

Curtis, J. J. 1959. The vegetation of Wisconsin. University of Wisconsin Press, Madison.

Gaudet, J. J. 1976. Nutritional relationships in the detritus of a tropical swamp. Arch. Hydrobiol. 78: 213–239.

Howard-Williams, C. 1975. Seasonal and spatial changes in the composition of the aquatic and semiaquatic vegetation of Lake Chilwa, Malawi. Vegetatio 30: 33–39.

Howard-Williams, C. 1977. The distribution of nutrients in Swartvlei, a Southern Cape Coastal Lake, Water S.A. 4: 213–217.

Howard-Williams, C. 1978. The growth and production of aquatic macrophytes in a south temperate saline lake. Verh. Internat. Verein. Limnol. 20: 1153–1158.

Howard-Williams, C. and W. Howard-Williams. 1978. Nutrient leaching from the swamp vegetation of Lake Chilwa, a shallow African lake. Aquat. Bot. 4: 257–267.

Howard-Williams, C. and W. J. Junk. 1977. The chemical composition of central Amazonian aquatic macrophytes with special reference to their role in the ecosystem. Arch. Hydrobiol. 79: 446–464.

Howard-Williams, C. and G. M. Lenton 1975. The role of the littoral zone in the functioning of a shallow tropical lake ecosystem. Freshwater Biol. 5: 445–459.

Howard-Williams, C. and T. G. Longman. 1976. A quantitative sampler for submerged aquatic macrophytes. J. Limnol. Soc. sth. Afr. 2: 31–33.

Howard-Williams, C. and B. H. Walker. 1974. The vegetation of a tropical African lake: classification and ordination of the vegetation of Lake Chilwa (Malawi). J. Ecol. 62: 831–854.

Junk, W. J. 1970. Investigations on the ecology and production biology of the floating meadows (Paspalo-Echinochloetum) on the middle Amazon. 1. The floating vegetation and its ecology. Amazoniana 2, 449–495.

Jupp, B. P., D. H. Spence and R. H. Britton. 1974. The distribution and production of submerged macrophytes in Loch Leven, Kincross. Proc. R. Soc. Edinb. (B) 74: 195–208.

Kvet, J. 1971. Growth analysis approach to the production ecology of reedswamp plant communities. Hidrobiologia, Bucur. 12: 15–40.

Lee, P. J. 1971. Multivariate analysis for the fisheries biology. Tech. Rep. Fish. Res. Bd Can. No. 244. Winnipeg, Canada.

Musil, C. F., J. O. Grunow and C. H. Bornman. 1973. Classification and ordination of aquatic

macrophytes in the Pongola River Pans, Natal. Bothalia 11: 181–190.
Pearsall, W. H. 1920. The aquatic vegetation of the English Lakes. J. Ecol. 8: 163–201.
Scheimer, F. and M. Prosser. 1976. Distribution and biomass of submerged macrophytes in Neusiedlersee. Aquat. Bot. 2: 289–307.
Sculthorpe, C. D. The biology of aquatic vascular plants. Edward Arnold, London.
Seddon, B. 1972. Aquatic macrophytes as limnological indicators. Freshwater Biol. 2: 107–130.
Shelford, V. E. 1918. Conditions of existence. In: Ward, H. B. and Whipple G. C. (Eds.) Freshwater Biology. John Wiley & Sons Inc. New York, pp. 21–60.
Spence, D. H. N. 1964. The macrophyte vegetation of freshwater lochs, swamps and associated fens. In: Burnett, J. H. (Ed.). The vegetation of Scotland, Oliver & Boyd, Edinburgh, pp. 306–425.
Swan, J. M. A. 1970. An examination of some ordination problems by use of simulated vegetational data. Ecology 51: 89–102.
Thompson, K. 1975. Swamp development in the headwaters of the White Nile. In: Rzoska J. (Ed.). The Nile, Biology of an ancient River. W. Junk, The Hague, pp. 177–196.
Westlake, D. F. 1975. Primary production of freshwater macrophytes. In: Cooper J. D. (Ed.). Photosynthesis and productivity in different environments. Cambridge University Press, Cambridge, pp. 189–206.
Wetzel, R. G. and R. A. Hough. 1973. Productivity and role of aquatic macrophytes in lakes. An assessment. Polskie Archwm Hydrobiol. 20: 9–19.
Wetzel, R. G. and N. Van Thuy. 1971. Distribution, production and role of aquatic macrophytes in a southern Michigan marl lake. Freshwater Biol. 1: 3–21.

Appendix 1

List of plant species mentioned in the text, with naming authorities.

Ceratophyllum demersum L.
Cyperus natalensis Hochst.
Cyperus papyrus L.
Dactyloctenium geminatum Hack.
Eleocharis sp.
Eragrostis chapelieri (Kunth.) Nees
Fuirena sp.
Hemarthria altissima Stapf. & C. E. Hubbard
Imperata cylindrica (L.) Beauv.
Ischaemum arcuatum (Nees) Stapf.
Juncus krausii Hochst.
Leersia hexandra Swartz
Myriophyllum spicatum L.
Najas pectinata (Parl.) Magnus
Nymphaea caerulea Savigny
Nymphaea capensis Thunb.
Panicum meyerianum Nees
Panicum repens L.
Phragmites mauritianus Kunth
Polygonum tomentosum Willd.
Potamogeton pectinatus L.

Potamogeton schweinfurthii A. Benn.
Pycreus polystachyus Beauv.
Scirpus littoralis Schrad.
Sporobolus virginicus (L.) Kunth
Typha latifolia L. susp. *capensis* Rohrb.
Utricularia inflexa var. *stellaris* P. Taylor

Appendix 2

Methods of analysis of environmental factors.

1) Water depth. As the depth was rising throughout the study, water depth was standardised to a particular date.

2) Exposure to wind and wave action. This was measured using the exposure rating described fully in Chapter 7.

3) pH. Water pH was measured with a glass electrode pH meter. Soil pH was determined on a 1:5 ratio of soil to distilled water. After shaking the mixture, the water was filtered off and pH was measured as above.

4) Conductivity. This was measured with an Electronic switchgear Model MC1 conductivity meter. Soil conductivity was measured on filtered water extracted from a 1:1 ratio of soil to distilled water as for pH.

5) Soil organic matter. Samples of known weight were ashed at 500°C for 3 hours and % loss on ignition was calculated. This was assumed to be equivalent to soil organic matter as the $CaCO_3$ content of the sands was very low.

7 The invertebrate communities: zooplankton, zoobenthos and littoral fauna

R. C. Hart

Introduction

In an aquatic ecosystem, the invertebrate fauna often represents the major group of secondary producers and provides an important, often essential pathway for energy flow into higher consumer levels. Notwithstanding this important ecological fact, there appears to have been a tendency in Africa to concentrate attention on the higher trophic levels which have direct economic importance, and invertebrate studies have generally lagged behind. Inspection of the index of Balon & Coche's (1974) account of Lake Kariba for citations to zooplankton, zoobenthos, or invertebrates illustrates this pre-occupation notwithstanding the existence of significant studies especially on benthic invertebrates in that system. Work on, for example, Lakes Kariba, Chad, George and Chilwa over the past decade shows that attention is being given to the invertebrate communities. Nevertheless, compared to the situation in north-temperate areas, the general biology and ecology of many, often major species of invertebrates of our African standing waters is poorly known. Thus our invertebrate studies on Lake Sibaya were initiated at a very basic descriptive level, additionally so in view of the occurrence of several estuarine species, the ecology of which was poorly known or unstudied even in their typical estuarine environments.

Basically, two themes may be identified in the research undertaken to date on the invertebrates of Lake Sibaya. Pioneering work by the late Dr R. E. Boltt (Boltt, 1969a,b; Boltt, *et al.*, 1969) on the benthos was concerned largely with the distributional ecology of dominant species. My own studies on zooplankton were also largely related to distributional ecology at the outset, but subsequently our interests shifted more to the productivity of the invertebrates, and this has been the major theme in later work which has progressed into the field of ecological energetics. Dictated by the available resources of manpower and funding, the studies undertaken to date have been in many cases autecological, with the underlying rationale of examining selected organisms in detail rather than attempting broader but necessarily more superficial community analyses.

What follows is an attempt to provide the reader with a general coverage of the known aspects of the biology of invertebrates in Lake Sibaya from published and unpublished sources. For reasons of convenience and clarity, the invertebrates are considered in three groups, – the zooplankton, the zoobenthos and the littoral fauna. A discussion of some attributes which appear to be of general relevance to the invertebrate fauna, together with a brief identification of major areas requiring further study concludes this section.

The Zooplankton

Species Composition

The limnetic zooplankton community of Lake Sibaya is fairly simple in terms of species diversity. The entomostracan zooplankton consists of one calanoid copepod (*Pseudodiaptomus hessei* (Mrázek)), four cyclopoid copepods (*Mesocyclops leuckarti aequatorialis* forma *micrura* (Kiefer), *Thermocyclops emini* (Mrázek), *Thermocyclops crassus consimilis* (Kiefer), *Tropocyclops brevis* Dussart) and two cladocerans (*Bosmina longirostis* (O. F. Muller) and *Moina* sp.). Several planktonic Rotifera are known and various Ostracoda and Amphipoda occur sporadically in the plankton. Apart from planktonic Hydracarina, the remaining macrozooplankton are mero-planktonic larval or juvenile stages. An indication of the structure of the macrozooplankton in summer and winter is given in Table 1, based on collections made with a coarse net (Allanson, Bruton & Hart, 1974).

The microzooplankton of the lake has been largely ignored, but as Table 2 shows, limited collections with finer nets have revealed the numerical importance of rotifers and of naupliar and copepodite stages of the often small cyclopoid copepods. While Table 3 shows that volumetrically, the single calanoid copepod clearly dominates the larger zooplankton, it can be

Table 1. Percentage frequency of occurrence of components of zooplankton community based on surface tows with a net of 158 μm mesh aperture, 1967–1968. After Allanson, Bruton & Hart (1974).

Taxa	% frequency	
	Summer	Winter
Rotifera (unsorted)	0.50	2.00
Ostracoda (unsorted)	0.25	0.00
Cladocera		
Moina sp.	2.00	4.50
Bosmina longirostris	0.80	3.50
Copepoda		
Cyclopoida (unsorted)	20.50	35.50
Calanoida		
Pseudodiaptomus hessei	69.50	52.00
Amphipoda (unsorted)	1.50	1.50
Brachyura		
Hymenosoma orbiculare (zoeae larvae)	3.00	0.50
Hydracarina (unsorted)	0.50	0.50
Insecta		
Ephemeroptera nymphs	0.25	0.00
Chironomidae larvae	0.20	0.00
Ceratopogonidae larvae	1.50	0.00
Pisces		
Gilchristella aestuarius (larvae)	0.50	0.00

Table 2. A comparison of the absolute and relative abundance of major components of the summer zooplankton community determined from vertical hauls with 'coarse' and 'fine' nets.

	Abundance (Numbers m^{-3})	
Taxa January 1977	158 μm mesh aperture net	86 μm mesh aperture net
Rotifera (unsorted)	15	3649
Cladocera (unsorted)	705	577
Calanoida (*P. hessei*)	700	383
Cyclopoida (unsorted)	533	3251
Copepod nauplii (unsorted)	127*	4172
Other taxa combined†	23	0

* exclusively calanoid nauplii
† includes zoeae larvae of *Hymenosoma orbiculare*, Hydracarina, Amphipoda and eggs and larvae of *Gilchristella aestuarius* (Pisces).

	Relative abundance	
February 1976	158 μm mesh aperture net	35 μm mesh aperture net
Rotifera (unsorted)	1	2146
Calanoida (*P. hessei*)	42	49
Cyclopoida (unsorted)	45	210
Copepod nauplii (unsorted)	22	806

Table 3. The relative abundance and biomass of major components of the zooplankton community based on summer collections made with a net of 158 μm mesh aperture.

Taxa	% occurrence	% biomass
Rotifera (unsorted)	0	0
Cladocera		
Moina sp	4.36	1.66
Bosmina longirostris	24.88	4.99
Copepoda		
Cyclopoida (unsorted)	24.39	8.02
Calanoida		
Pseudodiaptomus hessei	44.43	75.97
Ostracoda (unsorted)	0.32	}
Amphipoda (unsorted)	0.16	}
Hydracarina (unsorted)	0.16	}
Brachyura		} 9.36
Hymenosoma orbiculare zoeae	0.81	}
Pisces		}
Gilchristella aestuarius eggs	0.48	}

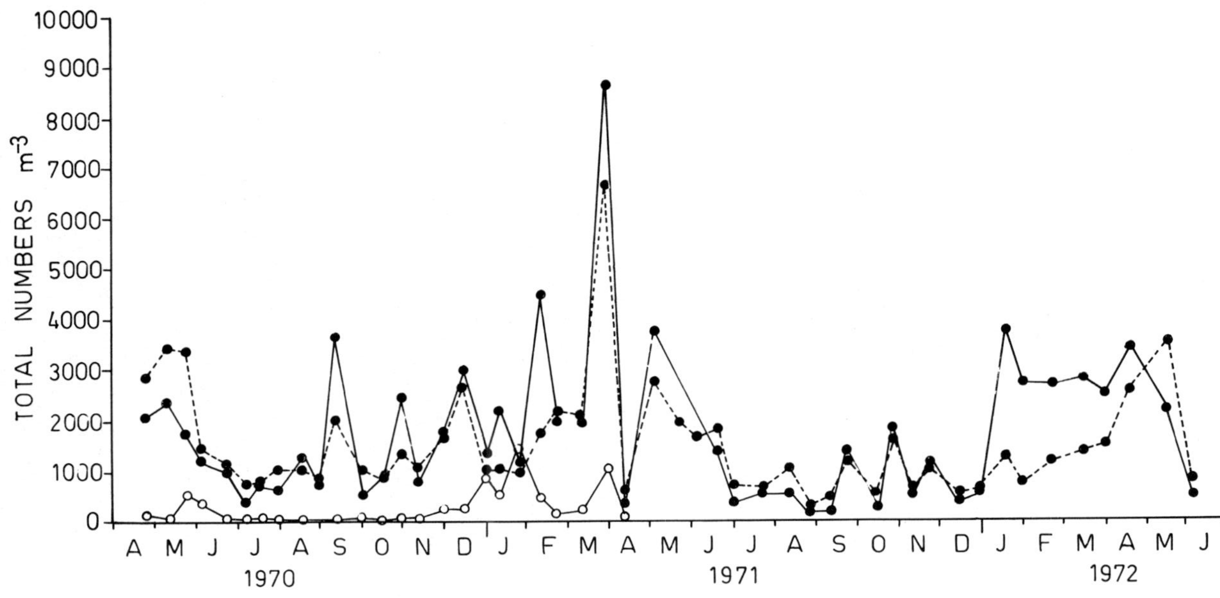

Fig. 1. Seasonal changes in abundance of cyclopoid copepods (●——●) *Pseudodiaptomus hessei* (●- - - -●) and cladoceran zooplankton (○——○) at a routine sampling site in water 30 m deep (Station III in Hart & Allanson (1975)). Based on vertical hauls with a net of 158 μm mesh aperture during 1970–1971.

expected that in view of their numerical abundance and especially in view of the higher turnover times which small forms generally exhibit, that these smaller zooplankters contribute significantly to the overall structure and functioning of the community.

Apart from the absence of larval insects in winter, seasonal variation in the taxonomic structure of the zooplankton community is not pronounced. As might be expected, however, the relative (Table 1) and absolute abundance (Figs. 1, 2b) of various components does change seasonally to some extent.

An interesting feature of the zooplankton community lies in the existence therein of several estuarine or marine forms. The dominant zooplankter,

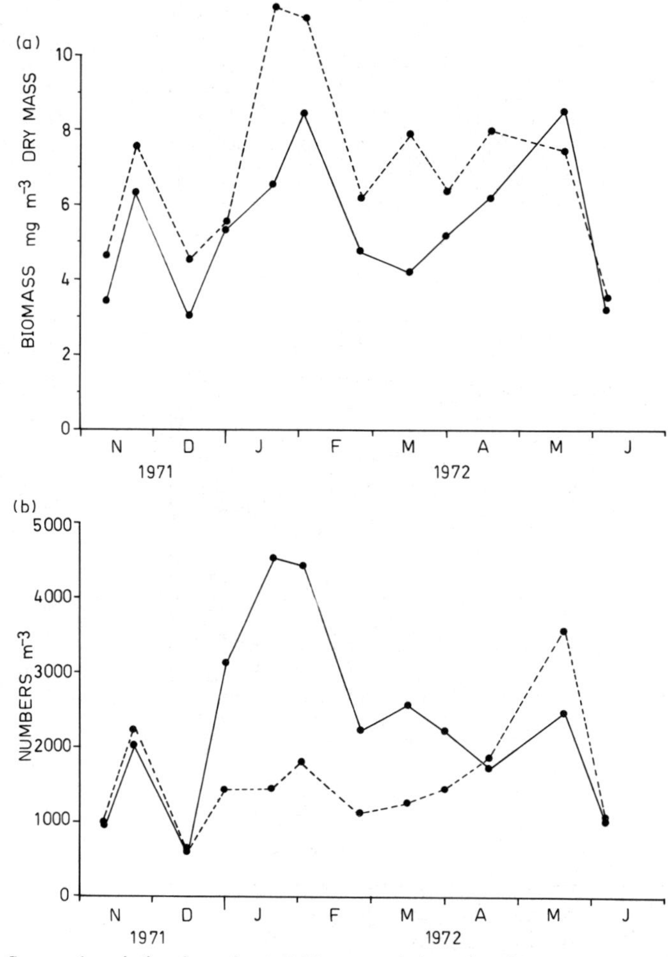

Fig. 2. (a) Seasonal variation in estimated biomass of the calanoid copepod *Pseudodiaptomus hessei* (●——●) and total zooplankton biomass (●- - - -●) weighed directly. Based on vertical hauls with a net of 158 μm mesh aperture. Values given are mean data for 5 routine sites (Stations I–V in Hart & Allanson (1975)) sampled during 1971–1972. (b) Corresponding changes in mean density of *Pseudodiaptomus hessei* (●- - - -●) and all cyclopoid copepods (●——●) at Stations I–V.

Pseudodiaptomus hessei which accounts for on average nearly 80% of the total biomass of zooplankton retained by 158 μm mesh aperture nets (Fig. 2a) is typically an estuarine or saline lagoon species. Other estuarine or marine organisms which contribute significantly to the zooplankton community include meroplanktonic zoeae larvae of the crown crab *Hymenosoma orbiculare*, and the eggs and larvae of the pelagic fish *Gilchristella aestuarius*. In addition the amphipods which occasionally venture into the plankton are estuarine forms.

Vertical Distribution and Diel Migration

The vertical distribution and migration of the calanoid *Pseudodiaptomus hessei* is the most thoroughly investigated aspect of zooplankton biology in Lake Sibaya. These and related aspects have been published elsewhere (Hart & Allanson, 1976; Hart, 1976, 1977, 1978) but are reconsidered briefly here. In addition, previously unpublished and less detailed information concerning the vertical distribution and migration of other components of the zooplankton is described below.

Interest in the vertical distribution and diel migration of zooplankton in Lake Sibaya arose largely from a consideration of the relatively homogeneous pelagic environment. The absence of thermal stratification and the resulting vertical uniformity of several major biotic and abiotic variables appeared to offer a unique opportunity to examine the perennial problem of vertical migration of zooplankton in a relatively simple environment.

Calanoida: The vertical distribution and diel migration of various life history stages of the copepod *P. hessei* has been studied intensively by Hart & Allanson (1976). In general its vertical migration (Fig. 3) conforms most closely to Hutchinson's (1967) 'nocturnal' pattern. The dusk ascent into surface and near-surface waters starts at low light intensities near sunset and continues until darkness. The timing and rates of ascent vary somewhat with stage of development – older stages migrating most rapidly and at lower light intensities than younger stages (Fig. 4). Vertical distribution is gradually depressed during the hours of darkness ('midnight sinking') (Fig. 3). Following a brief and inconspicuous dawn rise in some stages the population descends to its daydepth fairly soon after sunrise (Fig. 5).

The dusk ascent, and, less clearly, the dawn descent are related to rates of change of light intensity in nature but a component of endogenous rhythmicity has been revealed under laboratory conditions (Hart & Allanson, 1976). It appears likely that the vertical migration of *P. hessei* is cued endogenously while changes in light intensity merely provide the 'Zeitgeber' phasing the intrinsic rhythm.

One of the more remarkable aspects of the vertical distribution of *P. hessei* relates to the close association of this copepod with bottom substrates during daylight. The strength of this substrate association depends both upon stage of development, being most pronounced in older stages, and also upon depth of water (see Fig. 6). In shallow water (<10 m) even nauplii are to some extent benthic, while in deep water (40 m) only the adult

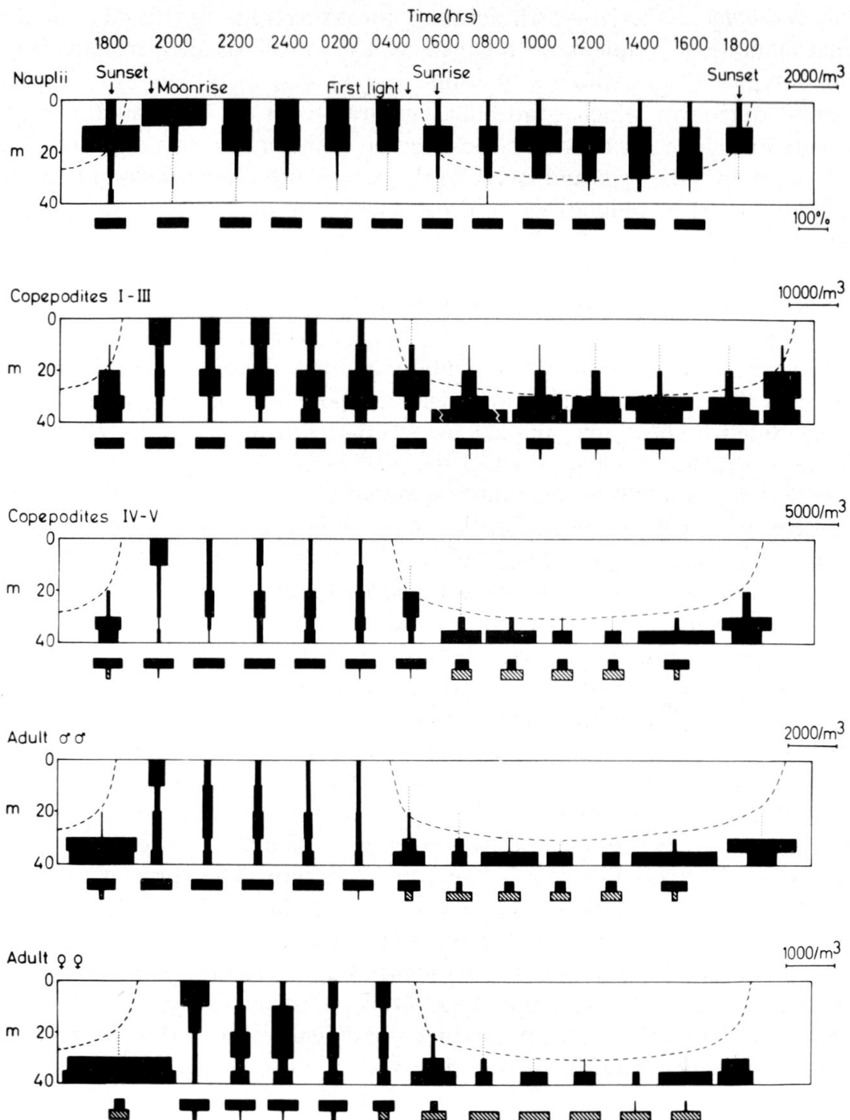

Fig. 3. Changes in vertical distribution of different life-history stages of the copepod *Pseudodiaptomus hessei* over 40 m of water, 1–2 March 1972. Scales of density per m^3 are shown for each panel. Dashed line shows depth of 1 lux isolume. Dotted sections of the depth frequency histogram represent catches too low for illustration on the appropriate scale. The depth-frequency histograms within each demarcated panel show the vertical distribution of pelagic animals. The histograms below each panel illustrate the relative percentages of the total population below a unit area of lake surface which were pelagic (solid black) or benthic (cross-hatched). Pelagic animals collected using 158 μm mesh aperture Nansen closing net. Benthic animals collected in substrate bins (Hart, 1976). After Hart & Allanson (1976).

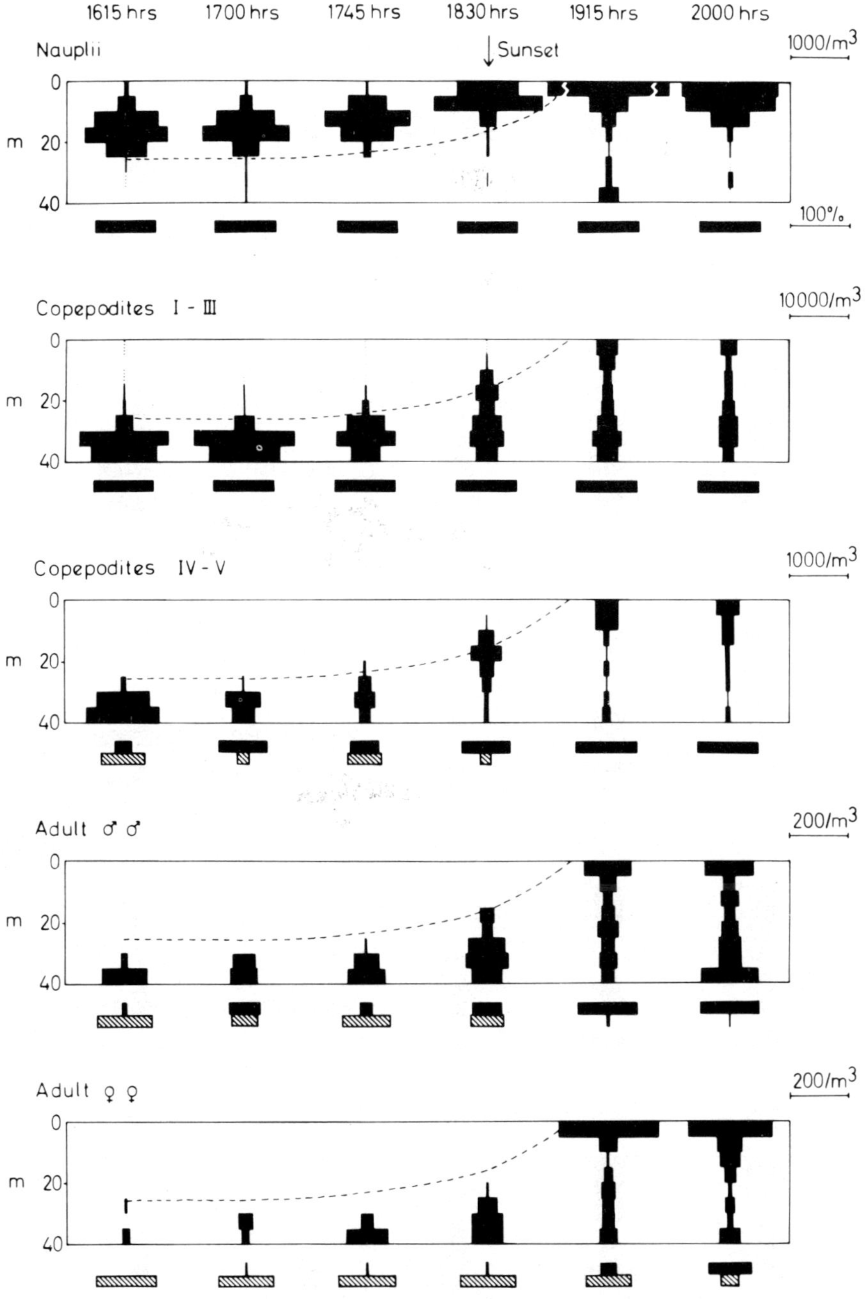

Fig. 4. Changes in vertical distribution of different life-history stages of *Pseudodiaptomus hessei* over 40 m of water at dusk, 15 February 1972. For further explanation see legend to Fig. 3. After Hart & Allanson (1976).

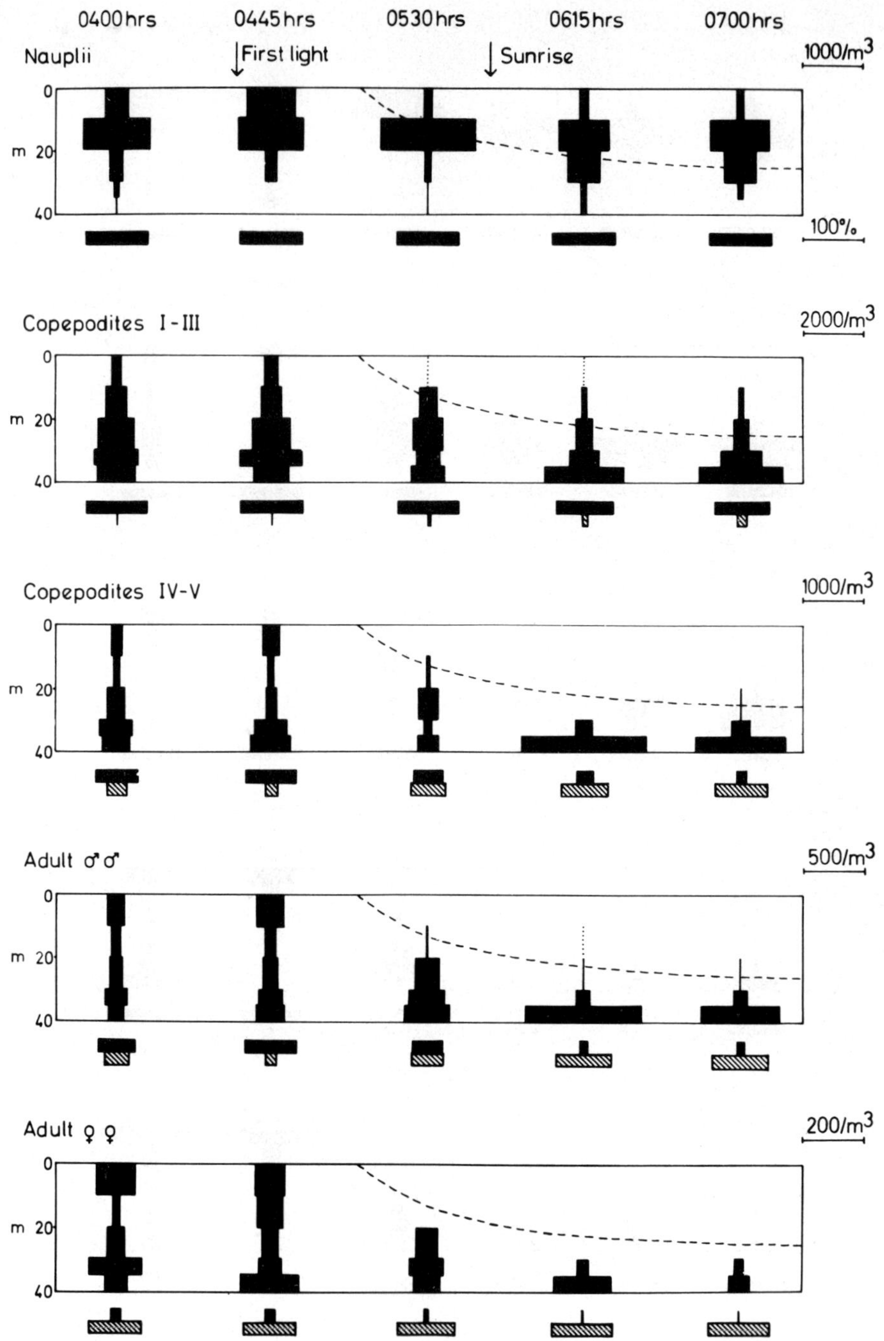

Fig. 5. Changes in vertical distribution of different life-history stages of *Pseudodiaptomus hessei* over 40 m of water at dawn, 19 February 1972. For further explanation see legend to Fig. 3. After Hart & Allanson (1976).

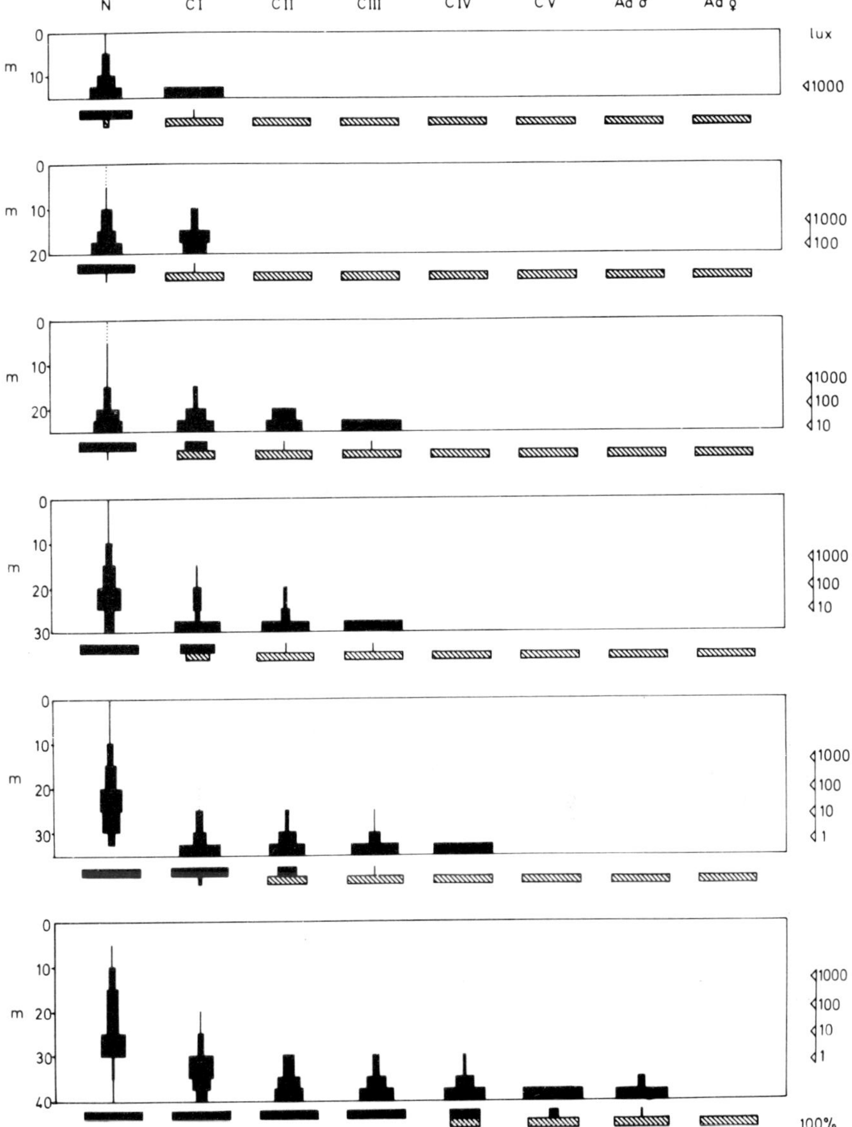

Fig. 6. Percentage day time vertical distribution of different life-history stages of *Pseudodiaptomus hessei* in water of various depth. Based on samples collected between 10.30 and 13.30 hours, 16 March 1972. Depths at which 1000, 100, 10 and 1 lux were recorded are shown. For further explanation see legend to Fig. 3. After Hart & Allanson (1976).

stages are predominantly benthic. Circumstantial evidence suggests that the copepods may bury themselves shallowly in the loose interface substrates and are thereby precluded from feeding (Hart, 1977). This benthic phase potentially confers obvious energy conserving opportunities by reducing metabolic costs of maintaining a pelagic distribution. In addition it may serve to reduce detection by visual predators.

Cyclopoida: The vertical distribution and migration of the total cyclopoid assemblage has been examined but data on specific cyclopoid species are not available. Considered as a group – the general pattern of vertical migration of the cyclopoids as shown in Fig. 7 is similar to that of *P. hessei.* Throughout the year their vertical distribution in the early evening (Fig. 8) corresponds broadly with that of *P. hessei.* In both groups, nocturnal vertical distribution appears to be influenced by lunar intensity. In the calanoid, this influence varies according to life history stage (Hart & Allanson, 1976). Comparable differences might be expected within the cyclopoids.

Specific attention has not been paid to the existence or otherwise of a benthic phase in the diel cycle of the cyclopoids. Analyses of samples collected from substratum bins (Hart, 1976) have indicated that some members of the cyclopoid assemblage are benthic during daylight.

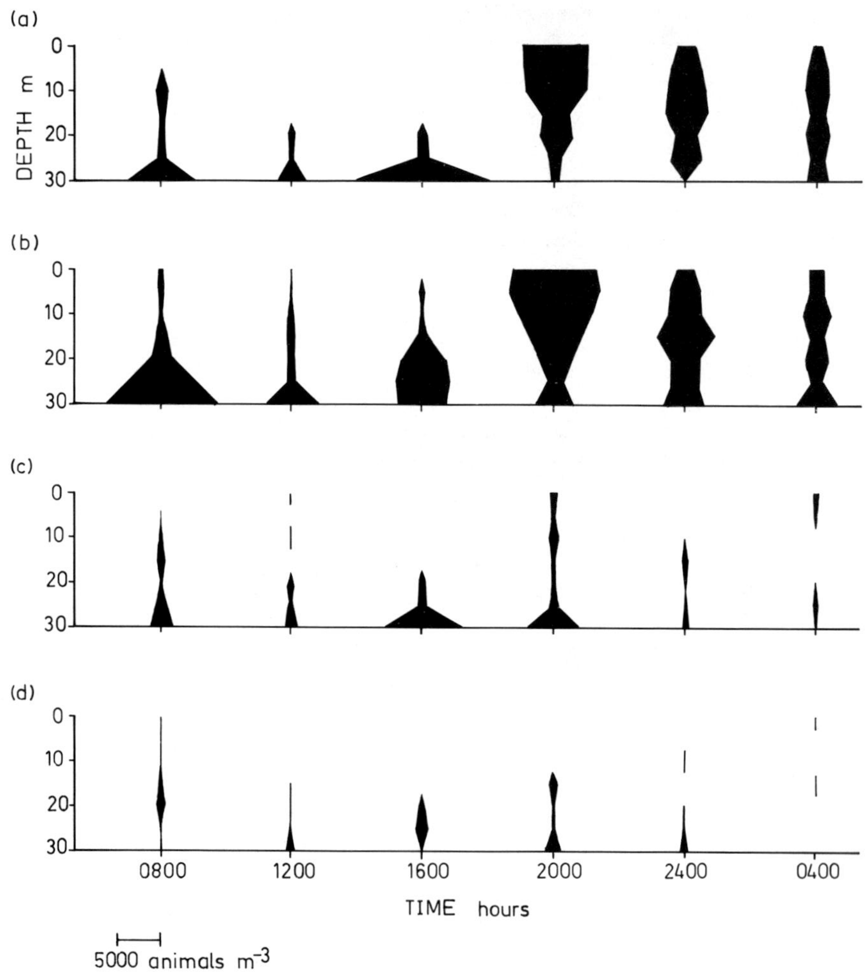

Fig. 7. Vertical distribution of (a) *Pseudodiaptomus hessei*, (b) cyclopoid copepods, (c) *Moina* sp. and (d) *Bosmina longirostris* through twenty-four hours at a 30 m station, 10–11 April 1970. Based on samples taken using a Friedinger water bottle and filtered on 158 μm mesh netting.

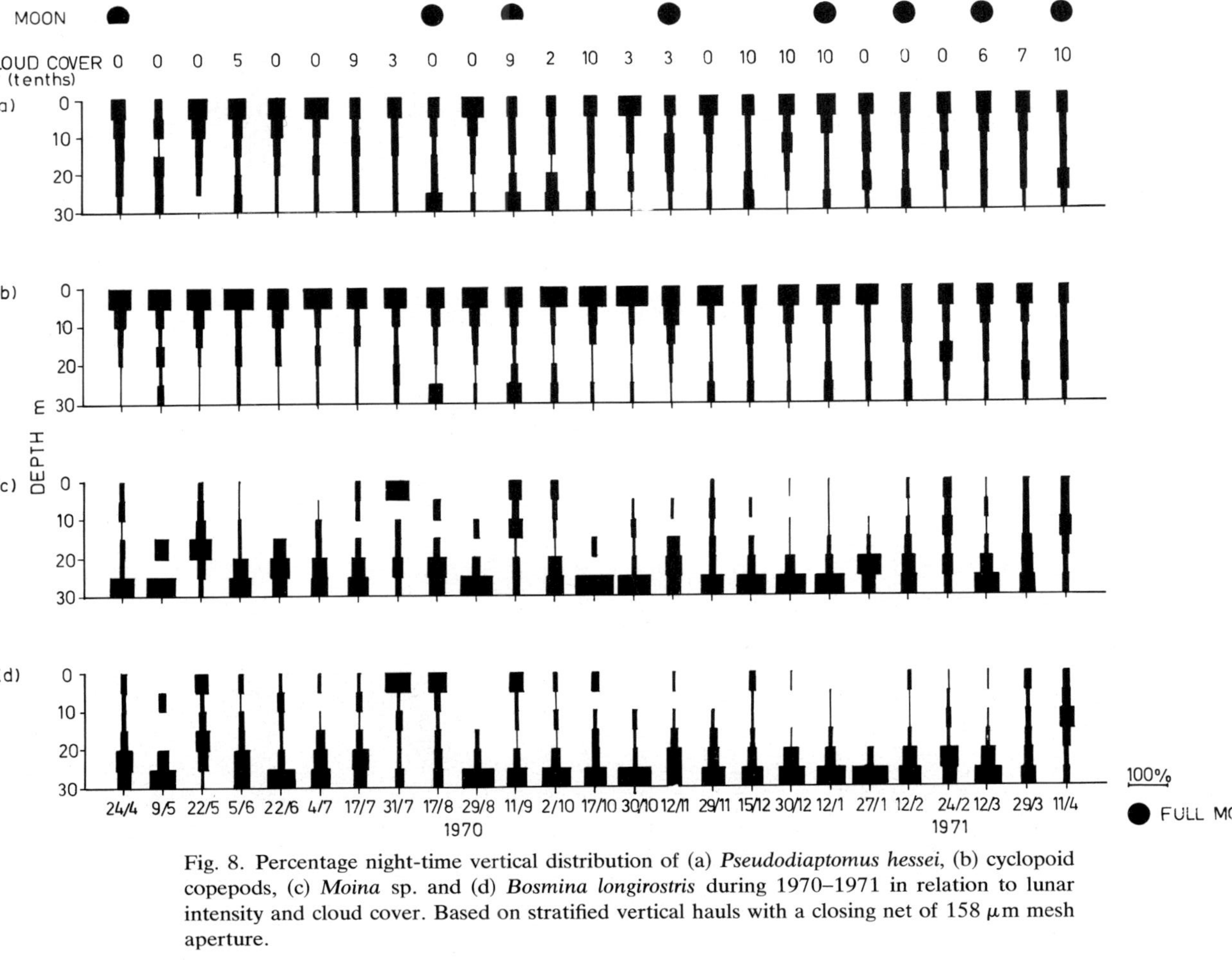

Fig. 8. Percentage night-time vertical distribution of (a) *Pseudodiaptomus hessei*, (b) cyclopoid copepods, (c) *Moina* sp. and (d) *Bosmina longirostris* during 1970–1971 in relation to lunar intensity and cloud cover. Based on stratified vertical hauls with a closing net of 158 μm mesh aperture.

Cladocera: Bosmina longirostris and *Moina* sp. generally occurred in relatively low numbers in most collections taken to describe vertical distribution and migration. Nevertheless as Fig. 7 shows it is possible to discern elements of migratory behaviour in these cladocerans.

The vertical distribution of the cladoceran zooplankton in the early evening (Fig. 7) contrasts with that of the copepod zooplankton. The cladocerans occurred deeper in the water column throughout the year than the copepods. This deeper nocturnal distribution may reflect a reduced amplitude of vertical migration in these cladocerans which accords with the relatively weak migrations shown by them in Fig. 8.

The difference in nocturnal distribution of the copepods (especially the cyclopoids) and the cladocerans is striking and would appear to represent some mechanism potentially reducing co-existence of these groups at night. Whether mutual competitive exclusions or predator-avoidance by the cladocerans is involved is not known. The deeper distribution of the cladocerans suggest that these zooplankters are at least to some extent excluded from utilizing food resources in the euphotic zone, although it has been shown that vertical differences in abundance of algae are marginal in this well mixed lake (Hart & Hart, 1977). Nevertheless, it may be that the cladoceran components of the zooplankton rely proportionately more upon bacterial and detrital food resources than the other microphagous zooplankton.

Horizontal Distribution

The horizontal distribution of zooplankton varies considerably with time in this turbulent system. Such variation is evident both in terms of total standing stock (Fig. 9) and within specific components of the zooplankton

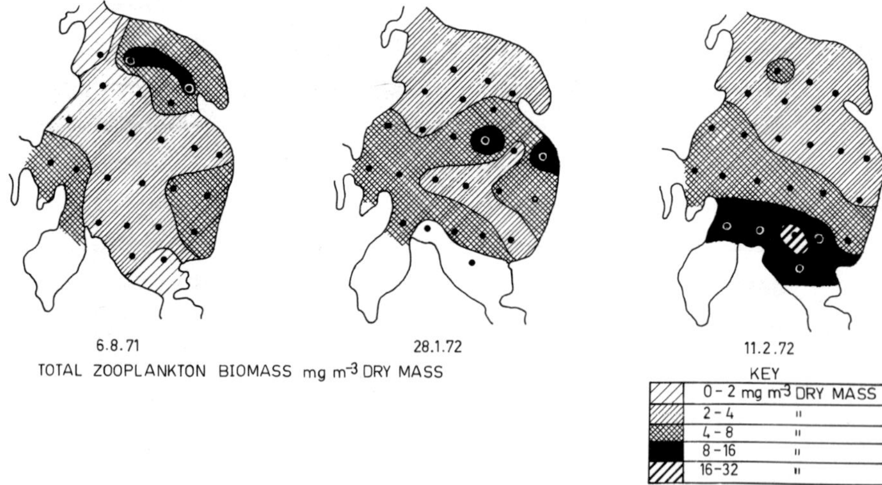

Fig. 9. Lakewide distribution of zooplankton biomass on three dates, based on bottom to surface hauls with a 158 μm mesh aperture net collected during 6–8 hours overnight. Total biomass determined directly. Dots represent sampling sites.

community. The nature and likely origins of lakewide distribution patterns of *P. hessei* have been described by Hart (1978), and it is clear that as in other lakes, wind-induced turbulence appears largely responsible for bringing about changes in distribution of this copepod over the lake. The nature and extent of these changes is determined in part by the vertical distribution of component life history stages: the younger stages which live in more turbulent water nearer the surface are affected most by wind-induced currents. Older, deeper-living stages are to some extent below the influence of strong directional currents and their distributions are accordingly less affected. Lakewide variations in heat content and food availability over this comparatively small but highly windswept lake are slight and unlikely to set up gradients of zooplankton abundance as reported for larger lakes.

The broad patterns of lakewide distribution of other components of the crustacean zooplankton shown in Fig. 10 appear to be similar to that of *P. hessei.*

Seasonal Changes in Abundance and Composition and Estimates of Production

No striking seasonal changes exist in the structure of the zooplankton community of Lake Sibaya, apart from an apparent absence of insect members in the cool season (Table 1). Additionally no extreme changes in abundance of entomostracan zooplankton occur (Figs. 1, 2) notwithstanding a considerable annual variation in lake temperature – a range of about 10°C (Fig. 13). On these grounds the entomostracan zooplankton community of this lake might be considered to be stable.

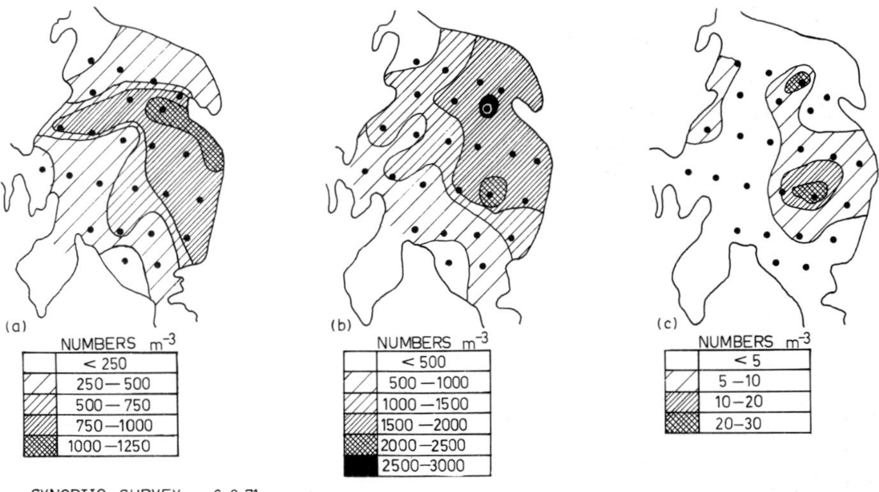

Fig. 10. Lakewide variation in density of (a) cyclopoid copepods, (b) *Pseudodiaptomus hessei* and (c) zoeae larvae of *Hymenosoma orbiculare* as determined by overnight sampling on 6–7 August 1971, using a 158 μm mesh aperture net hauled vertically from lake bottom to surface. Dots represent sampling sites.

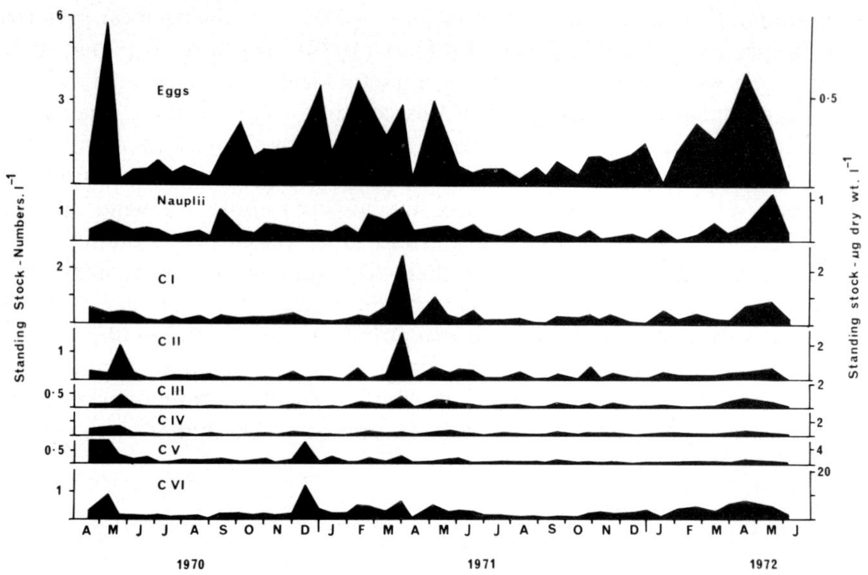

Fig. 11. Seasonal changes in abundance and estimated biomass of different life-history stages of the copepod *Pseudodiaptomus hessei* at a 30 m sampling site, based on collections made with a net of 158 μm mesh aperture. After Hart & Allanson (1975).

The most detailed information on aspects of seasonality relate to the calanoid copepod *P. hessei*. Changes in population structure, standing stock and production over 27 months are described by Hart & Allanson (1975). This copepod breeds throughout the year (on average >60% of adult females were ovigerous) and all life-history stages are present at all times of the year (Fig. 11). As reflected in Fig. 12, the progression of clearly discernible cohorts through the population is not evident even on the basis of short-interval sampling at a single site. To some extent, any such analysis of cohort progression is complicated by variations in copepod density seemingly reflecting the movement of 'patches' (and therefore different sub-populations) of high or low density past the fixed sampling site. Nevertheless the evidence for continuous recruitment is compelling.

Seasonal variation in total biomass and production of *P. hessei* is shown in Fig. 13. Dry biomass values estimated from counts of life history stages and the known stage-mass relationships for *P. hessei* (Hart & Allanson, 1975) varied from below 2 mg m^{-3} dry mass to somewhat above 20 mg, with an average standing stock of 5.7 mg m^{-3}. Much of the variation in standing stock apparent in Fig. 13 may be attributed to random and sampling error associated with spatial variation in standing stock (*e.g.* see Fig. 9). Where mean data are available for several sites the extent of variation is reduced (Fig. 2). Strong annual peaks in biomass of *P. hessei* occurred at the routine station between March and May – when lake temperatures were dropping.

Changes in average biomass of *P. hessei* (estimated values) and total zooplankton (weighed after 5 years storage) at 5 sampling sites between November 1971 and June 1972 have been shown in Fig. 2a. The time-

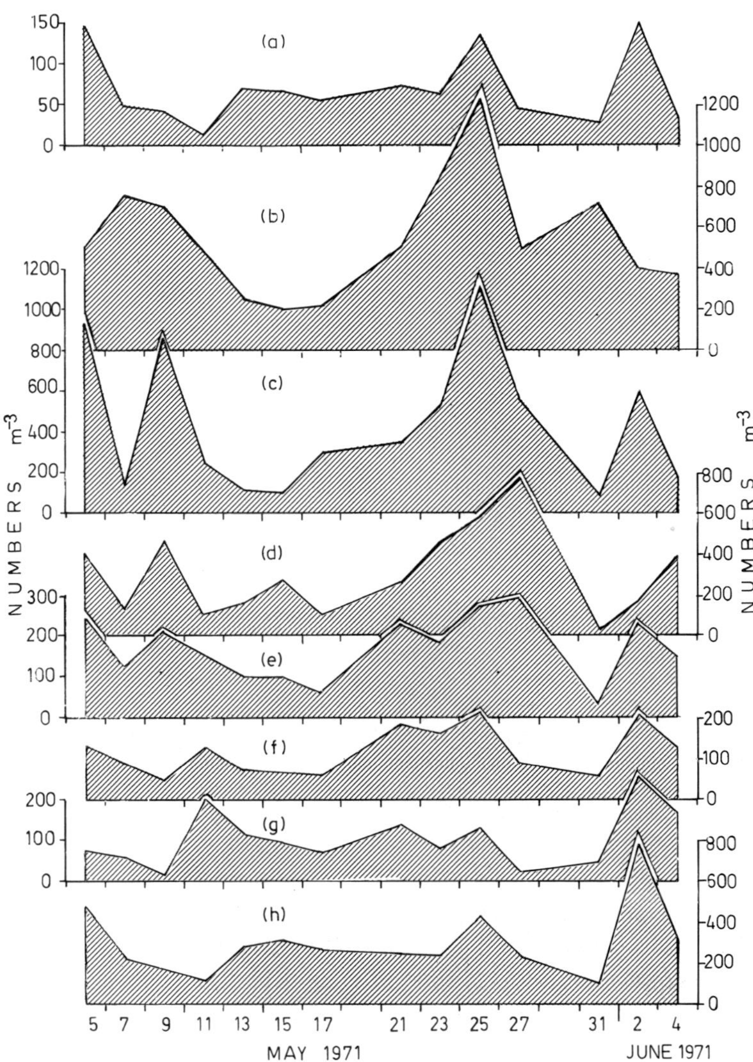

Fig. 12. Short term variation in population structure and abundance of the copepod *Pseudodiaptomus hessei* at a routine sampling site over 30 m of water, during May–June 1971. Based on vertical hauls with a net of 158 μm mesh aperture. (a) Ovigerous ♀♀, (b) Nauplii, (c) to (h) successive copepodite stages C_I to C_{VI}.

course of changes in these components is very similar possibly indicating that common or related causal factors regulate zooplankton standing stocks. However, there are indications in Figs. 1 and 2 that the April–May peak in calanoid numbers was associated with a drop in abundance of cyclopoid copepods, which showed highest densities in January–February although no comparable change occurred during the preceding summer of 1970–1971.

While average standing stocks of zooplankton are low, analyses of production reveal that turnover-times are comparatively high, certainly in *P.*

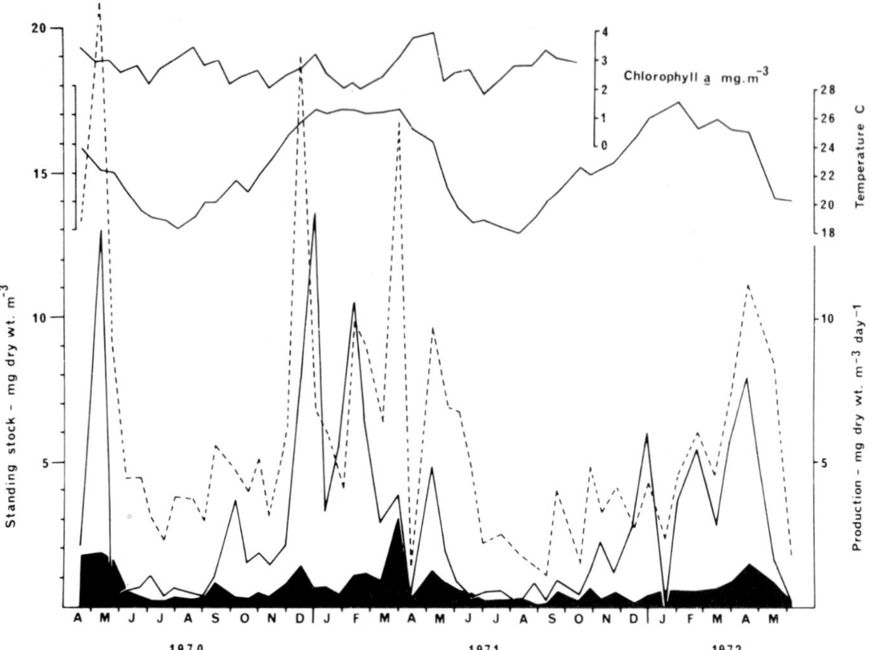

Fig. 13. Seasonal changes in daily production ($mg\,m^{-3}$ dry mass) and biomass ($mg\,m^{-3}$ dry mass) of the copepod *Pseudodiaptomus hessei* in relation to surface temperature and chlorophyll *a* content at a routine 30 m sampling site. Shaded area – production calculated by summing daily biomass increments over all life history stages (Winberg, 1971). Solid line – production calculated from finite birth rate (egg ratio method – Elster, 1954; Edmondson, 1960). Broken line – estimated biomass of *P. hessei*. After Hart & Allanson (1975).

hessei. For example, the more conservative estimate of average daily production for this copepod of $0.634\,mg\,m^{-3}$ dry mass (Hart & Allanson, 1975) leads to a Production/Biomass ratio (P/B) of 0.11. This represents a biomass replacement time of 9 days or 40 replacements per annum, based upon Winberg's (1971) method of summing growth increments over all stages per unit of time to estimate production. By contrast, use of the egg-ratio method (Elster, 1954; Edmondson, 1960) which essentially describes turnover time of numbers rather than biomass yields an estimate of average production of $2.46\,mg\,m^{-3}\,day^{-1}$ dry mass, with a corresponding P/B ratio of 0.43. This represents an unacceptably high turnover rate of biomass, *viz*. of the order of once every 2 days, although it may be more realistic in relation to the replacement time of numbers. As shown in Fig. 13, the discrepancy between Winberg's and Elster's methods is greatest during summer, when the egg stock is largest.

Estimates of average biomass, production and P/B ratios over the main basin of Lake Sibaya in summer and winter are given in Table 4. Notwithstanding significant differences in estimates of production based upon the two methods, it is clear that summer production is two to three times as great as in the cool season.

Table 4. Estimates of average biomass and production of *P. hessei* over the main basin of Lake Sibaya. *n* denotes the number of sites sampled on each date. After Hart & Allanson (1975).

	Biomass mg m^{-3} dry mass		Egg ratio method (Elster)		Growth increment method (Winberg)		
	Mean	±S.E.	Production mg m^{-3} day^{-1} dry mass	Daily P/B	Production mg m^{-3} day^{-1} dry mass Mean	±S.E.	Daily P/B
6 Aug. 1971 $n = 24$	3.95	±0.41	0.57	0.144	0.40	±0.035	0.101
28 Jan. 1972 $n = 25$	4.20	±0.44	1.91	0.455	0.77	±0.085	0.183
11 Feb. 1972 $n = 25$	5.84	±0.79	1.52	0.260	1.28	±0.219	0.219

Feeding Biology

The feeding biology of the zooplankton of Lake Sibaya has been severely neglected, largely through manpower limitations. Some aspects of the feeding biology of *P. hessei* have been examined in relation to the vertical distribution and migration of this copepod. In addition, assessments of the major categories of food items filtered by this copepod have been made.

A diel feeding rhythmicity associated with the diel alternation between the benthic and pelagic phases of *P. hessei* has been described (Hart, 1977). Under natural conditions feeding activity (as reflected by the proportion of animals with food in the gut) of adults, and to a lesser extent the pre-adult stages is largely restricted to the hours of darkness and is reduced or terminated during the day while the animals are benthic. Hart (*loc. cit.*) has suggested that this rhythmicity arises through the assumption of an albeit marginally 'infaunal' rather than merely 'epifaunal' existence during their benthic phase such that physical constraints preclude filter-feeding. Only during the resumption of their nocturnal planktonic or pelagic phase does it become possible for these copepods to resume normal filter-feeding.

Examination of the fore-gut contents of *P. hessei* suggests that it feeds largely upon nannoplankton and bacterioplankton. Comparatively few net phytoplankton are apparently ingested, although they may provide something of a nutritional windfall when taken. Measurements of filtration rate, while requiring refinement and extension seemingly confirm the relative importance of nanno- and bacterio-plankton as food items. By contrast filtration rates obtained for net phytoplankton were neglibible.

Comparisons with Other Southern African Coastal Lakes

Little published information exists regarding the zooplankton of other coastal lakes, and indeed, apart from check-listings, the zooplankton has been generally neglected. Many of the coastal lakes are influenced in varying degree by periodic or aperiodic sea-water ingress with the consequent introduction of estuarine-marine forms. The longevity of such components in the recipient coastal lakes must depend largely upon their survival and reproductive capabilities within the resulting salinity regimes to which they are exposed. Scope exists for experimental work in this direction.

In Swartvlei an interesting example of the modifying influences of this sea-water mass upon the distribution of the copepod *Pseudodiaptomus hessei* has been revealed by Grindley & Wooldridge (1973). Swartvlei is a meromictic lake with a strong halocline present between 5 and 9 m for much of the year. The hypolimnion is anaerobic and supports large populations of a cyclopoid copepod *Acartia* sp. The calanoid *P. hessei* does not occur in the hypolimnion, but neither is it found to any great extent in the epilimnion over deep water, although in the shallow fringes where the epilimnion extends to the bottom substrates it is present in large quantities. This may be indicative that the diurnal benthic phase of the adults is an obligate feature of their biology.

Hemens *et al.* (1971) captured few *Pseudodiaptomus* during daylight in Lake Nhlange, whereas nocturnal sampling some years earlier (Hart, unpublished) resulted in large catches. In Lagoa Poelela, Boltt (1975) collected specimens of a related species *P. charteri* (= *P. stuhlmanni*, Grindley, pers. comm.) in his grab sampling of the benthos. These observations provide limited circumstantial evidence that *Pseudodiaptomus* possesses a benthic phase in these coastal systems as well as in Lake Sibaya.

The coastal lakes with very different but nevertheless most comparable physico-chemical similarities to Lake Sibaya, *viz.* Lakes Nhlange, Piti and Poelela are, unfortunately, probably the poorest known in terms of their zooplankton. The better known systems, Lake St. Lucia and Swartvlei have strong marine influences, rendering comparisons of limited relevance. The copepod *Pseudodiaptomus stuhlmanni* is an important component of the zooplankton in St. Lucia (Grindley, Blaber, pers. comm.) but the zooplankton of Swartvlei is dominated at least seasonally by spat of the bivalve *Musculus virgiliae* (Coetzee, pers. comm.) which in later life festoons the dense macrophyte beds which fringe that system. The other major species in Swartvlei include the cyclopoids *Halicyclops* sp. and *Acartia* sp. which judging from their distribution are tolerant of the generally inhospitable anaerobic conditions within the hypolimnion. These species are also confined to higher salinities.

The Zoobenthos

Species Composition

Many benthic organisms have been recorded from Lake Sibaya and a checklist of the known forms is given in Table 5 which indicates broadly the distribution of some of the animals. Where possible, species which are apparently important numerically or volumetrically are identified, as are those species which are typically estuarine in habitat. Undoubtedly the most important species occur within the Crustacea and the Mollusca with the former strongly represented by estuarine-marine forms and the latter by freshwater forms. As a further generalization it may be noted that the major 'infaunal' components of the benthos are estuarine species, while with the exception of the marine crown crab *Hymenosoma orbiculare* (which may bury itself, particularly during daylight) the major large 'epifaunal' elements are typical freshwater species.

The quantitative composition of the benthos is difficult to describe in average lakewide terms because of the restricted depth ranges and substratum preferences of at least some of its components (Boltt, 1969a). In addition, larger motile epibenthic organisms undoubtedly have been undersampled in collections for which data are available for the infauna. Ignoring these and other limitations it is clear, nevertheless, from Table 6 that the smaller infaunal organisms numerically dominate the benthos overall. Table 6 shows that the three amphipods and tanaid comprise over 90% of the community

Table 5. Distribution of benthic invertebrates recorded from Lake Sibaya, showing any known vertical restrictions and apparent preferences for substratum type. Data from Boltt (1969a), Allanson, Bruton & Hart (1974) and unpublished records.

		Inshore excluding marginal vegetated areas	Submerged macrophytes and stems	Sandy bottoms	Silty bottoms	Apparent depth restriction	Common or important species	Uncommon or rare species	Estuarine species
Coelenterata									
	Hydra? sp.		+	+				+	
Annelida									
Polychaeta	*Ceratonereis keiskama*		+	+		<20 m	+		+
Oligochaeta	unsorted	+				No	+		
Crustacea									
Ostracoda	*Cypria* sp.								
	Darwinula sp.								
	Heterocypris sp.								
	Loxoconcha sp.								
	Perissocytheridae estuaria								
Branchiura	*Argulus*? sp.							+	
Tanaidacea	*Heterotanais* nov. sp.?		+					+	
	Leptochelia sp.								
	Apseudes digitalis		+	+	+	No	+		+
Isopoda	*Cyathura carinata*			+	+	No	+		+
	Pseudosphaeroma barnardi		+			<5 m		+	+
	Paramunna laevifrons		+					+	+
Amphipoda	*Grandidierella lignorum*	+	+	+	+	No	+		+
	Corophium triaenonyx	+	+	+	+	No	+		+
	Afrochiltonia capensis							+	
Decapoda	*Hymenosoma orbiculare*	+	+	+		No	+		+
	Caridina nilotica		+	+	+	No	+		
Insecta									
Ephemeroptera	*Povilla adusta*		+						
	Cloeon? *crassi*		+						
Trichoptera	*Dipseudopsis capensis*		+						
	Dipseudopsis simplex		+						
	Ecnomus thomasetti		+						
	Homilia knysnaensis		+						
	Leptocerus inflatus		+						
	Oecetis sibayiensis		+						
Diptera									
(Chironomidae)	Chironomid larvae	+	+				+		
	Chironomus spp.		+						
	Polypedilum spp.		+						
	Tanytarsus sp.		+						
	Micropsectra sp.		+						
	Corynoneura sp.		+						
	Cricotopus sp.		+						
	Tanypus sp.		+						
	Prociadius sp.		+						
	Ablabesmyia spp.		+						
(Culicidae)	Chaoborid larvae							+	
Dytiscidae	*Nymphula*? *capensis*		+						
Mollusca									
Prosobranchia	*Melanoides tuberculatus*	+	+	+	+	No	+		
	Bellamya capillata	+	+	+	+		+		
Pulmonata	*Lymnaea natalensis*		+				+		
	Bulinus (*B.*) *natalensis*	+	+	+		<7 m			
	Bulinus (*P.*) *globosus*		+						
	Biomphalaria pfeifferi		+						
	Gyraulus costulatus		+						
	Anisus natalensis		+						
	(≡ *Ceratophallus* sp)								
	Burnupia sp.		+						
	Succinea striata								
Lamellibranchia	*Corbicula africana*			+		No		+	
	Sphaerium capense		+						
	Eupera ferruginea		+						

Table 6. Approximate 'average' composition of more important species of invertebrate zoobenthos in Lake Sibaya over entire lake. Calculations ignore unequal distributions with depth and substrate type, and seasonal changes in density. Likewise unequal sampling emphasis on small sessile and large motile forms is ignored.

	Grandidierella lignorum	*Corophium triaenonyx*	*Apseudes digitalis*	*Cyathura carinata*	*Ceratonereis keiskama*	Oligochaeta	Chironomid larvae	*Hymenosoma orbiculare*	*Bellamya capillata*	*Melanoides tuberculatus*	*Bulinus natalensis*
Mean density m^{-2}	1220‡	812*	5700‡	168*	40*	247*	147*	0.3*·‖	4.0*·¶	32.2*	0.6*
% frequency	14.6	9.7	68.1	2.0	0.5	3.0	1.8	0.004	0.05	0.4	0.007
Mean mass animal^{-1} (mg)	0.0608‡	0.0665§	0.0417‡	0.0614§	0.1090†	0.0113†	0.0581†	48.0‖	99.0¶	54.0*	—
Mean biomass mg m^{-2} (dry)	74.2	54.0	237.7	10.3	4.4	2.8	8.5	14.4	396.0	1738.8	—

* Boltt (1969a),
† Boltt (1975),
‡ Boltt (unpublished ms.),
§ Batchelor (1972),
‖ Forbes & Hill (ms.),
¶ Rudd (unpublished).

by number. Notwithstanding their small individual biomass, these components also contribute substantially to the total biomass of the benthos.

Seasonal variation in taxonomic *composition* of the benthos appears to be slight on the basis of hot-cool season comparisons reported by Boltt (1969a), certainly as regards the infaunal Crustacea. Limited information is available regarding the seasonality of other components. However, there is some evidence to suggest that *densities* of infauna were higher in winter than in summer* with corresponding seasonal changes recorded for *Hymenosoma orbiculare.* In contrast the mollusc *Bellamya capillata* was less numerous in winter.

Distribution of Benthos in Relation to Depth and Substrata Types

The distribution or abundance of several benthic invertebrates in Lake Sibaya is affected by substrate type or depth. The substrata of Lake Sibaya are predominantly sandy (Boltt, 1969a), with silty substrata occurring over probably no more than 10 to 15% of the lake bottom. The sandy substrata throughout the lake are extremely similar with respect to particle size distribution with a mean median phi (ϕ) value of 3.203. However, the amount of extremely fine (<0.04 mm) flocculent mainly organic material (≡subsieve fraction, Boltt, 1969a) which settles on and covers the sand grains increases with depth and may comprise up to a third by volume of the superficial substrata. Silty substrata consisting of oozy grey mud comprised of particles <0.04 mm are confined largely to submerged valleys leading from the northern and western arms of the lake and are not specifically confined to deepest water areas.

In general the densities of the infaunal benthos are dramatically lower in silty substrata than in sandy substrata (Table 7) but in contrast, the density of the two major epifaunal molluscs is greater on the silty substrata.

The influence of depth on the distribution of the benthos is more complex. Boltt (1969a) recognized two types of distribution. He considered the amphipods, tanaid and polychaete *Ceratonereis keiskama* to be 'discontinuously' distributed, *i.e.* occurring in different densities over different depths. By contrast the isopod *Cyathura carinata*, oligochaetes and chironomid larvae were considered to be evenly or continuously distributed although interspecific differences within the latter two groups could be obscured by combining taxa. Boltt (1969a) attempted to establish whether depth *per se* or the proportionate increase with depth of subsieve material in sandy substrata was important in regulating the distribution of benthic animals. He concluded that at least as far as sandy substrata were concerned, many of the observed changes in distribution appeared to be related to depth alone.

A particularly interesting example of discontinuous distribution was shown for the amphipod *Grandidierella lignorum* (Boltt, 1969a). Changes in abundance of this amphipod with depth are compared with those shown by

* The reality of this difference was subsequently questioned following further sampling (Boltt, unpublished ms.).

Table 7. A comparison of the densities of the major components of the benthic community in and on sandy and silty substrata in Lake Sibaya. Data largely from Boltt (1969a).

	Grandidierella lignorum	*Corophium triaenonyx*	*Apseudes digitalis*	*Cyathura carinata*	*Ceratonereis keiskama*	Oligochaeta	Chironomidae larvae	*Hymenosoma orbiculare*	*Bellamya capillata*	*Melanoides tuberculatus*
Sandy substrata										
Density m^{-2}	2329	1542	9093	280	80	493	244	0.7	0.5	8.9
Silty substrata										
Density m^{-2}	111	82	2300	56	—	—	45		7.4	56

another amphipod *Coraphium triaenonyx* and the tanaid *Apseudes digitalis* in Fig. 14. While common above 20 m, *G. lignorum* is sparse in deeper water while the tanaid *A. digitalis* reaches its highest densities below 20 m. The factors responsible for the distribution of *G. lignorum* formed the basis of a detailed eco-physiological investigation (Boltt, 1969b, unpublished ms.).

Having excluded the effects of pressure, oxygen tension, pH, hydrogen sulphide and phototactic behaviour or substratum choice in determining the distribution of *G. lignorum*, Boltt (*loc. cit*) was able to account for this amphipod's distribution in terms of its sensitivity to dissolved CO_2, and more specifically to rates of change of $p\,CO_2$. At low rates of increase of $p\,CO_2$ (<14 mm Hg $p\,CO_2\,hr^{-1} \equiv 31\,mg\,l^{-1}\,hr^{-1}\,CO_2$) the animals remained within their burrows where they became anaesthetized and eventually died. At rates of increase of $p\,CO_2 > 20$ mm Hg $p\,CO_2\,hr^{-1}$ ($44\,mg\,l^{-1}\,hr^{-1}\,CO_2$) the amphipods evacuated their tubes, and in addition reversed their photo responses, becoming positively phototactic. Measurements in the field revealed that during temporary microstratification at the substratum-water interface significant increases in $p\,CO_2$ occurred at depths of 30 to 32 m but not in shallower water (19 to 20 m). Although the maximum CO_2 tension recorded immediately at the substratum-water interface was only 6.3 mm Hg $p\,CO_2$($14\,mg\,l^{-1}\,CO_2$) it is likely that rates of change of $p\,CO_2$ within the animals' burrows may have exceeded the critical levels eliciting evacuation behaviour. In any case, individuals of *G. lignorum* are frequently planktonic at night: this is probably an important dispersal phase. The animals are normally photonegative and swim to and burrow in the substratum by day. Dr R. E. Boltt considered that individuals reaching a microstratification near the substratum during this 'settling' behaviour would register a gradient of increasing $p\,CO_2$ which would reverse the phototaxis and keep them in the plankton away from the substratum. Only on reaching a substratum where no microstratification existed could they settle and this largely precluded them from occupying deeper waters. Even deepest areas of the lake are presumably periodically without microstratification, however, thus allowing a few individuals to settle and remain there as within certain limits they are able to adapt to changes in $p\,CO_2$.

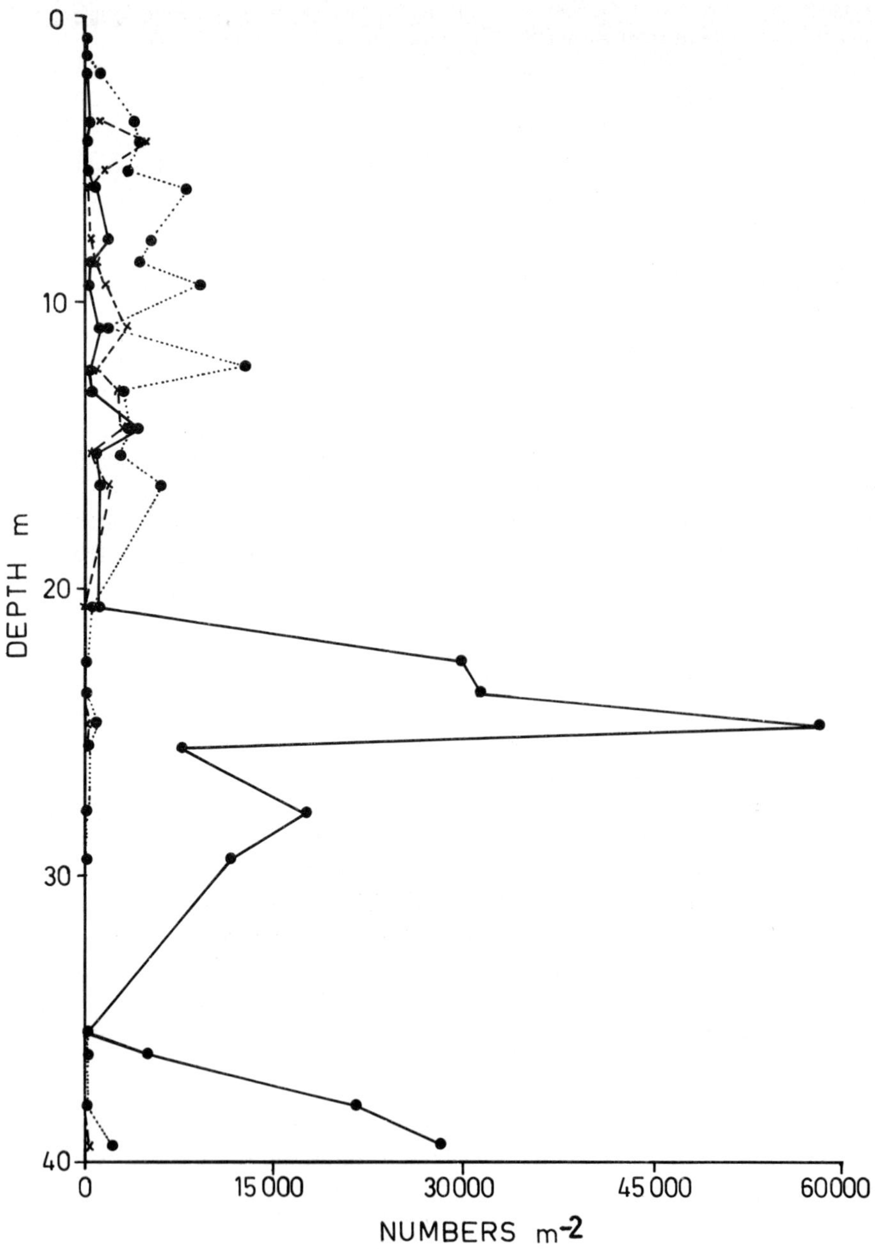

Fig. 14. Variation in abundance of three infaunal Crustacea with depth in Lake Sibaya. Based on samples collected from sandy substrata only. *Grandidierella lignorum* (●- - - - - - -●), *Apseudes digitalis* (●——●) and *Corophium triaenonyx* (X- - - -X). Data from Boltt (1969a).

By contrast, the tanaid *A. digitalis* is not sensitive to CO_2. Subjected to increasing CO_2 tension it eventually becomes anaesthetized and dies, but only at CO_2 levels greatly in excess of those experienced in nature. Accordingly its depth distribution is not limited in the same fashion as that of *G. lignorum*. The striking increase in density of *A. digitalis* below 20 m suggests that some strong competitive interaction may be involved between this tanaid and *G. lignorum*, with *G. lignorum* having the advantage above 20 m where its occurrence is not CO_2 limited.

Benthic Fauna of Submerged Macrophyte Beds

Table 5 shows the great diversity of benthic invertebrates associated with submerged macrophytes, with molluscs and insects particularly well represented. Significant observations on the distribution and abundance of the molluscan component of submerged macrophyte beds have been made by Appleton (1976, 1977a) who also considers aspects of molluscan zoogeography within Tongaland (1977a). Appleton recorded 12 species of aquatic Mollusca in sheltered bays within Lake Sibaya, including *Ceratophallus* sp. (≡ *Anisus natalensis* in Table 5) and *Bulinus* (*Ph.*) *globosus*, attributed to *B.* (*Ph.*) *africanus* by Allanson *et al.* (1974). In perennial ponds adjacent to Lake Sibaya only 11 of these 12 species occurred, and only 4 were present in perennial pans adjacent to the lake. Appleton (1977a) suggested that the decline in number of species in lentic habitats associated with the lake corresponded to the increasing severity of the thermal regimes within these habitats (see also Appleton, 1977b,c).

On the basis of a comparative analysis of the malacofauna in one sheltered and one exposed submerged macrophyte bed within Lake Sibaya, Appleton (1976, 1977a) suggested that pulmonate snails in general were able to penetrate to greater depths within sheltered than exposed weed beds and that this difference was conceivably related to differences in wave action. His findings of egg capsules and juvenile individuals of various species up to 5 m below the surface indicated that breeding could occur on macrophyte vegetation well below the surface. The occurrence of *Biomphalaria pfeifferi* up to 5 m below the surface has implications regarding the respiratory physiology of this snail.

Continuous recording of water temperature within macrophyte beds during his study revealed the thermal stability of these habitats, where temperatures seldom exceeded 25°C or varied daily by more than 2°C.

Table 8 shows the density of snails per kg of macrophyte (wet mass) recorded along a transect of progressively deeper water in Guguswana Bay (see Fig. 16). The six pulmonate species were most abundant between 1.4 and 2.8 m at the time of the investigation (July 1973), but their occurrence in deeper water led Appleton to suggest they were able to colonize deeper macrophyte beds. Little can be said regarding the insects and other faunal components of the submerged weed beds apart from Boltt's (1969a) observations of a greater species diversity in dense *Myriophillum* than in sparse *Potamogeton* beds. The chironomids, ecologically important in numerous

Table 8. Distribution and abundance (snails kg^{-1} macrophyte (wet mass)) of snails in a submerged macrophyte bed (mostly *Myriophyllum spicatum*) as a function of depth in Guguswana Bay. From Appleton (1977a).

Depth (m)	*Biomphalaria pfeifferi*	*Bulinus* (*P.*) *globosus*	*Bulinus* (*B.*) *natalensis*	*Ceratophallus* sp.	*Lymnaea natalensis*	*Burnupia* sp.	*Melanoides tuberculata*
0.5	5.0	1.8	3.6	0	1.1	0	1.4
1.0	4.0	1.6	7.3	0.3	0.7	0.3	3.0
1.4	7.7	2.3	21.0	0.5	6.4	0	4.9
2.8	3.3	0	43.9	0.7	1.8	2.3	7.9
3.2	1.2	0	29.6	0.7	0	1.1	13.2
3.4	1.1	0	10.7	0	0	0	3.6
3.7	1.0	0	16.0	0	0.5	0.5	4.0
4.5	2.8	0	26.7	0	0.3	0.7	3.9
4.7	0	0	17.5	0	0	0.7	1.4
5.0	0	0	6.0	0	0	0	6.7

aquatic habitats, are scarcely known even at a taxonomic level in Lake Sibaya. While the submerged macrophyte beds are somewhat limited in this lake (see Howard-Williams, this volume) their fauna is obviously important in the totality of the system, and requires additional study. Strongly motile forms such as the shrimp *Caridina nilotica* which occur in large numbers in some macrophyte beds but not others (Hart, unpublished) pose considerable sampling difficulties.

Benthic Standing Stocks

A rigorous assessment of spatial and temporal variation in benthic standing stocks is not presently possible. The two major divisions of the benthos, *viz.* infauna and epifauna have not been adequately sampled concurrently, and indeed sampling emphasis has been largely biased towards the small infaunal components. Information on the proportion of sandy and silty bottoms is not available. Notwithstanding these and related difficulties (not least of which relate to possible long term cycles and seasonal phenomena within the benthos) an attempt has been made to assemble all the information available to permit a broad overview of the standing stocks and biomass of the major species. This analysis, which incorporates data collected over nearly a decade, is given in Table 9.

From this analysis, however approximate, some patterns emerge. Firstly, the biomass of the numerically dominant infauna and small 'sessile' epifauna together is approximately equivalent to that of the larger 'motile' epifauna. Furthermore, in both categories, a few species make up the bulk of the biomass. Within the infauna, the estuarine amphipods and tanaid jointly

Table 9. Mean densities (numbers m^{-2}) and standing stocks (mg m^{-2} dry mass) of major invertebrate benthos considered in 10 m depth increments. Calculations ignore seasonal differences and no account has been taken of the differing densities observed on sandy and silty substrata in deriving these average figures. Data from Boltt (1969a, unpublished ms.), Forbes & Hill (ms.), and Rudd (unpublished). Standing stocks estimated using the average biomass values given in Table 6.

		Grandidierella lignorum	*Apseudes digitalis*	*Corophium triaenonyx*	*Ceratonereis keiskama*	*Cyathura carinata*	Oligochaeta	Chironomidae larvae	Total microbenthos	*Hymenosoma orbiculare*	*Bellamya capillata*	*Melanoides tuberculatus*	Total macrobenthos	Total biomass gm per m^2	Area of bottom km^2	Lake total Tonnes
0 to	Density	6023	329	1708	180	131	305	216		0.19	2.5	8.5			45.48	57.0
9.9 m	Biomass	366.2	13.7	113.6	19.6	8.0	3.4	12.5	537.0	9.1	247.5	459.0	715.6	1.253	(58.74%)	
10 to	Density	4933	2075	1570	89	291	316	173		0.25	0.48	0.15			18.15	10.9
19.9 m	Biomass	299.9	86.5	104.4	9.7	17.9	3.6	10.1	532.1	12.0	47.5	8.1	67.6	0.600	(23.44%)	
20 to	Density	2114	17,092	1594	0	319	653	199		0.36	2.5	2.9			10.80	15.2
29.9 m	Biomass	128.5	712.7	106.0	0	19.6	7.4	11.6	985.8	17.3	247.5	156.6	421.4	1.407	(13.95%)	
30 to	Density	897	15,193	104	0	274	126	193		0.15	37.6	22.2			3.00	17.0
40 m	Biomass	54.5	633.5	6.9	0	16.8	1.4	11.2	724.3	7.2	3722.4	1198.8	4928.4	5.653	(3.87%)	
Average over all	Density	5028	3652	1598	127	200	349	203		0.23	3.4	6.3			77.43	
depths (weighted)	Biomass	305.7	152.3	106.2	13.8	12.3	3.9	11.8	606.0	10.9	335.1	339.8	685.8	1.292		
																100.1

dominate. The epifauna is co-dominated by two freshwater molluscs although the freshwater shrimp *Caridina nilotica* which has defied numerical analysis to date may actually be a co-dominant with the molluscs. The apparent dominance of the molluscs in the epifauna is concluded on the basis of a severely limited sampling programme. While only further sampling will improve quantitative information about these molluscs,* this general conclusion may well remain valid.

The total biomass per unit area of the microbenthos is remarkably similar over the upper 20 m, and reaches its maximum in the depth range 20 to 30 m, where it is approximately double that found in the upper 20 m. Intermediate values occur below 30 m. The three estuarine crustaceans, *G. lignorum*, *A. digitalis* and *C. triaenonyx* for which the most rigorous data are available, jointly show the same pattern. In contrast, within the epifaunal macrobenthos the variation in biomass of molluscs between the four depth intervals is far greater. To some extent this must be attributable to limited sampling, but the markedly higher standing stocks recorded for water deeper than 30 m probably have some foundation in reality. The most thoroughly researched epifaunal benthic invertebrate, the crab *Hymenosoma orbiculare* shows significant changes in abundance and biomass per unit area with changes in depth, with highest standing stocks recorded around 30 m (Forbes & Hill, ms.). As shown in Table 10 changes with depth occur both in sex ratio and proportion of females in berry. Additional information is required concerning the abundance and distribution of immature crabs, as these were not collected by the sampling technique used (manual capture following visual sighting by SCUBA divers). Similar considerations apply to the molluscan epifauna which was most extensively sampled by SCUBA divers who would have tended to overlook small individuals.

Table 10. Differences in density, biomass, energy content, sex ratio and proportion of berried females of the crown crab *Hymenosoma orbiculare* in relation to depth. Data from Forbes & Hill (unpublished ms.)

Depth interval m	Area searched m^2	No. crabs collected	Mean density m^{-2}	% age ♀♀	% age ♀♀ in berry	Mean biomass $mg\ m^{-2}$ dry mass	Mean energy content $J\ m^{-2}$
3 to 4	360	48	0.13	31.2	6.7	3.56	45
7 to 9	360	89	0.25	37.0	45.5	10.68	133
10 to 12	360	91	0.25	50.5	50.0	11.42	143
13 to 17	360	91	0.25	42.8	53.8	11.97	150
22 to 24	480	144	0.30	54.8	65.0	14.35	176
26 to 29	240	113	0.47	49.6	48.2	21.59	206
33 to 35	240	36	0.15	58.3	57.1	7.38	92
39 to 40	90	15	0.17	46.7	14.3	6.42	109

* For example, the extremely high density of *Bellamya capillata* between 30 and 40 m may be attributable to a bias in sampling effort in favour of silty substrata which support considerably higher densities of this mollusc than the sandy substrata (see Table 7).

The production of two infaunal crustaceans, *Grandidierella lignorum* and *Apseudes digitalis* at summer temperatures has been examined by Boltt (unpublished ms.) on the basis of turnover time estimated from analyses of egg stocks and embryonic durations. Mean numerical stocks of these animals in four consecutive 10 m depth intervals were calculated from samples taken over several years, disregarding the seasonal differences in abundance observed between hot and cool seasons. The silty substrata were excluded

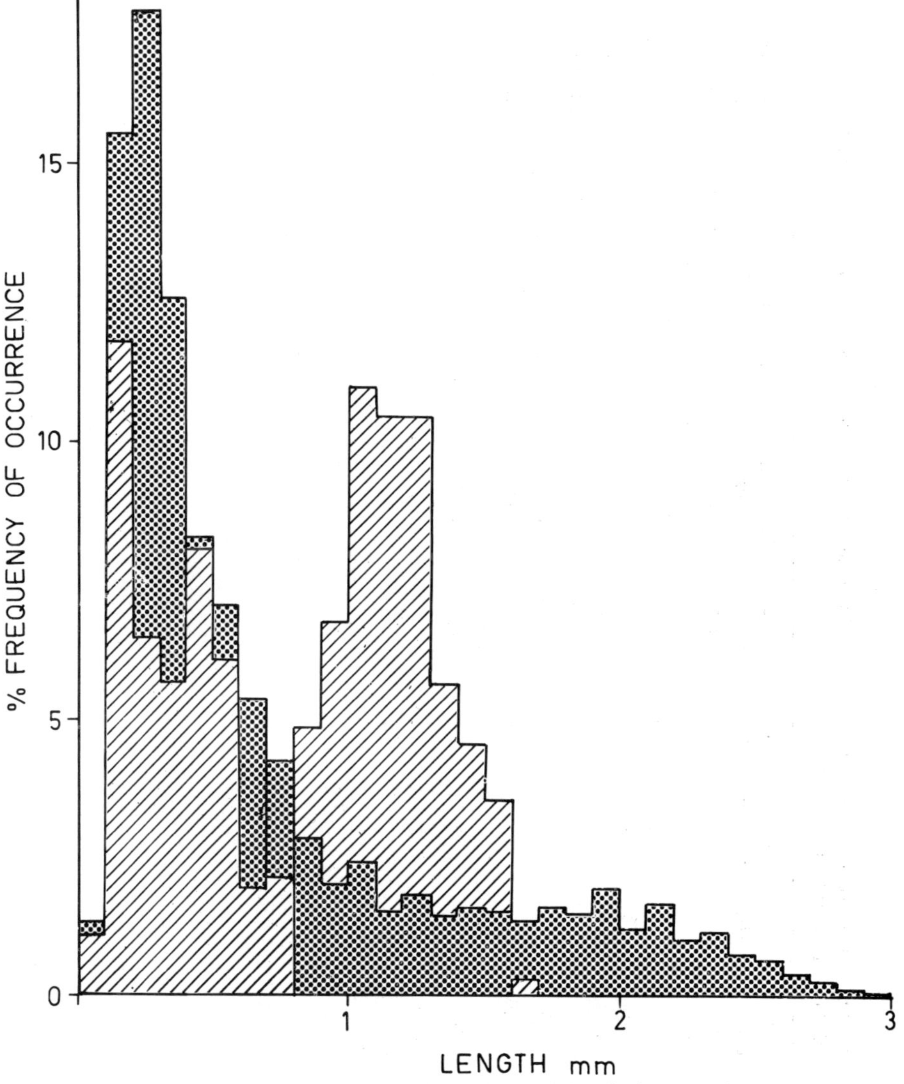

Fig. 15. Percentage size structure of populations of *Grandidierella lignorum* (stippling, sample size = 5500) and *Apseudes digitalis* (cross hatching, $n = 3810$) in sandy substrata during summer 1973–74. Data from Boltt (unpublished ms.).

Table 11. Calculations of summer production of *Grandidierella lignorum* and *Apseudes digitalis* in Lake Sibaya. Incomplete data provided for *Corophium triaenonyx*. Data from Boltt (unpublished ms.).

	Depth interval m	Density m^{-2}	No. ♀♀ with eggs m^{-2}	Mean clutch size	Egg stock m^{-2}	Embryonic duration days	Recruitment rate eggs d^{-1}	Finite birth rate B	Turnover time 1/B (days)	Biomass mg m^{-2} dry mass	Production mg m^{-2} d^{-1} dry mass	P/B
Grandidierella lignorum	0–9.9	6023	325	7.46	2425	4.5	539	0.089	11.18	366.3	32.77	0.089
	10–19.9	4933	266	7.46	1984	4.5	441	0.089	11.19	300.0	26.82	0.089
	20–29.9	2114	35	2.65	93	4.5	21	0.010	104.02	128.6	1.24	0.010
	30–40	897	15	2.65	40	4.5	9	0.010	101.55	54.5	0.54	0.010
Apseudes digitalis	0–9.9	328	38	8.38	318	10.0	32	0.097	10.30	13.7	1.33	0.097
	10–19.9	2057	240	8.38	2011	10.0	201	0.098	10.23	85.8	8.39	0.098
	20–29.9	17,092	1343	9.70	13,027	10.0	1303	0.076	13.12	713.0	54.34	0.076
	30–40	15,193	1194	9.70	11,582	10.0	1158	0.076	13.12	633.8	48.32	0.076
Corophium triaenonyx	0–9.9	1708	231*	5.89								
	10–19.9	1570	212*	5.89								
	20–29.9	1594	118*	5.12								
	30–40	104	8*	5.12								

* based on % of females in population during Dec–Jan 1973–1974.

from the analysis because of the paucity of animals in them, and their small contribution to the total area of lake bottom. Average volumetric standing stocks of *G. lignorum* were calculated using the length-mass relationship (Mass = 0.0053 Length$^{2.44}$) established for this species and applied to analyses of length-frequency distribution observed in collections made during the summer of 1973–74 and shown in Fig. 15. An average individual mass was calculated for *A. digitalis* from a large random sample and this value (0.0417 mg) was multiplied by observed densities of the tanaid to provide an estimate of its biomass per unit area. The resulting average numerical and volumetric standing stocks are those given in Table 9. Egg development times of 4.5 and 10.0 days were established in the laboratory at 28°C for *G. lignorum* and *A. digitalis* respectively. Field data necessary for the application of the egg-ratio method of estimating production were collected largely during intensive sampling in the summer of 1973–74 and the resulting summer production estimates are given in Table 11. In general the two species which carry a similar number of eggs on average show remarkably similar turnover times notwithstanding a twofold difference in embryonic development rate which is counterbalanced by interspecific differences in the proportion of females carrying eggs. Disparately long turnover-times are apparent for *G. lignorum* populations living below 20 m. These are attributable to the low proportion of egg-bearing females which Boltt (unpublished ms.) considered to reflect the adverse $p\,CO_2$ conditions experienced by *G. lignorum* in deeper water, as discussed above. In addition, recruitment to this deep-water population may be attributable predominantly to immature non-breeding animals which would result in disproportionately low values of egg production and therefore finite birth rates.

Estimates of production of *G. lignorum* based upon changes in biomass were determined from laboratory estimates of post-embryonic growth rate at two temperatures. These values were applied to the population structure and abundance in the field and give production estimates which are generally about half those obtained from calculations of finite birth rates (Table 12). However, the growth method provides considerably elevated and presumably more realistic estimates of production for the deep water sediments than the recruitment method for reasons which will be apparent

Table 12. Comparative estimates of production of *Grandidierella lignorum* calculated on the basis of summation of biomass increments (growth method) with estimates based upon calculations of finite birth or recruitment rates. Data from Boltt (unpublished ms.).

	Summer Production (mg m^{-2} d^{-1} dry mass)		Winter Production (mg m^{-2} d^{-1} dry mass)	
Depth interval	Growth method	Recruitment method	Growth method	Recruitment method
0 to 9.9	16.0	32.7	13.9	24.6
10 to 19.9	13.0	26.8	11.4	20.1
20 to 29.9	5.6	1.2	4.9	0.9
30 to 40	2.1	0.5	2.4	0.4

from the preceding paragraph. The absence of growth data preclude a comparable analysis for *Apseudes digitalis.*

Considered jointly, the production of *G. lignorum* and *A. digitalis* based on recruitment amounts to 34, 35, 56 and 49 mg m^{-2} day^{-1} dry mass within the 0–10, 10–20, 20–30 and 30–40 m depth intervals respectively. This is of the same order of magnitude as that attributable to the copepod *Pseudodiaptomus hessei* when converted to production per m^2 of lake surface (*viz.* about 1.7 mg m^{-3} day^{-1} dry mass or 22 mg m^{-2} day^{-1} dry mass based on growth, and assuming an average depth of 13 m for Lake Sibaya).

The increase in standing stocks and production of benthos below 20 m is notable. Whether this is related to nutritional factors or to a lowered incidence of predation by fishes is not known. The swim-bladder physiology of various cichlid fishes regulates in varying degree their ability to penetrate deep water (>20 m) although this is modified by temperature and age (Caulton, 1975; Caulton & Hill, 1973). Blaber & Whitfield (1977) found that the burrowing goby *Croilia mossambica* which feeds extensively on benthic invertebrates in Lake Sibaya, is restricted to water shallower than about 16 m, apparently because it relies upon visual detection of prey. The more common goby in Lake Sibaya, *Glossogobius giurus,* which also feeds on invertebrates, is apparently not limited in its depth distribution. On balance there is some evidence for decreased predation by fish in deeper water.

Comparisons with Other Southern African Coastal Lakes

Better comparative data are available on the zoobenthos of several coastal lakes than exist for the zooplankton, but the relevance of the comparative approach for the purposes of this account is, as for the zooplankton, limited by the great differences in the physico-chemical characteristics of the systems under discussion. The benthos of two lakes showing some similarity to Lake Sibaya, i.e. Lake Nhlange and Lagoa Poelela, have fortunately been examined in some detail, while the most comparable system, Lake Piti, has not.

Lake Nhlange (3 to 5‰ salinity) is part of the Kosi lake system which opens to the sea. It is notable that notwithstanding its salinity, this lake lacks two of the small estuarine crustaceans which dominate the infaunal benthos of Lake Sibaya – viz. *Grandidierella lignorum* (*G. bonnieri* occurs instead) and *Apseudes digitalis.* Likewise the isopods *Cyathura carinata* and *Pseudosphaeroma barnardi* which are estuarine forms occurring in Lake Sibaya are absent from Lake Nhlange (Boltt & Allanson, 1975). The absence of these four crustaceans from Lake Nhlange cannot be accounted for simply in terms of the general chemistry of the system. During periods of exceptional rainfall, flushing of Lake Nhlange is a significant event. This may be accompanied by the introduction of large amounts of organic material from the upstream swamps and surrounding catchment area leading to de-oxygenation of deeper waters, with correspondingly profound influences on the benthos. From this viewpoint, Lake Sibaya presents a considerably more stable environment in which the impact of known long-term climatological cycles (Allanson, Hill, this volume) upon the benthos is seemingly less drastic than such aperiodic events as flooding is to the benthos of Lake Nhlange. Conversely, Boltt & Allanson (1975, p. 261) suggested that the benthos of Lake Sibaya was intrinsically unstable.

In general the epifaunal component of the benthos of Lake Sibaya is far richer than that in Lake Nhlange. In Lake Nhlange, one of the few important epibenthic organisms is the estuarine bivalve *Musculus virgiliae.* The freshwater prosobranch molluscs *Bellamya capillata* and *Melanoides tuberculatus* which dominate in Lake Sibaya are understandably absent. Boltt & Allanson (1975) commented that the paucity of epibenthos in Lake Nhlange might be attributable to the predation pressure of large immigrant marine fish. A second macro-epibenthic species, the crab *Hymenosoma orbiculare* occurred only in small numbers notwithstanding the fact that this crab readily buries itself shallowly in the bottom substrata, leaving only its eye-stalks and eyes exposed.

Lagoa Poelela, a coastal lake in Mozambique, shows some interesting features apparently in common with both Lake Nhlange and Lake Sibaya. This lake was fairly saline (8‰) during the benthic investigation reported by Boltt (1975), at which time its benthos was dominated by the estuarine bivalve *Musculus virgiliae.* However, shells of the freshwater molluscs *Melanoides tuberculatus* and *Bulinus natalensis* were found in the bottom substrata, possibly reflecting a prior freshwater phase. Since the Rio Inharrime, which is presumably inhabited by such freshwater components flows into Lagoa Poelela, there exists the alternative possibility that these shells were of extraneous origin.

Data on the total biomass of benthos are not available for Lake Nhlange, but direct estimates are available for Lagao Poelela. In that system, unlike Sibaya, the biomass of benthos dropped strikingly below 13 m apparently as a result of the depth restriction of *Chara,* a substratum colonized by the dominant, *Musculus.* Thus for the depth zones 0 to 3 m, 5 to 12 m and 13 to 23 m, Boltt (1975) reported total standing stocks of 0.5, 2.6 and 0.007 g m^{-2} dry mass. The values reported for the 5 to 12 m depth interval are approximately double those estimated for the 0–10 m interval in Lake Sibaya.

Littoral Invertebrates

Species Composition

The division between the zoobenthos and the littoral fauna is somewhat artificial since the littoral and benthic habitats intergrade and several important species are not exclusively confined to one habitat or the other. Nevertheless, it provides a convenient separation of those organisms associated with the marginal areas from those which occur in offshore benthic habitats. In this section it is intended to deal with the organisms which occur in the littoral or marginal areas, whether or not they are present elsewhere.

Table 13. The numbers of species of invertebrates within various taxonomic groups recorded from littoral habitats in Lake Sibaya. A+ signifies that additional species may be involved as the present taxonomic knowledge is incomplete.

Phylum	Class	Number of taxa recorded
Porifera		1+
Annelida	Polychaeta	1
	Oligochaeta	2+
	Hirudinea	3
Arthropoda	Crustacea	18
	Insecta	36+
Mollusca	Gastropoda	10
	Lamellibranchiata	1

The fauna of the littoral of Lake Sibaya is diverse, with the Mollusca, Crustacea and Insecta strongly represented. A checklist of the organisms known to occur in littoral habitats is given by Allanson, Bruton & Hart (1974). This information is condensed in Table 13 which indicates the numbers of organisms recorded within higher taxonomic groupings. From this analysis it is apparent that the greatest species diversity exists within the insects. Granted that many of these are transitory aquatic larval stages of aerial adults (some possibly more representative of the wetlands immediately adjacent to the lake than to the lake itself) – several perennial aquatic insects are also represented, particularly within the Hemiptera.

The majority of the littoral invertebrates are typical freshwater species, with only some estuarine Crustacea present. The latter are seemingly not numerically important in the littoral, unlike their conspecifics which may dominate offshore benthic habitats.

Distribution and Standing Stocks of the Littoral Fauna

Discrete localized habitats exist along restricted areas of the lake shore, but much of the shore-line is characterized by a continuum of intergrading habitats, differing largely in the amount of rooted vegetation developed (see Howard-Williams, Chapter six). The substrates of the littoral are predominantly sandy. The littoral habitats are extremely labile, with classical botanical successions related to the cycle of receding and advancing shore-line associated with fluctuations in lake level.

Wide differences in both species composition and abundance of littoral invertebrates exist over the range of littoral habitats found around the shores of Lake Sibaya. While no detailed analysis of this faunal variation has been made, it is clear that many of the differences observed are attributable to the degree of wind and wave exposure to which given sections of shore-line are subjected. The development of marginal vegetation is affected by the degree of exposure. This in turn affects the availability of food and shelter for the littoral fauna. Allanson *et al.* (1974) recorded only 29 taxa from the exposed sparsely vegetated sandy shore-lines, while 83 were recorded from more sheltered and more densely vegetated 'quiet' shores.

An indication of the degree of exposure of shore-lines around Lake Sibaya, based upon the length of wind fetch facing particular sections of the shore-line during any of the predominant winds is given in Fig. 16. This analysis takes no account of differences in intensity of particular wind regimes, nor of differences in their frequency. Notwithstanding this and other limitations (see below) Fig. 16 shows that most of the shore-line around the main basin of the lake is very exposed. Considering the lake as a whole, about 24% of its shore-line faces a fetch of more than 5 km. Only in the northern and western arms of the lake are there extensive 'quiet' shores.

The only quantitative information on the standing stocks of the major invertebrate groups in the littoral of Lake Sibaya was collected during January–February 1977 – a period of extremely high lake level. Some indication of the density and biomass of the larger invertebrates at that time is given in Table 14 which illustrates the dominance of the shrimp *Caridina nilotica.* The mean standing stock of this shrimp exceeds by an order of magnitude that of the crab *Potamon sidneyi* which, although numerically sparse, ranks second in terms of biomass.

While exposure is clearly a major factor affecting faunal diversity and abundance in littoral areas, the analysis of its influence is complex. It is clear from Table 15 that the average biomass of invertebrates is greatest along the most sheltered shores. While standing stocks are lower along exposed

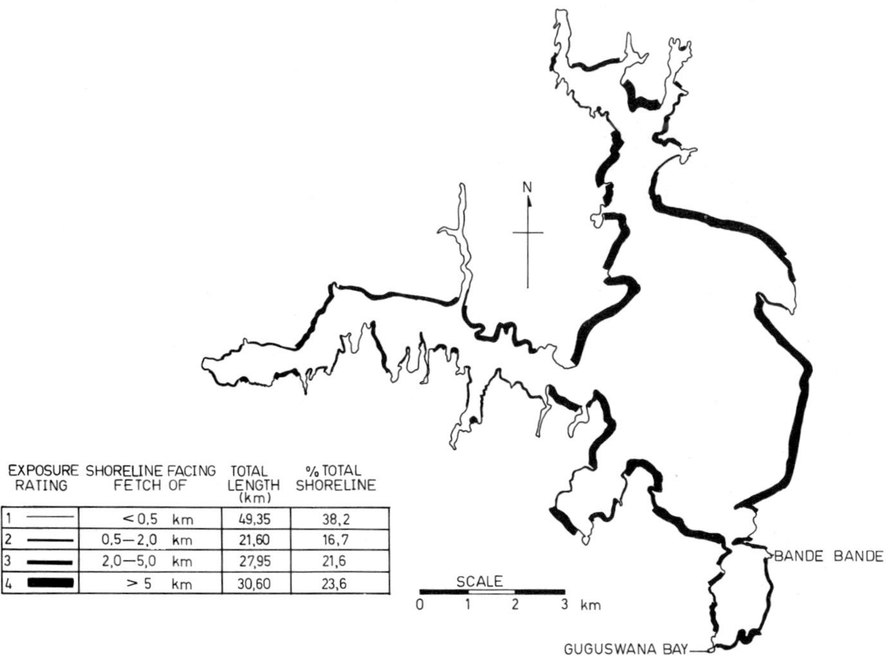

EXPOSURE RATING	SHORELINE FACING FETCH OF	TOTAL LENGTH (km)	% TOTAL SHORELINE
1	<0,5 km	49,35	38,2
2	0,5–2,0 km	21,60	16,7
3	2,0–5,0 km	27,95	21,6
4	>5 km	30,60	23,6

Fig. 16. Outline of Lake Sibaya (1964 Trigonometrical survey) showing the degree of exposure of its shorelines to the predominant winds.

Table 14. Average standing stocks of littoral invertebrates in Lake Sibaya, January–February 1977. Samples (5 replicates of $0.1\ m^2$ each) were collected at 22 sites around the lake, using a scoop net of 0.410 mm aperture mesh operated within the confines of an open-ended rectangular box delimiting the given area.

Taxa	Mean density Numbers m^{-2}	Mean Biomass mg m^{-2} (dry mass)
Caridina nilotica	695.7	2,845.0
*Potamon sidneyi**	0.6	285.0
Coleoptera	6.5	149.0
Mollusca†	10.2	119.0
Odonata	6.0	35.6
Hemiptera	1.6	16.6
Ephemeroptera	46.9	16.0
Amphipoda	23.5	1.5
Diptera	1.2	0.8
Hirudinea	0.5	0.7
Isopoda	2.5	0.4
Trichoptera	0.2	0.1

* *P. sidneyi:* ash free dry mass = ~65% of total dry mass.
† Dry mass not corrected for shell content.

shores, the decrease is not progressively related to increased exposure. Using the same data given in Table 15, rank correlation between average standing stock and exposure index is greatly improved by the use of a subjective exposure rating, based on familiarity with conditions at the sampling sites. Average standing stocks of *Caridina nilotica* along shores designated as sheltered, gently wave-washed, moderately wave-washed, and extremely exposed were 4.0, 2.6, 2.3 and 0.9 g m^{-2} dry mass. This improved correspondence between standing stock and exposure undoubtedly reflects the consideration made in a subjective analysis of such factors as the damping effect of offshore vegetation, modifying effects of adjacent under-

Table 15. Average biomass of *Caridina nilotica* and other invertebrates (excluding burrowing forms) in relation to exposure ratings of shore-line given in Fig. 16. 1, fetch <0.5 km; 2, fetch 0.5–2.0 km; 3, fetch 2–5 km; 4, fetch >5 km. Data based on five replicate samples collected at each of 22 sites, Jan.–Feb. 1977. Biomass of invertebrates other than *C. nilotica* based on pooled samples from each site.

Exposure rating	No. of sites	*C. nilotica* biomass (g m^{-2} dry mass) Mean	(range)	Other invertebrates Biomass (g m^{-2} dry mass) Mean	(range)
1	3	6.363	(1.60–10.53)	0.897	(0.15–1.88)
2	3	0.334	(0.20–0.54)	0.095	(0.04–0.20)
3	7	2.336	(0.51–5.20)	0.773	(0.01–2.86)
4	9	2.906	(0.33–7.26)	0.581	(0.005–2.44)

Table 16. The frequency of samples containing given densities or biomass of the shrimp *Caridina nilotica* in the littoral of Lake Sibaya, January–February 1977. Five replicate samples (each of 0.1 m^2) were collected at 22 sites around the lake, using a scoop net of 0.410 mm aperture mesh operated within a physically delimited area. Smallest shrimps which may be numerically important but contribute minimally to total biomass are undersampled by this technique.

Density m^{-2}	No. of samples	Biomass g dry mass m^{-2}	No. of samples
0	2	0	2
1–250	43	<1	44
251–500	24	1–2	22
501–750	12	2–3	12
751–1000	5	3–4	7
1001–1250	7	4–5	5
1251–1500	3	5–6	1
1501–1750	4	6–7	4
1751–2000	2	7–8	3
2001–2500	3	8–9	1
2501–3000	1	9–10	2
3001–4000	2	10–12	1
4001–5000	0	12–14	2
5001–6000	1	14–16	1
6001–7000	0	16–18	2
7001–8000	0	18–20	0
8001–9000	1	20–22	1

water topography and local intensity of prevailing winds, which are ignored in the analysis of exposure provided by Fig. 16. However, the dynamic nature of the shore-line, particularly as a result of fluctuating water levels, precludes reasonable exposition of exposure rating based on subjective assessments (cf. Fig. 16).

The shrimp *Caridina nilotica*, in addition to dominating the littoral fauna, occurs in almost every marginal habitat in Lake Sibaya, apart from barren sandy beaches. As Table 16 shows, it was present in all but 2 of 110 samples collected from 0.1 m^{-2} quadrats in January–February 1977. The absence of this shrimp along unvegetated sandy shores appears to be related to the absence of shelter (from physical effects of wave action or from predators) rather than to limitations of food resources. Large numbers of *C. nilotica* occur within the tangled 'rope' of vegetation (see Plate 1) which arises during periods of increasing lake level when the root material of terrestrial vegetation (largely monocotyledonous) is exposed and subsequently consolidated by wave wash. This tangled mass of vegetation generally occurs at the water's edge, and from it, large numbers of shrimps can be seen to move offshore at dusk to feed on the apparently 'clean' sandy terraces which are rich in diatoms, organic detritus and often faecal ropes left by the shoals of cichlid fish which feed upon such terraces during calm periods (Bruton, 1973; Bowen, 1976).

Other factors in addition to exposure *per se* would be expected to influence the distribution and abundance of the littoral fauna. While biomass

Plate 1. Consolidated plant material at the lake's edge providing a refuge for *Caridina nilotica* in an otherwise inhospitable area.

of *C. nilotica* was not correlated with soil organic content, hydrophyte biomass or water depth at the sampling sites, observations suggest that standing stocks of this shrimp were greatest where periphytic communities were conspicuous on the littoral hydrophytes. Clearly much scope exists for more critical analyses of factors controlling the distribution and abundance of the littoral fauna and particularly the smaller species and burrowing forms which the present analysis has largely ignored.

Temporal Events

The littoral habitats and particularly their associated flora are conspicuously affected by the fluctuations in water level which have been recorded during the past decade (see Chapter 3). Less conspicuous but equally significant changes must occur within the littoral fauna. Appleton (1977a,b) has discussed some effects of water level changes on the ecology of the snail *Biomphalaria pfeifferi.* Intuitively, it would appear that the effects of the hydrological regime in this lake are more marked upon the littoral communities than upon the benthic and planktonic communities.

This is particularly so for the population of the polymitarchid mayfly, *Povilla adusta.* Its larvae at low lake level occur sporadically in *Typha* stems. Following inundation of the fringing thorn-tree community the population rises sharply in response to the increase in suitable larval substratum.

Obviously competition for larval habitat space is reduced and the ensuing population explosion is primarily responsible for the collapse of the thorn-tree canopy into the water. Continued utilization of this newly submerged material by the larval stages aids in fragmentation of the trees and contributes to the allochthonous organic input to the lake. These changes are illustrated in Chapter 2.

The effect of fluctuating water levels is alternatively to flood marginal areas (resulting in a marked increase of allochthonous organic matter into the aquatic system) and then, during receding water levels to allow the return of more typical psammosere communities. The dynamic and striking changes in habitat structure of the littoral areas have not been quantified (but see Howard-Williams, this volume) and neither have the accompanying faunal changes been investigated except at a somewhat anecdotal level. The only quantitative data on faunal abundance relate to the changes in standing stocks of the shrimp *C. nilotica* recorded at three sites on the shore-line during 1975–1976, a period of flooding (Hart, unpublished). Some of these observations are given below.

Average standing stocks fluctuated between January 1975 and March 1976 with an apparently consistent decline in the population over the last 6 to 8 months of the study as shown in Fig. 17. This decline coincided with rising water levels the effect of which was to inundate the previously terrestrial habitats bordering the lake. The decrease in shrimp population at this time seemingly results from the dispersion of animals into the newly-available habitats, with the consequent localized reduction in population density rather than from a population decline. Sampling has shown that shrimps are common in the inundated marginal grasslands and, in fact, move quickly into freshly-inundated areas. Grasslands in which the grass *Imperata cylindrica* is abundant appear to be particularly favourable, seemingly as a result of the rich periphytic community (especially diatoms) which develops upon this grass after inundation (see Plate 2) and upon which *C. nilotica* in turn feeds. Likewise, the leaf-litter beneath flooded stands of the 'm' Doni' tree *Syzygium cordatum* also provides an apparently favoured habitat for the shrimps.

Reproductive output of *C. nilotica*, reflected in egg stock per unit area, while fluctuating widely at one site, remained fairly stable at a second, and showed no consistent seasonal trends (Fig. 17) suggesting that conditions for breeding did not alter drastically through the year. The number of eggs carried per female was higher on average in winter than in summer, as a result of the greater proportion of larger females which carry correspondingly larger clutches. This potentially favourable influence on recruitment potential is largely offset, however, by the proportionately slower embryonic development rates at winter temperatures (Hart, in press *a*).

Population size and sex structure also remained fairly stable during the study, apart from the presence both of much larger females, and of a greater proportion of large females in winter than in summer. However, differences in population size structure between two routine sampling sites (see Fig. 16) were in evidence; more specifically, the proportion of the smallest size class

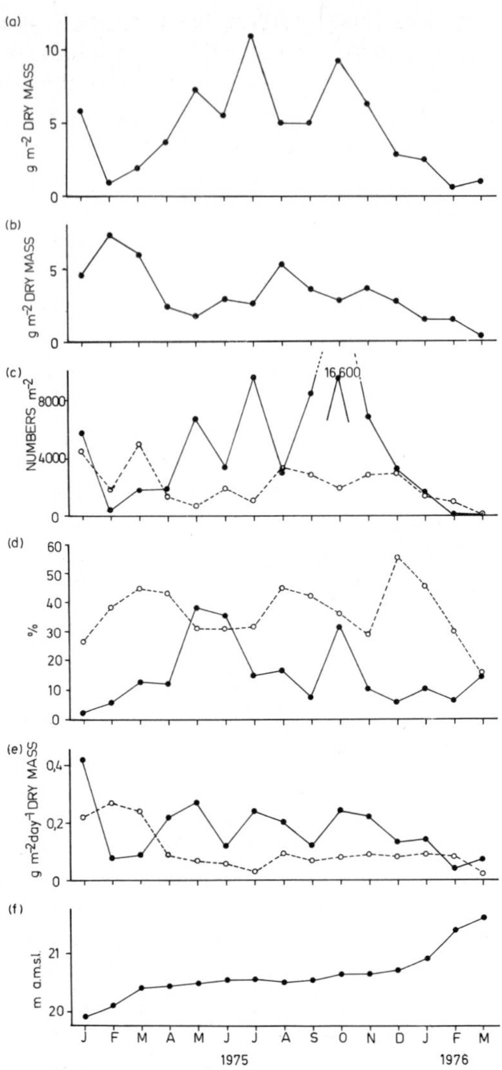

Fig. 17. Selected population attributes of *Caridina nilotica* at two shoreline sites on Lake Sibaya during 1975–1976: (a) Standing stocks of shrimp at Guguswana Bay and (b) Bande Bande; (c) Egg stocks at the two sites; (d) Percentage of population at each site comprising animals of less than 0.83 mm carapace length; (e) Estimates of daily production at each site, based upon summation of daily growth increments for the population; (f) Changes in lake level during the study period. For panels c, d and e:– Guguswana Bay (●——●) and Bande Bande (○- - -○).

of animals (<0.83 mm carapace length) was much lower at Guguswana Bay, a site characterized by higher standing stocks and egg stocks than Bande Bande, the second sampling locality (Fig. 17). On average, the proportion of these small shrimps at Guguswana Bay was half that recorded at Bande Bande, notwithstanding the fact that the average egg stock at Guguswana Bay was approximately twice that at Bande Bande. Differences in the

Plate 2. Scanning Electron Micrograph of the surface of a blade of *Imperata cylindrica* collected from an inundated stand of this grass, showing the dense periphytic diatom community of *Mastogloia elliptica* (×195).

pattern of survivorship would appear to be involved, although their origin remains speculative. Guguswana Bay appears to be a more diverse habitat, and may support greater diversity and density of predators than Bande Bande, with correspondingly higher predation pressure which might increase juvenile mortality especially.

Production of Littoral Invertebrates

Estimates of production exist only for the shrimp *Caridina nilotica*, and these are summarized in Table 17. Some idea of the month-to-month variations is given in Fig. 17. Widely differing estimates are obtained depending upon the method of estimation used, with the growth increment method yielding more conservative and presumably more realistic values than those based upon egg ratio methods, *i.e.* population turnover times. The two sites studied show rather different patterns of production based upon growth increments. At Bande Bande, average summer production was approximately twice that in the cool season, whereas in Guguswana Bay, mean summer and winter production was similar. The egg ratio method suggests very different results, with stable production at Bande Bande, but

Table 17. Estimates of production of *Caridina nilotica* at two sites on Lake Sibaya, based on monthly sampling (5 replicates and 3 replicates at Bande Bande and Guguswana Bay respectively) from January 1975 to March 1976. The distinction between summer and winter was made on the basis of monthly average temperatures. Summer – temperatures exceeding 24.0°C; *viz.* November to May, 10 months of study period. Winter – temperatures below 24.0°C *viz.* June to October, 5 months of study period.

		Growth increment method			Egg ratio method		
		Production ($g\,m^{-2}\,d^{-1}$) Mean	Range	Mean Daily P/B	Production ($g\,m^{-2}\,d^{-1}$) Mean	Range	Mean Daily P/B
Bande Bande	Summer	0.13	0.02–0.27	0.04	0.30	0.02–0.79	0.17
	Winter	0.07	0.03–0.09	0.02	0.28	0.09–0.48	0.22
Guguswana Bay	Summer	0.17	0.04–0.42	0.06	0.40	0.01–0.99	0.10
	Winter	0.18	0.12–0.24	0.03	1.17	0.21–2.58	0.16

high winter production at Guguswana Bay. On several counts, the egg ratio method yields apparently anomalous results. These call into question the appropriateness of this rather widely-used technique.

While the estimates of production, based on summation of growth increments over the population are currently being refined, the values given here are unlikely to be seriously misrepresentative. From them it is clear that as with the other invertebrates studied in this subtropical lake, production is a continuous process. Other members of the littoral invertebrate community, particularly various crustaceans are likely to show patterns of continuous production. It is probable that discontinuous production is a feature only of some of the more strongly seasonal-breeding insects whose aquatic larvae are important in the littoral areas. The dynamics of production of such groups may throw greater insight into the control and regulation of secondary productivity in the littoral habitats.

Comparisons with Other Southern African Coastal Lakes

The fauna of littoral margins appears not to have been examined to any appreciable extent in other southern African coastal lakes apart from limited taxonomic and distributional studies. Significant comparisons on the biology or ecology of littoral fauna between coastal lakes are seemingly precluded by lack of information.

General Discussion

Having attempted some description of the invertebrate communities it is appropriate to discuss some of the more general considerations which seem to apply to these communities.

One of the most consistent features of the invertebrates, although by no means exclusively confined to this group, is the generally low standing stocks supported by the system. Likewise levels of production are low. Taken together, it would appear that the system is oligotrophic, and the paucity of available P and N in the system (Allanson, this volume) apparently provides an explanation for this oligotrophy. What appears to set this oligotrophic system apart from other, for instance temperate oligotrophic waters, relates to the recycling rate of organic matter through the invertebrate populations, as reflected by the relatively high turnover times of the organisms studied. The more conservative estimate of the mean daily P/B ratio for the copepod *Pseudodiaptomus hessei* throughout the year indicates a population turnover-time of about nine days, which would permit 40 numerical population replacements (but not generations) per year. While the significance of annual P/B ratios for predictive purposes has been questioned (Waters, 1977) the use of daily P/B ratios is unlikely to be seriously misrepresentative and in this instance appears to provide a useful indication of turnover rates within the population.

As might be expected, the productivity of those perennially-present invertebrates studied is maintained at the same order of magnitude throughout the year in this subtropical system. Clear parallels in this regard exist with equatorial Lake George. Interesting contrasts with Lake George do exist, however. Although the productivity of *Thermocyclops hyalinus*, the dominant zooplankter in Lake George, is two orders of magnitude higher than that of *P. hessei* in Lake Sibaya, its mean daily P/B coefficient is considerably lower, reflecting a turnover-time of 12.5 days or about 29 turnovers per year. The causal factors for such differences can only be surmised. In how far they reflect differing nutritional conditions, life-history or other genetical strategies, or other factors is unknown.

Metabolic processes within poikilotherms are generally temperature dependent and rates normally increase positively with temperature. Over the annual temperature range experienced by the Sibaya invertebrate fauna, such temperature-related effects apply. For example, embryonic durations of the copepod *P. hessei* and the shrimp *C. nilotica* both vary with temperature (Hart & Allanson, 1975; Hart, in press a), and the respiratory demands of *C. nilotica* are also temperature dependent (Hart, in press b). However, there are indications that this sub-tropical fauna has thermally adapted, such that the metabolic processes are not as rapid proportionately as might be expected on the basis of comparisons with rates reported for temperate fauna, over *their* seasonal thermal range. The post-embryonic development rate of *P. hessei* at the annual average temperature which it experiences is not dissimilar from that recorded for comparable temperate calanoids at their much lower annual average temperature. Such thermal adaptation would appear to be energetically advantageous. Its generality requires confirmation of the sort provided for the embryonic development of marine copepods by McLaren *et al.* (1969).

Notwithstanding thermal adaptation, the metabolic costs of maintaining populations at high temperatures are high. The mechanism(s) by which these energetic costs are balanced in warm-water invertebrates require further examination for a proper understanding of warm-water ecosystems. Energetic studies on *Caridina nilotica* (Hart, in press a, b, and unpublished) have suggested that for this organism at least the energetic costs of the vital production processes (growth, reproduction, respiration) are not proportionately dissimilar from those recorded for quasi-comparable temperate invertebrates. However, it remains to be seen whether or not the ecological efficiency of warm-water invertebrates is generally improved or impaired by high temperatures, relative to that of their temperate counterparts, or whether physiological homeostases result in some measure of universal uniformity.

From the preceding sections, it will be obvious that a major gap exists in knowledge of the invertebrate communities, namely the feeding biology of representative members. While it has been demonstrated or inferred that certain animals are variously filter-feeders, surface scrapers, substrate ingestors or predators, the precise nature of the food resources and its utilization has not been examined to any significant extent. Observations on the copepod *Pseudodiaptomus hessei* have suggested that it filters nannoplankton and bacterioplankton at greater rates than net phytoplankton. From this it has been inferred that it relies largely upon the former categories of food items. It remains to be shown that this inference is correct, and it is fundamentally important to be able to describe the primary origin of the heterotrophic food resources. In other words, the relative importance of autochthonous and allochthonous particulate organic matter in the energetic functioning of the system requires examination.

The feeding biology of several benthic invertebrates which scrape sand grains and gyttja on the lake bed is equally poorly understood. The importance of organic detritus with its associated microflora in supporting the sometimes vast standing stocks of benthic invertebrates below the euphotic zone can only be inferred. Various questions are pertinent in this context. Is this material primarily phytoplanktonic in origin? To what extent are heterotrophic recycling processes associated with this organic floc responsible for the maintenance of this benthic fauna, or is this fauna dependent largely upon a continual input of 'new' organic matter from the euphotic zone or allochthonous sources? The analysis of such problems poses serious methodological difficulties, especially at an isolated field station. The difficulty is compounded by the sometimes catholic diets of component species. The shrimp *Caridina nilotica* provides a good example. This animal feeds by scraping surfaces with its highly modified setose chelipeds (Fryer, 1960). In those littoral areas supporting extensive periphyton communities, its diet is mixed, consisting of autotrophic and heterotrophic organisms. In other littoral areas, it forages extensively on decaying plant material, and it largely ingests detritus with its associated heterotrophic microflora. Its diet in submerged hydrophyte stands is presumably a mixture of autotrophic and heterotrophic periphytic organisms, but must include a largely heterotrophic component associated with decaying plant fragments. On profundal bottoms it is surmised that autotrophic organisms largely disappear from its diet. In addition, the shrimp readily scavenges when the opportunity arises. The feeding ecology of this opportunistic 'detritivore' is apparently geared to the food resources available, and attempts to quantify the relative importance in its diet of auto- and allochthonous organics will have to take into account spatial and temporal variations in its food environment, considerably complicating the analysis. Similar complications undoubtedly apply to other species.

In the descriptive accounts of the zooplankton and zoobenthos, especially the latter, the existence and often importance of marine-estuarine species in this essentially fresh-water system was mentioned (see also Allanson *et al.* 1966). Thus another interesting facet to the biology of invertebrates of Lake Sibaya relates to the osmoregulatory physiology of these species. Forbes & Hill (1969) investigated experimentally the ability of the crab *Hymenosoma orbiculare* to survive and breed in freshwater. They showed that the crabs possessed considerable osmoregulatory abilities, but revealed that the zoeae larvae hatched from eggs carried by females of marine stock could not survive salinities as low as those found in Lake Sibaya. They suggested that the calcium enrichment in the waters of Lake Sibaya (Allanson & van Wyk,1969) could be important in permitting survival and development of zoeae in Lake Sibaya, by reducing the permeability of the body surface in the classical manner. On the basis of observations that crabs from Lake Sibaya were able to withstand transfer back to sea water for at least six days, they suggested that the Sibaya population of this crab did not constitute a distinct physiological race.

A similar conclusion was drawn by Boltt (1969b) regarding the amphipod *Grandidierella lignorum.* Estuarine stocks of this species were able to tolerate freshwater conditions, given some acclimation in diluted sea water. Their natural occurrence in Lake Sibaya, but absence from truly freshwaters at the heads of estuaries in which they occurred was ascribed to the marginally elevated NaCl concentrations in Lake Sibaya.

The ionic-osmotic physiology of other marine-estuarine species living in Lake Sibaya has not been investigated experimentally. The copepod *Pseudodiaptomus hessei* appears to be extremely euryhaline, occurring elsewhere both in waters considerably less saline and in waters more saline than sea water, although the salinities it experiences in Lake Sibaya (~0.5‰) appear to be the lowest from which it is known.

More problematic in a sense is the occurrence of various estuarine benthic invertebrates in Lake Sibaya which are not found in the considerably more saline Lake Nhlange, a mere 60 km north of Lake Sibaya (Boltt & Allanson, 1975). The origin of these differences does not appear to lie within the realm of ionic-osmotic physiology, but rather in other aspects of the animals' biology (Boltt & Allanson, 1975).

In conclusion, it may be suggested that while some of the invertebrates of Lake Sibaya may be among the better known of those in freshwater environments in Africa, much remains to be discovered about their precise functional role in the ecosystem. Much of the groundwork upon which such analyses depend has been provided, and it remains a personal hope that the studies on this pristine sub-tropical lake may be extended before adverse demophoric developments occur.

Acknowledgements

It is with great pleasure that I acknowledge the help, advice and stimulus given me by my colleagues, the late Dr R. E. Boltt, Dr B. J. Hill, and especially Professor B. R. Allanson during my studies on Lake Sibaya. My hope is that their rigorous scientific discipline and intuition, so freely shared, has rubbed off, resulting in a better contribution to this present volume. I am grateful for the opportunity afforded me of examining unpublished data and information in drafted manuscripts of the late Dr R. E. Boltt which I have incorporated in the text, and to Drs B. J. Hill, A. T. Forbes and C. Howard-Williams for placing unpublished information at my disposal. Mrs Pat Eva patiently and meticulously typed and retyped the various drafts; Robin Allanson and my wife, Rei, prepared the figures. To these people, as well as Professor B. R. Allanson and Dr C. Howard-Williams who helped in various other ways, I offer my sincere thanks.

References

Allanson, B. R., M. N. Bruton & R. C. Hart. 1974. The plants and animals of Lake Sibaya, KwaZulu, South Africa: A checklist. Rev. Zool. afr. 88 (3): 507–533.

Allanson, B. R. & J. D. van Wyk. 1969. An introduction to the physics and chemistry of some lakes in northern Zululand. Trans. R. Soc. S. Afr. 38: 217–240.

Allanson, B. R., B. J. Hill, R. E. Boltt & V. Schultz. 1966. An estuarine fauna in a freshwater lake in South Africa. Nature, Lond. 209: 532–533.

Appleton, C. C. 1976. The influence of abiotic factors on the distribution of *Biomphalaria pfeifferi* (Krauss, 1848) (Planorbidae: Mollusca) and its life-cycle in south-eastern Africa. Unpublished M.Sc. dissertation, Rhodes University.

Appleton, C. C. 1977a. The fresh-water Mollusca of Tongaland, with a note on molluscan distribution in Lake Sibaya. Ann. Natal Mus. 23: 129–144.

Appleton, C. C. 1977b. The influence of temperature on the life-cycle and distribution of *Biomphalaria pfeifferi* (Krauss, 1948) in south-eastern Africa. Int. J. Parasitol. 7: 335–345.

Appleton, C. C. 1977c. The influence of above-optimal constant temperatures on South African *Biomphalaria pfeifferi* (Krauss) (Mollusca: Planorbidae). Trans. R. Soc. trop. Med. Hyg. 71: 140–143.

Balon, E. K. & A. G. Coche. 1974. Lake Kariba, a man-made tropical ecosystem in central Africa. Monographiae Biologicae 24: xii+767. W. Junk, The Hague.

Batchelor, G. R. 1972. A preliminary investigation of the benthos of West Kleinemond Estuary. Honours thesis, Rhodes University, Grahamstown.

Blaber, S. J. & A. K. Whitfield. 1977. The biology of the burrowing goby *Croilia mossambica* Smith (Teleostei, Gobiidae). Environ. Biol. Fishes 1: 197–204.

Boltt, R. E. 1969a. The benthos of some southern African lakes. Part II. The epifauna and infauna of the benthos of Lake Sibayi. Trans. R. Soc. S. Afr. 38: 249–269.

Boltt, R. E. 1969b. A contribution to the benthic biology of some southern African lakes. Unpublished Ph.D. dissertation, Rhodes University.

Boltt, R. E. 1975. The benthos of some southern African lakes. Part IV: The benthos of Lagoa Poelela. Trans. R. Soc. S. Afr. 41: 273–281.

Boltt, R. E. & B. R. Allanson. 1975. The benthos of some southern African lakes. Part III: The benthic fauna of Lake Nhlange, KwaZulu, South Africa. Trans. R. Soc. S. Afr. 41: 241–262.

Boltt, R. E., B. J. Hill & A. T. Forbes. 1969. The benthos of some southern African lakes. Part I. Distribution of aquatic macrophytes and fish in Lake Sibayi. Trans. R. Soc. S. Afr. 38: 241–248.

Bowen, S. H. 1976. Feeding ecology of the cichlid fish *Sarotherodon mossambicus* in Lake

Sibaya, KwaZulu. Ph.D. dissertation, Rhodes University.
Bruton, M. N. 1973. A contribution to the biology of *Tilapia mossambica* Peters in Lake Sibaya, South Africa. M.Sc. dissertation, Rhodes University.
Caulton, M. S. 1975. The ability of the cichlid fishes *Tilapia rendalli* Boulenger, *Tilapia sparrmanii* A. Smith and *Hemihaplochromis* (= *Pseudocrenilabrus*) *philander* (M. Weber) to enter deep water. J. Fish Biol. 7: 513–517.
Caulton, M. S. & B. J. Hill. 1973. The ability of *Tilapia mossambica* (Peters) to enter deep water. J. Fish Biol. 5: 783–788.
Edmondson, W. T. 1960. Reproductive rates of rotifers in natural populations. Memorie Ist. ital. Idrobiol. 12: 21–77.
Elster, H. J. 1954. Über die populationsdynamik von *Eudiaptomus gracilis* Sars und *Heterocope borealis* Fischer im Bodensee-Obersee. Arch. Hydrobiol. 20: 546–614.
Forbes, A. T. & B. J. Hill. 1969. The physiological ability of a marine crab *Hymenosoma orbiculare* Desm. to live in a subtropical freshwater lake. Trans. R. Soc. S. Afr. 38: 271–283.
Fryer, G. 1960. The feeding mechanism of some Atyid prawns of the genus *Caridina*. Trans. R. Soc. Edinb. 64: 217–244.
Grindley, J. R. & T. Wooldridge. 1973. The plankton of the Wilderness lagoons. Unpublished report – Port Elizabeth Museum.
Hart, R. C. 1976. The substrate bin – a new sampling device for studying diel vertical migratory movements on to and off lake sediments. Freshwater Biol. 6: 155–159.
Hart, R. C. 1977. Feeding rhythmicity in a migratory copepod (*Pseudodiaptomus hessei* (Mrázek)). Freshwater Biol. 7: 1–8.
Hart, R. C. 1978. Horizontal distribution of the copepod *Pseudodiaptomus hessei* in a subtropical Lake Sibaya. Freshwater Biol. 8: 415–421.
Hart, R. C. (in press a) Embryonic duration, and post-embryonic growth rates of the tropical freshwater shrimp *Caridina nilotica* (Decapoda: Atyidae) under laboratory and experimental field conditions. Freshwater Biol.
Hart, R. C. (in press b). Oxygen consumption in *Caridina nilotica* (Decapoda: Atyidae) in relation to temperature and size. Freshwater Biol.
Hart, R. C. & B. R. Allanson. 1975. Preliminary estimates of production by a calanoid copepod in subtropical Lake Sibaya. Verh. int. Ver. Limnol. 19: 1434–1441.
Hart, R. C. & B. R. Allanson. 1976. The distribution and diel vertical migration of *Pseudodiaptomus hessei* (Mrázek) (Calanoida: Copepoda) in a subtropical lake in southern Africa. Freshwater Biol. 6: 183–198.
Hart, R. C. & R. Hart. 1977. The seasonal cycles of phytoplankton in subtropical Lake Sibaya: A preliminary investigation. Arch. Hydrobiol. 80: 85–107.
Hemens, J., H. Ketz, W. D. Oliff & D. E. Simpson. 1971. Natal coast estuaries: Environmental surveys. Survey No. 3 – The Kosi Bay estuarine lakes. C.S.I.R.–N.I.W.R. report.
Hutchinson, G. E. 1976. A treatise on Limnology. Vol. 2. Introduction to Lake Biology and the Limnoplankton. xi + 1115. John Wiley & Sons, Inc. New York.
McLaren, I. A., C. J. Corkett & E. J. Zillioux. 1969. Temperature adaptation of copepod eggs from the arctic to the tropics. Biol. Bull. mar. biol. Lab., Woods Hole 137: 486–493.
Waters, T. F. 1977. Secondary production in inland waters. Adv. Ecol. Res. 10: 91–164.
Winberg, G. G. (Ed.). 1971. Methods for the estimation of production of aquatic animals. xii + 175. Academic Press, London.

Appendix

Checklist of Invertebrate fauna of Lake Sibaya (modified from Allanson, Bruton & Hart, 1974).

PORIFERA

Freshwater sponge

COELENTERATA

Hydra sp.

TURBELLARIA (unsorted)

ANNELIDA

POLYCHAETA

Ceratonereis keiskama Day, 1953

OLIGOCHAETA

Tubificidae (unsorted)

Naididae (unsorted)

HIRUDINEA

Glossiphoniidae

Placobdella unita Moore, 1958

Batracobdella tricarinata (Blanchard, 1897)

Hirudidae

Limnatis fenestrata Moore, 1939

MOLLUSCA

PROSOBRANCHIA

Thiridae

Melanoides tuberculatus (Müller, 1774)

Viviparidae

Bellamya capillata (Frauenfeld, 1865)

PULMONATA Basommatophora

Lymnaeidae

Lymnaea natalensis Krauss, 1848

Planorbidae

Biomphalaria pfeifferi Krauss, 1848

Bulinus (*Bulinus*) *natalensis* (Küster, 1841)

Bulinus (*Physopsis*) *globosus* (Morelet, 1866)

Gyraulus costulatus Krauss, 1848

Ceratophallus sp. Brown & Mandahl-Barth, 1973.

Ancylidae

Burnupia sp.

PULMONATA Stylommatophora

Succineidae

Succinea patentissima Pfeiffer, 1853

Succinea striata Krauss, 1848

LAMELLIBRANCHIA

Cyrenidae

Corbicula africana (Krauss, 1848)

Sphaeriidae

Sphaerium capense (Krauss, 1848)

Eupera ferruginea (Krauss, 1848)

CRUSTACEA

CLADOCERA

Bosmina longirostris (O. F. Müller, 1785)
Simocephalus vetulus (O. F. Müller, 1776)
Moina sp.

OSTRACODA

Cypria sp.
Darwinula sp.
Heterocypris sp.
Loxoconcha sp.
Perissocytheridae sp. aff. *estuaria* Benson & Maddocks, 1964

COPEPODA

Cyclopoida

Mesocyclops leuckarti aequatorialis forma *micrura* (Kiefer, 1928)
Thermocyclops crassus consimilis (Kiefer, 1934)
Thermocyclops emini (Mrázek, 1895)
Tropocyclops brevis Dussart, 1972

Calanoida

Pseudodiaptomus hessei (Mrázek, 1895)

BRANCHIURA

Dolops ranarum (Stuhlmann 1891)

TANAIDACEA

Heterotanais nov. sp.
Leptochelia sp.

Apseudidae

Apseudes digitalis Brown, 1956 (? *Apseudes cooperi*)

ISOPODA

Anthuridae

Cyathura carinata (Kröyer, 1847)

Euridicidae

Pontogeloides latipes Barnard, 1914

Munnidae

Paramuna laevifrons Stebbing, 1910

Sphaeromidae

Pseudosphaeroma barnardi Monod, 1931

AMPHIPODA

Aoridae

Grandidierella lignorum Barnard, 1935

Corophiidae

Corophium triaenonyx Stebbing, 1904

Talitridae

Afrochiltonia capensis (Barnard, 1916)
Orchestia ancheidos Barnard, 1916

DECAPODA

Hymenosomatidae

Hymenosoma orbiculare Desmarest, 1825

Potamonidae

Potamon (*Potamonautes*) *sidneyi* Rathbun, 1937

Atyidae

Caridina nilotica (P. Roux, 1833)

ARACHNIDA

Hydracarina (unsorted)
Pisauridae
Thalassius spenceri Pickard-Cambridge, 1898.

INSECTA

EPHEMEROPTERA
Polymitarchidae
Povilla adusta Navás
Baetidae
Baetis? bellus
Cloeon? crassi
Austrocaenis? capensis

ODONATA Anisoptera
Aeshnidae
Aeshna minuscula McLachlan, 1895
Anax imperator mauricianus Rambur, 1842
Gomphidae
Ictinogomphus ferox Rambur, 1842
Paragomphus hageni (Sélys, 1870)
Libellulidae
Acisoma panorpoides ascalaphoides Rambur, 1842
Brachythemis leucosticta (Burmeister, 1839)
Chalcostephia flavifrons Kirby, 1889
Orthetrum farinosum Förster, 1898
Palpopleura lucia (Drury, 1773)
Pantala flavescens (Fabricius, 1798)
Philonomon luminans (Karsch, 1893)
Rhyothemis semihyalina (Desjardins, 1835)
Trapezostigma basilaris Beauvais, 1805
Trapezostigma limbatum (Desjardins, 1832)
Trithemis arteriosa (Burmeister, 1839)
Trithemis sp.
Urothemis edwardsi Sélys, 1849

ODONATA Zygoptera
Agrionidae
Phaon iridipennis iridipennis (Burmeister, 1839)
Coenagriidae
Agriocnemis exilis Sélys, 1872
Ceriagrion glabrum Burmeister, 1839
Enallagma nigridorsum Sélys, 1876
Ischnura senegalensis Rambur, 1842
Pseudagrion massaicum Sjöstedt, 1909
Pseudagrion nigerrimum Pinhey, 1950
Lestidae
Lestes tridens McLachlan, 1895

HEMIPTERA
Belostomatidae
Sphaerodema capensis (Mayr, 1871)
Gerridae
Limnogonus hypoleucus (Gerstaecker, 1873)
Mesovelliidae
Mesovelia vittigera Horvath, 1895
Naucoridae
Laccocoris limigenus (Stål, 1865)

Pleidae
Plea pullulà Stål, 1855
Ranatridae
Ranatra parvipes Signoret, 1860

TRICHOPTERA
Dipseudopsis capensis Walker, 1852
Dipseudopsis simplex Ulmer, 1906
Ecnomus thomasetti Mosely, 1932
Homilia knysnaensis Barnard, 1940
Leptocerus inflatus Kimmins, 1962
Oecetis sibayiensis Scott, 1968
Parasetodes maguira (Mosely, 1948)
Orthotrichia sp.

DIPTERA
Chironomidae
Chironomus spp.
Cryptochironomus spp.
Polypedilum spp.
Tanytarus sp.
Micropsectra sp.
Tanypus sp.
Procladius sp.
Ablabesmyia spp.
Corynoneura sp.
Cricotopus sp.
Ceratopogonidae
Tipulidae

COLEOPTERA
Dytiscidae
Canthydrus notula (Erichson, 1843)
Cybister marginicollis Bohman, 1848
Hydrophillidae } unsorted
Gyrinidae }

LEPIDOPTERA
Nymphula? capensis

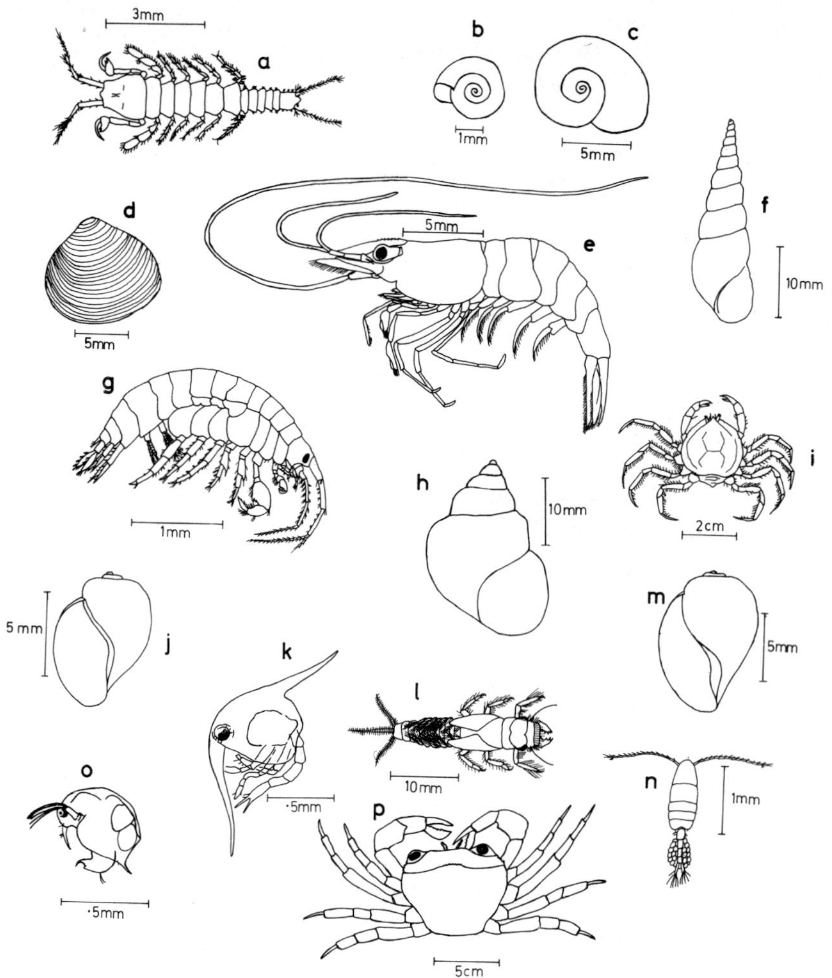

Appendix Figure. Outline sketches of some dominant or important invertebrates encountered in Lake Sibaya. (a) *Apseudes digitalis* (after Brown, 1956); (b) *Ceratophallus* sp. (after Brown & Mandahl-Barth, 1973); (c) *Biomphalaria pfeifferi* (after Oberholzer & van Eeden, 1967); (d) *Corbicula africana* (after Crowley, Pain & Woodward, 1964); (e) *Caridina nilotica* (from Fryer, 1960 and from Kensley, 1972); (f) *Melanoides tuberculatus* (after Oberholzer & van Eeden, 1967); (g) *Grandidierella lignorum* (from Griffiths, 1976); (h) *Bellamya* sp. (after van Eeden, 1960); (i) *Hymenosoma orbiculare* (after Broekhuysen, 1955); (j) *Bulinus* (*B.*) *natalensis* (after Oberholzer & van Eeden, 1967); (k) *Hymenosoma orbiculare* zoea larva (after Broekhuysen, 1955); (l) *Povilla* sp. (after Agnew, in press); (m) *Bulinus* (*P.*) *globosus* (after Oberholzer & van Eeden, 1967); (n) *Pseudodiaptomus hessei* (from Grindley, 1971); (o) *Bosmina longirostrus* (from Scourfield & Harding, 1966); (p) *Potamon sidneyi* (original).

Appendix References

Agnew, J. (in press). Ephemeroptera. In: Textbook of sub-Saharan entomology.
Allanson, B. R., M. N. Bruton & R. C. Hart. 1974. The plants and animals of Lake Sibaya, KwaZulu, South Africa: A checklist. Rev. Zool. afr. 88: 507–532.
Broekhuysen, G. J. 1955. The breeding and growth of *Hymenosoma orbiculare.* Ann. S. Afr. Mus. 41: 313–343.
Brown, A. C. 1956. Additions to the genus *Apseudes* (Crustacea: Tanaidacea) from South Africa. Ann. Mag. nat. Hist., ser. 12 9: 705–709.
Brown, D. S. & G. Mandahl-Barth. 1973. Two new genera of Planorbidae from Africa and Madagascar. Proc. malac. Soc. Lond. 40: 287–302.
Crowley, T. E., T. Pain & F. R. Woodward. 1964. A monographic review of the Mollusca of Lake Nyasa. Ann. Mus. r. Afr. Cent. 8°, Zool. 131: 1–58.
Fryer, G. 1960. The feeding mechanism of some Atyid prawns of the genus *Caridina.* Trans. R. Soc. Edinb. 64: 217–244.
Griffiths, C. L. 1976. Guide to the benthic marine amphipods of southern Africa. 106 pp. South African Museum Publication, Cape Town.
Grindley, J. R. 1971. Zooplankton. In: Animal life in Southern Africa. Nasau Ltd., Cape Town.
Kensley, B. 1972. Shrimps & Prawns of Southern Africa. 65 pp. South African Museum Publication, Cape Town.
Oberholzer, G. & J. A. van Eeden. 1967. The freshwater molluscs of the Kruger National Park. Koedoe 10: 1–42.
Scourfield, D. J. & J. P. Harding. 1966. A key to the British species of freshwater Cladocera. Freshwater Biological Association. Scientific Publication No. 5. 3rd Edition 55 pp.
van Eeden, J. A. 1960. Key to the genera of South African freshwater and estuarine gastropods (Mollusca). Ann. Transv. Mus. 24: 1–17.

8 The fishes of Lake Sibaya

M. N. Bruton

Previous Research

No scientists are known to have visited Lake Sibaya until the beginning of the present century. In 1906 and 1907 Dr E. Warren, then of the Natal Museum, collected *Tilapia rendalli*, *Sarotherodon mossambicus* and *Pseudocrenilabrus philander* from Lake Sibaya and dispatched them to the British Museum (Natural History) in London. These specimens were incorporated in Boulenger's (1915) catalogue of the freshwater fishes of Africa, but the *S. mossambicus* specimens were attributed to Kosi.

In August 1935 twenty-nine large-mouth bass *Micropterus salmoides*, 7–15 cm long, were introduced into Lake Sibaya by Mr B. Nicholson (Harrison, 1936). These fishes had been transported from Mbabane in Swaziland and were introduced into the lake to provide a sports fishery. The bass have apparently not survived as none have been caught since their introduction. Harrison (1936) mentioned that large bass fingerlings were introduced into the lake due to the presence there of the tiger-fish *Hydrocynus lineatus* (=*H. vittatus*). This erroneous locality record for *H. vittatus* was repeated by Barnard (1948, who voiced some sceptism), Smith (1951) and Jubb (1952), but corrected by Crass (1960) and Jubb (1967).

The Tongaland expeditions of 1947, 1948 and 1949, organized by the Natal Society for the Preservation of Wildlife and Natural Resorts, worked mainly at Kosi but small fish collections were made at Lake Sibaya. *Aplocheilichthys myaposae* and *A. johnstonii*, the latter possibly confused with *A. katangae*, were reported from the lake by Campbell & Allanson (1952) and *Tilapia*, 'gobies' and 'barbel' by Campbell (1947, 1969). In 1958 a Natal Parks Board Game Ranger, K. L. Tinley, surveyed aspects of the ecology of Lake Sibaya and collected eleven species of fish from the main lake and Skombane stream (Tinley, 1958, 1976; also mentioned by Crass, 1960, 1964). These papers represent the first record of the presence in Lake Sibaya of two fish species of marine origin (*Gilchristella aestuarius* and *Hepsetia breviceps*). The traditional methods used by Africans to capture fishes in Lake Sibaya and other coastal lakes and rivers in northern Zululand and southern Mozambique were also reviewed by Tinley (1964).

In 1965 an expedition from Rhodes University under the direction of B. R. Allanson visited the lake, thus beginning a 12 year association. In an early note Allanson *et al.* (1966) noted the presence of many estuarine and marine organisms, both vertebrate and invertebrate, in the lake, and Boltt *et al.* (1969) reported the discovery of two interesting gobies–*Croilia mossambica*, an elongate burrowing form previously known only from Lagoa Poelela in Mozambique, and *Silhouettea sibayi*, a small cryptic species described by Farquharson (1970) as new. The presence of *Barbus natalensis* in Lake Sibaya (Boltt *et al.*, 1969) has not been confirmed. Subsequent collections by Pike (1969), Allanson *et al.* (1974) and Bruton (1974) led to the final listing of 18 fish species from the main lake.

The establishment of a research station at Lake Sibaya in 1968 led to the first ecological research on the fishes. The pioneering work of Pike (1969) and Minshull (1968, 1969, 1970) on the distribution, food and condition of Sibaya fishes laid the foundation for more detailed studies on growth, breeding, environmental tolerances, resource partitioning and predation, which are reviewed here. Comparative studies have also been made with the ichthyofauna of other coastal lakes on the Mozambique plain.

Faunal Diversity

The fish fauna of Lake Sibaya consists of 18 species (Table 1, Figs. 1 to 5) and is dominated by cichlids (4 species) and gobiids (3). When this ichthyofauna is compared with that of other lakes, estuarine lagoons and floodplain rivers of the southern Mozambique plain, (Table 2), several trends are evident:

a) the large estuarine lagoons with direct openings to the sea, such as Lakes St Lucia and Kosi, have a high ichthyofaunal diversity comprised mainly of euryhaline/marine species which inhabit their brackish or estuarine sections. (The very high number of fishes in the Kosi system is due largely to the presence of 43 marine reef species around a rocky outcrop in the estuary (Blaber, 1978a).)

b) lakes with a tenuous connection to the sea, such as Poelela, Mgobezeleni and Piti have a relatively low diversity of fish. Lagoa Poelela, which is brackish (8‰, Hill *et al.*, 1975), has more euryhaline/marine fishes than primary freshwater forms, whereas Mgobezeleni and Piti which are fresh, are dominated by freshwater species.

c) Lake Sibaya, which is isolated and fresh, has a more diverse ichthyofauna than the other freshwater coastal lakes. Although it has mainly freshwater fishes, the euryhaline/marine component (33%) is important, as in the other coastal lakes.

d) the three large rivers which cross the southern Mozambique plain (Pongola, Limpopo and Incomati) have diverse fish populations dominated by freshwater species.

The total number of primary and secondary freshwater fish species in the rivers, lakes and lagoons of the southern part of the Mozambique plain (49, Bruton & Cooper, 1979) represents an extremely diverse fauna which comprises 23% of all freshwater fish species known from southern Africa (212; Jackson, 1975). This diversity can be explained in terms of the great variety of

Table 1. Checklist of the fishes of Lake Sibaya.

CLUPEIDAE	
Gilchristella aestuarius (Gilchrist & Thompson, 1916)	estuarine round-herring
MORMYRIDAE	
Marcusenius macrolepidotus (Peters, 1852)	bulldog
CYPRINIDAE	
Barbus paludinosus Peters, 1852	straightfin barb
Barbus viviparus Weber, 1897	bowstripe barb
Labeo molybdinus du Plessis, 1963	leaden labeo
CLARIIDAE	
Clarias gariepinus (Burchell, 1822)	sharptooth catfish
Clarias theodorae Weber, 1897	snake catfish
CYPRINODONTIDAE	
Aplocheilichthys katangae (Boulenger, 1912)	striped minnow
Aplocheilichthys myaposae (Boulenger, 1908)	Natal topminnow
ATHERINIDAE	
Hepsetia breviceps (Cuvier in C. & V., 1835)	Cape silverside
CICHLIDAE	
Pseudocrenilabrus philander (Weber, 1897)	southern mouthbrooder
Sarotherodon mossambicus (Peters, 1852)	Mozambique tilapia
Tilapia rendalli swierstrae Gilchrist & Thompson, 1917	southern redbreast tilapia
Tilapia sparrmanii Smith, 1840	banded tilapia
GOBIIDAE	
Croilia mossambica Smith, 1955	burrowing goby
Glossogobius giurus (Hamilton-Buchanan, 1822)	tank goby
Silhouettea sibayi Farquharson, 1970	Sibayi goby
ANABANTIDAE	
Ctenopoma multispinis Peters, 1844	manyspined climbing perch

habitats available on the southern Mozambique plain (from floodplain rivers and shallow swamps to deep coastal lakes), the relatively warm waters of the region, and the presence of invasion routes down the low-lying Mozambique plain and along inland drainages from the diverse tropical ichthyofaunae to the north.

Origin and Zoogeography

It is generally considered that central tropical Africa, especially the Zaire river, is the source of origin of most freshwater fishes in southern Africa (Bowmaker *et al.*, 1978). On the basis of the similarity of the ichthyofaunae

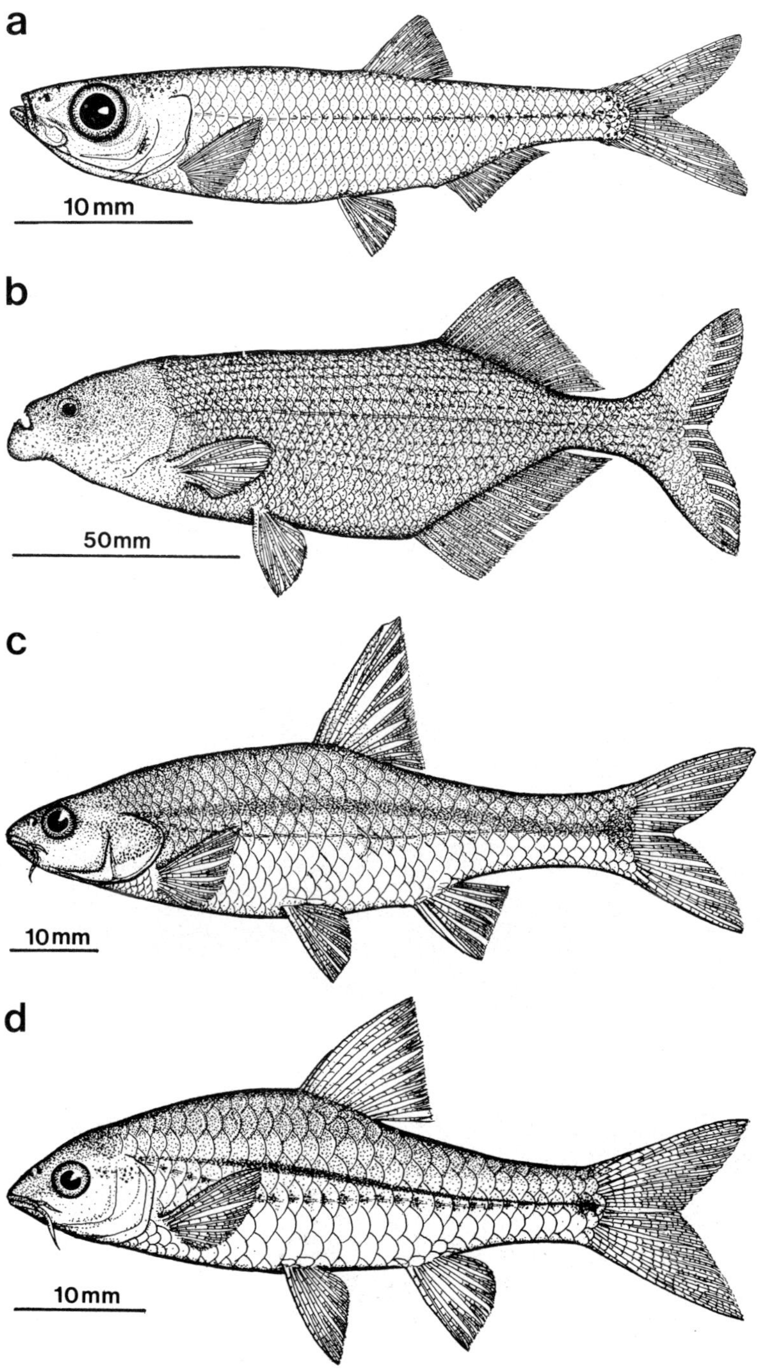

Fig. 1. A. *Gilchristella aestuarius* (Gilchrist &Thompson, 1916) B. *Marcusenius macrolepidotus* (Peters, 1852) C. *Barbus paludinosus* Peters, 1852 D. *Barbus viviparus*, M. Weber, 1897. Drawn by E. M. Tarr.

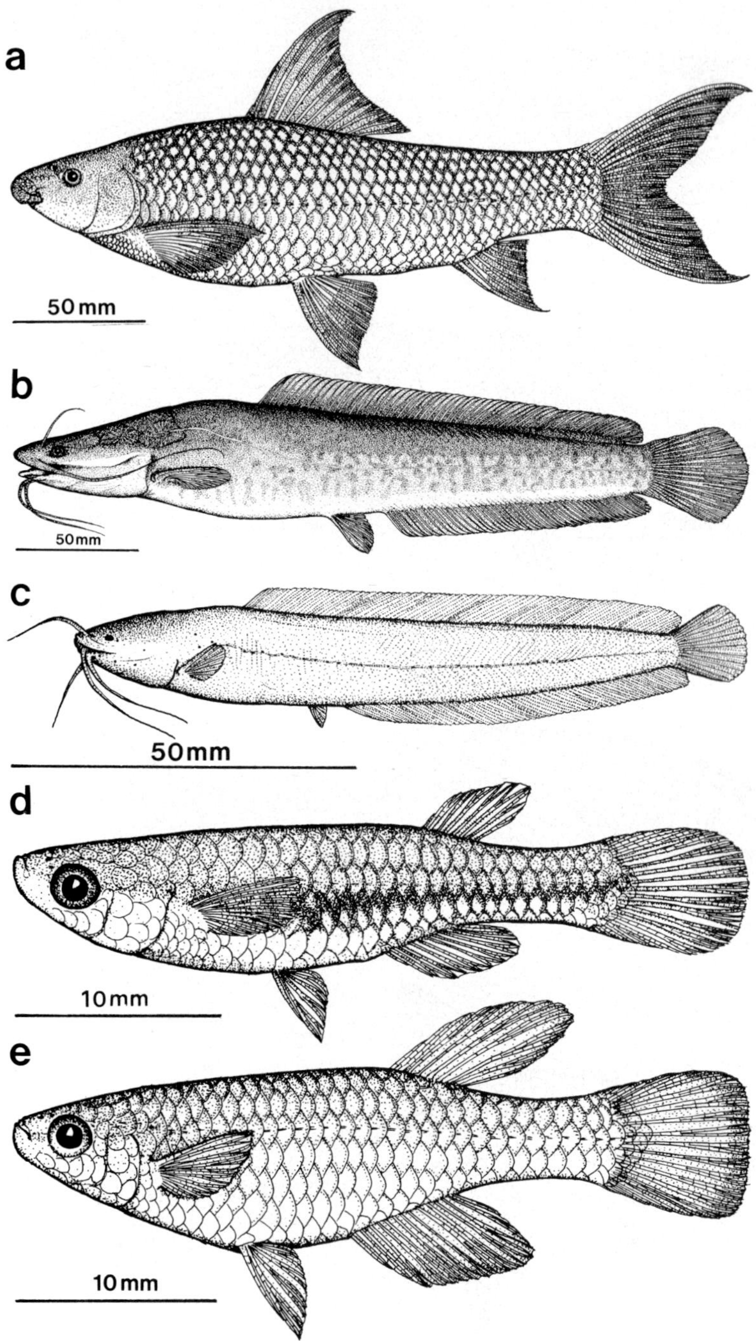

Fig. 2. A. *Labeo molybdinus* du Plessis, 1963 B. *Clarias gariepinus* (Burchell, 1822) C. *Clarias theodorae* M. Weber, 1897 D. *Aplocheilichthys katangae* (Boulenger, 1912) E. *Aplocheilichthys myaposae* (Boulenger, 1908). Drawn by E. M. Tarr.

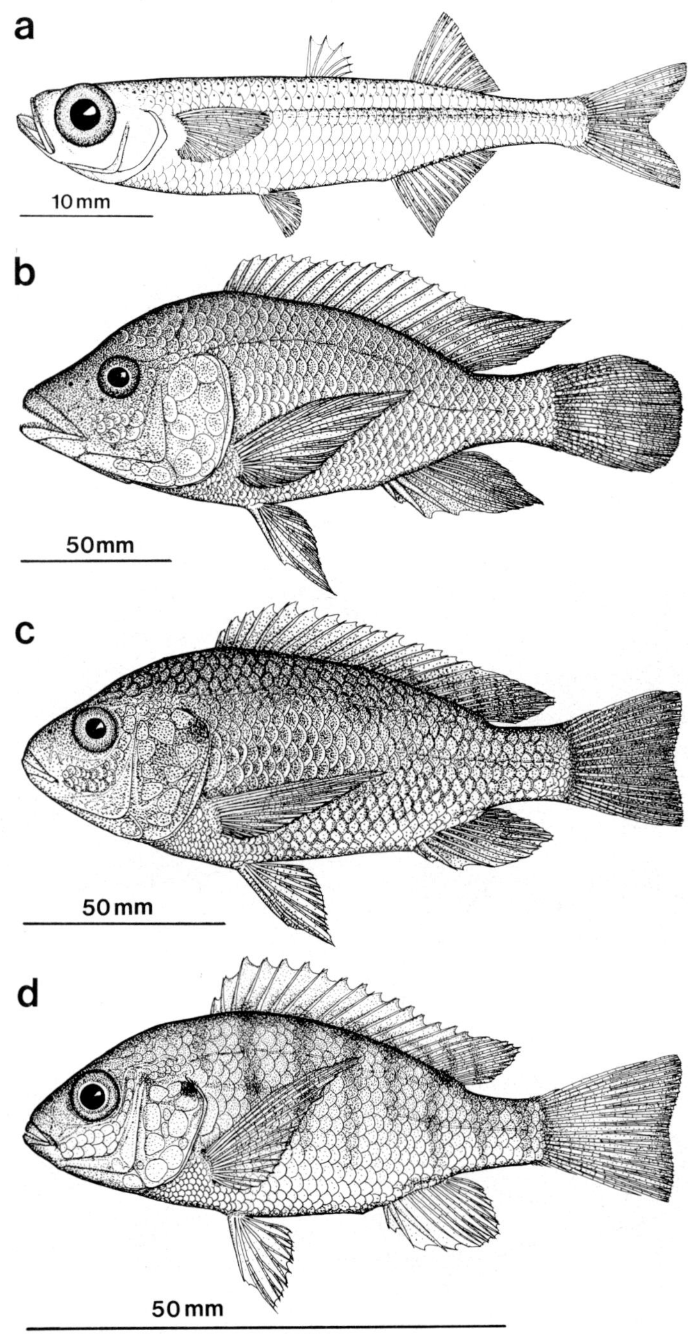

Fig. 3. A. *Hepsetia breviceps* Cuv., in C & V, 1835 B. *Sarotherodon mossambicus* (Peters, 1852) Mature male C. *Sarotherodon mossambicus* (Peters, 1852) Mature female D. *Sarotherodon mossambicus* (Peters, 1852) Juvenile. Drawn by E. M. Tarr.

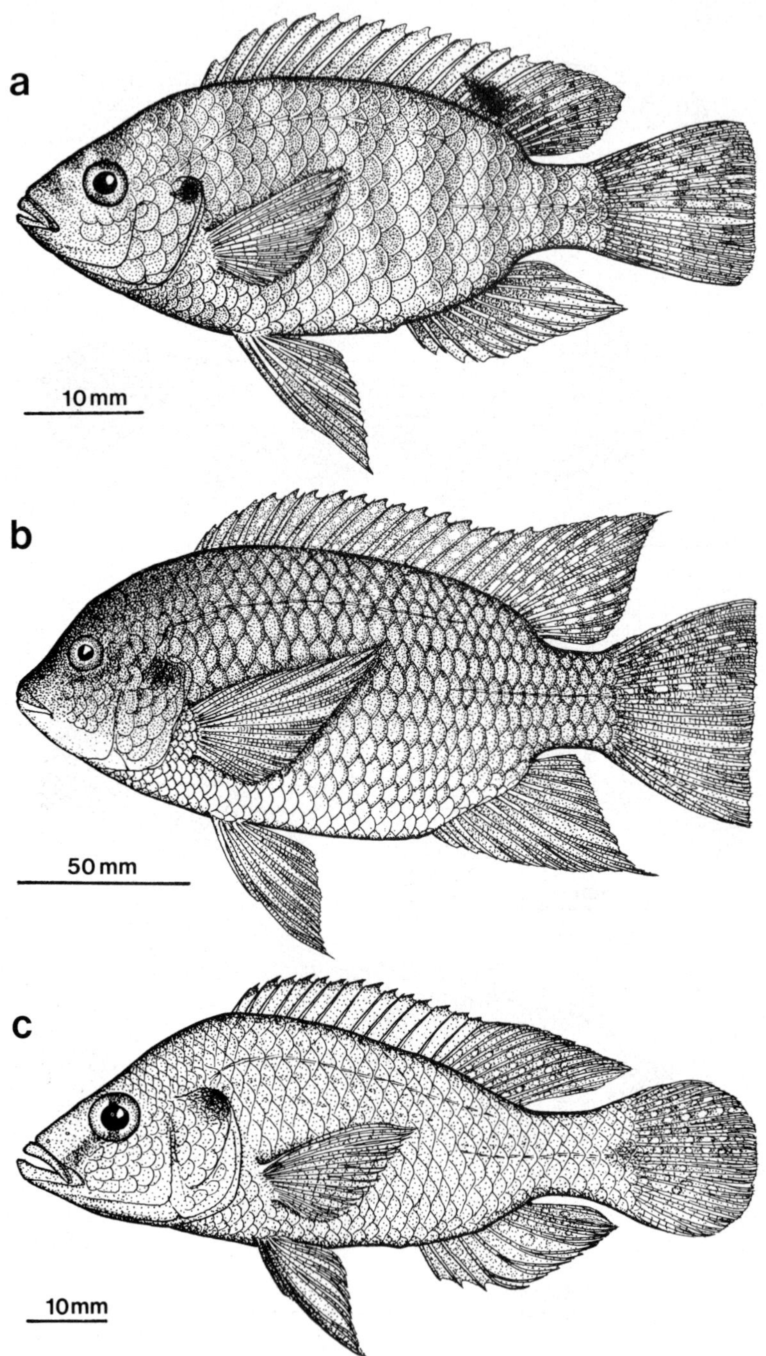

Fig. 4. A. *Tilapia sparrmanii* A. Smith, 1840 B. *Tilapia rendalli swierstrae* Gilchrist & Thompson, 1917 C. *Pseudocrenilabrus philander* (M. Weber, 1897). Drawn by E. M. Tarr.

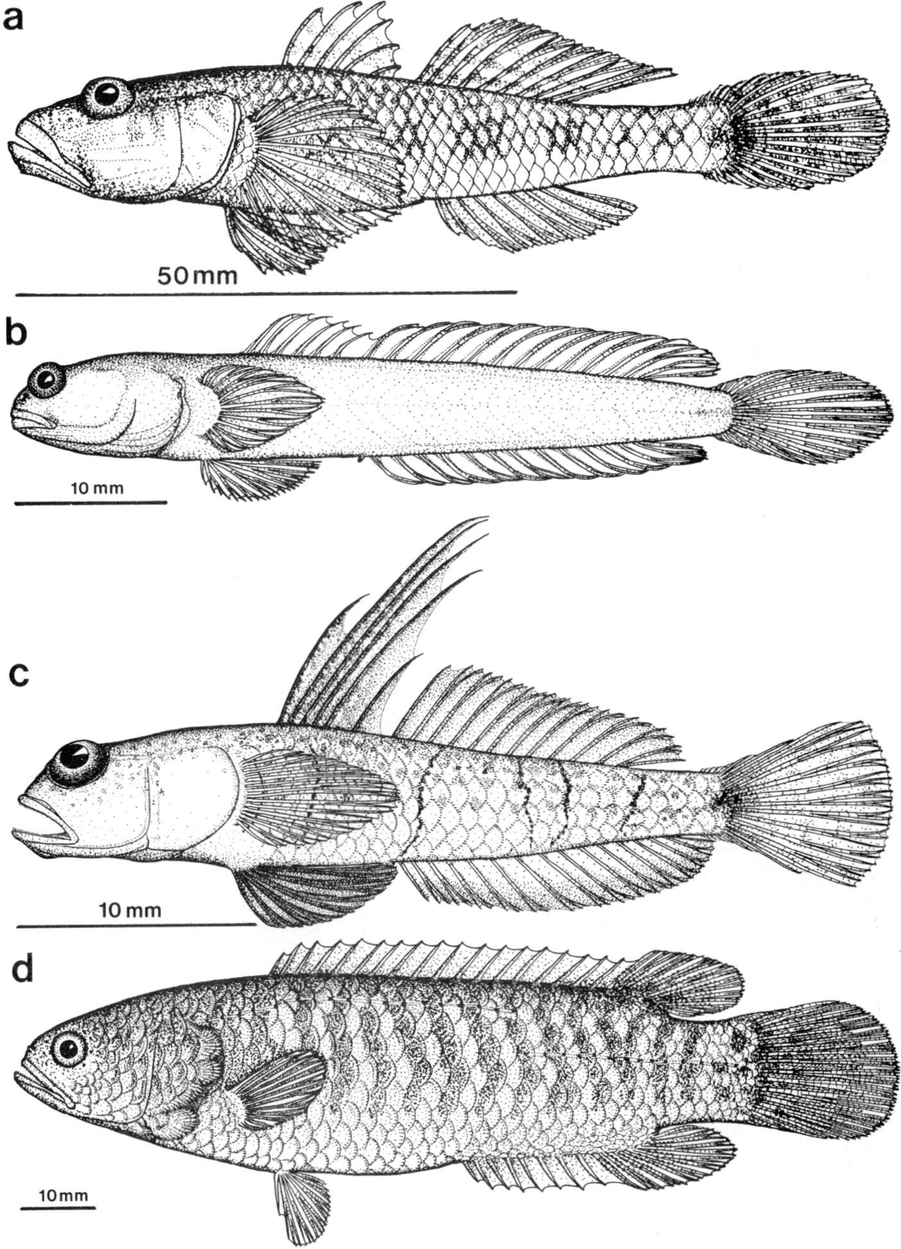

Fig. 5. A. *Glossogobius giurus* (Hamilton-Buchanan, 1822) B. *Croilia mossambica* J. L. B. Smith, 1955 C. *Silhouettea sibayi* Farquharson, 1970 D. *Ctenopoma multispinis* Peters, 1844. Drawn by E. M. Tarr.

Table 2. The diversity of primary, secondary and peripheral fishes in the coastal lakes, estuarine lagoons and lowland rivers of the southern Mozambique plain. Information from Bruton (1974, 1975), Bruton & Appleton (1975), Hill *et al.* (1975), van der Elst *et al.* (1976), Blaber (1978a), Gaigher, Jubb, Kok (pers. comm.) and unpublished data of the author.

	Surface area	Salinity	Connection to	Fishes								Total
				Primary		Secondary		Peripheral		Marine		
	km^2	‰	the sea	No.	%	No.	%	No.	%	No.	%	No.
Lake Sibaya	65	0	Isolated	5	27.8	8	44.4	5	27.8	0	0	18
L. Mgobezeleni	1.8	0.3	2 km swampy connection	1	6.6	8	53.3	6	40.0	0	0	15
Lake Piti		3	shallow, swampy, intermittent	2	15.4	6	46.2	4	30.8	1	7.7	13
Lake Poelela (+Quissico)	65	8	tenuous 75 km connection	0	0	5	38.5	6	46.2	2	15.4	13
Lake St Lucia		0–90	estuary usually open	2	2.4	3	3.7	10	12.2	67	81.7	82
Kosi system	35	Estuary 25–35 Nhlange 0–5 Amanzimnyama 0	estuary usually open	1	0.7	8	7.0	13	8.7	112	83.6	134
Pongola river & pans		Variable	100 km down Maputo river	24	53.3	14	31.1	5	11.1	2	4.4	45
Limpopo river (lower reaches and delta)		Variable	open estuary	16	59.3	8	29.6	3	11.1	0	0	27
Incomati river (lower reaches)		Variable	open estuary	16	55.2	8	27.6	5	17.2	0	0	29
Maputo river		Variable	open estuary	7	50	5	35.7	2	14.3	0	0	14
Small pans		0–0.3	isolated or tenuously connected	14	48.3	11	37.9	4	13.8	0	0	29

of different river basins the most likely invasion route of north-eastern Zululand is postulated to be along the Luangwa valley, lower Zambezi river, the coastal lakes on the Mozambique plain and the lower Limpopo river during the Pleistocene epoch (Crass, 1966; Gaigher & Pott, 1973; Bowmaker *et al.*, 1978; Fig. 6). Species which could not invade Maputaland via coastal lakes (possibly due to salinity barriers) may have entered the area by way of an inland route. Fish movements across the water-sheds of the Orange/Vaal and east coast river systems are considered probable by Gabie (1965), Gaigher & Pott (1973) and Bowmaker *et al.* (1978).

Roberts (1975) divided inland fishes into three groups according to their ability to tolerate salt water during distributional movements: peripheral, secondary and primary. Peripheral division fishes live in both fresh and sea water and readily disperse along a marine coast (e.g. the gobiid, clupeid and atherinid fishes in Lake Sibaya). Secondary division fishes normally live in freshwater but they are sufficiently euryhaline to disperse along the marine coast. The best example in Lake Sibaya is *Sarotherodon mossambicus*, but Roberts includes all cichlids, clariids and cyprinodontids in this category, as well as other fishes. Primary division fishes are intolerant of salt water which therefore presents a formidable barrier to their distribution *e.g.* mormyrids and cyprinids in Lake Sibaya.

All the primary freshwater species in Lake Sibaya also occur in the Pongola River, which is their most likely source of origin. The secondary species could have invaded the lake by either route, whereas the peripheral

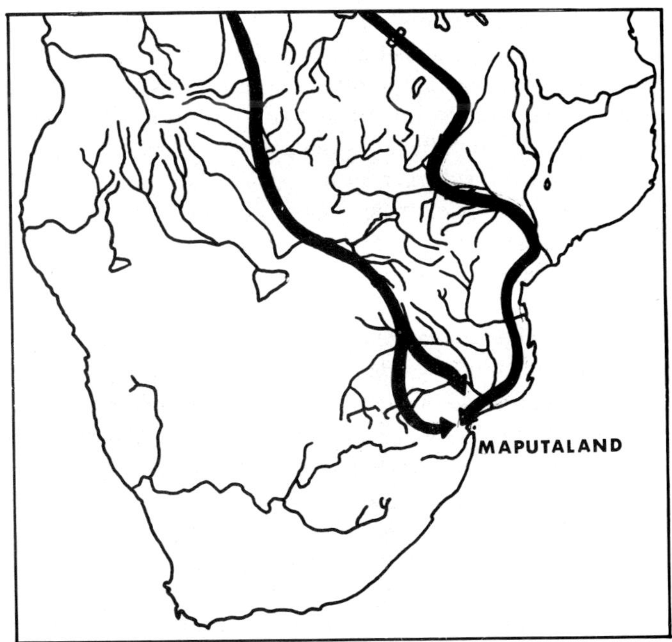

Fig. 6. Proposed coastal and inland invasion routes from tropical Africa to Maputaland of 'Zambezian zoogeographical province' freshwater fishes (after Bowmaker *et al*, 1978).

species have almost certainly distributed themselves southwards along the coast.

The invasion of Lake Sibaya from the Pongola floodplain in the west and the Kosi system in the north is not difficult to envisage today, and was certainly possible during past pluvial periods when large parts of the low-lying plain would have been flooded. In 1975, after several years of above average rainfall, the water level of Lake Sibaya and other lakes and rivers rose sharply and in many places the watertable was exposed above ground level to form extensive swamps. In early 1976 shallow water connections existed between Sibaya and Kosi via Skombane stream, Vasa pan and the Sihadla river, and in the area of Pelindaba flooded grasslands were almost contiguous with swamps in the area of Kosi, Musi and northwards to the Maputo River (forming a west-east, invasion route). At this time *Clarias ngamensis*, a catfish previously recorded only from the Pongola-Maputo river in Maputaland was found within 6 km of Lake Sibaya, and various other clariid as well as cichlid, cyprinodont, anguillid, cyprinid and anabantid fishes were widespread in the inundated areas. Many of these species are well-adapted to living in shallow anaerobic swamps *e.g.* the clariids and anabantids have aerial breathing organs and the cyprinodontids are known to extract oxygen from the water film (Lewis, 1972 in Roberts, 1975).

The coastal plain of Mozambique and Zululand (referred to here as the Mozambique Plain) is of considerable zoogeographical interest. It is one of the few extensive low-lying areas in sub-Saharan Africa (Roberts, 1975) and it has been subject to a recent marine transgression. Furthermore, the warm Mozambique and Agulhas currents flowing southwards along this coast result in a southern extension down the lowland plain of a tropical climate. The 18°C effective temperature isoline, a useful indicator of the outer boundary of the tropics, intersects the south-east African coast a few kilometres south of Lake Sibaya (Stuckenberg, 1969). This delimitation corresponds well with the subtraction of tropical terrestrial vertebrates (see next chapter) and freshwater fishes (see below). The subtraction zone for plants occurs further south, near St Lucia estuary (28°S, Acocks, 1975) and that for marine fishes even further south, probably in southern Natal and Pondoland (Smith, 1965).

The fish fauna of Lake Sibaya clearly reflects the marine origin of the lake as well as its location at the southern tip of this 'tropical peninsula'. Five (28%) of the eighteen fish species have marine affinities (*Gilchristella aestuarius, Hepsetia breviceps, Croilia mossambica, Glossogobius giurus, Silhouettea sibayi*) and one is euryhaline (*Sarotherodon mossambicus*). Only one species (*Aplocheilichthys myaposae*) has its centre of distribution further south. The other eleven species (61%) are part of the Zambezian zoogeographical province (Roberts, 1975) and extend as far north as the Zambezi, Kafue and Zambian Congo river systems, or further. These trends are also reflected in the composition of the freshwater ichthyofauna of the southern Mozambique plain as a whole *i.e.* including the lower reaches, floodplain pans and delta lakes of the Pongola, Maputo, Incomati and Limpopo rivers, and the coastal lakes and estuarine lagoons of St Lucia, Bangazi-south, Sodwana, Piti and Poelela. This fauna is comprised of 69 species of which 20 (29%) have marine affinites and 40 (58%), all primary and secondary freshwater species, belong to the Zambezian ichthyofaunal province to the north. Only two Maputaland freshwater fishes have their centre of distribution to the south (*Barbus natalensis* and *Aplocheilichthys myaposae*, which are both centred in and largely confined to Natal). In addition, two fish species have been introduced into Maputaland: *Micropterus salmoides*, an exotic from the northern hemisphere, and *Barbus gurneyi*, from central Natal. Both species were introduced in June 1970 into a small separate pan which lies to the west of Nhlange lake in the Kosi system.

Five species (7.3%) are considered here to be endemics of the tropical transition zone on the low-lying southern end of the Mozambique plain. These include two mochocids (*Chiloglanis swierstrae* and *C. paratus*) which are confined to the floodplains of the Pongola and Limpopo rivers, and three gobiids. *Mugiligobius pongolensis* is known only from the Pongola floodplain pans and Lake St Lucia, *Silhouettea sibayi* only from Lake Sibaya and *Croilia mossambica* from Lakes Sibaya, Nhlange, Sifungwe, Piti, Poelela and the coastal delta pans of the Limpopo river.

The freshwater ichthyofauna of Maputaland therefore consists of four elements:

a) a large Zambezian province component which is mainly distributed to the north (40 species, with 12 species in Lake Sibaya). Some of the fishes in this category are also widely distributed in South Africa, *e.g. Barbus paludinosus, Clarias gariepinus, Pseudocrenilabrus philander, Sarotherodon mossambicus* and *Tilapia sparrmanii.*

b) a component with marine affinities, but resident in freshwater (20 species, 5 in Lake Sibaya).

c) a group of transitional lowland endemics (5 species, 2 in Lake Sibaya; three of these species are also included in category (b) above).

d) a small number of Natal endemics (2 species, one in Lake Sibaya).

Similar elements comprise the populations of molluscs (van Bruggen & Appleton, 1977) and terrestrial vertebrates in Maputaland (see next chapter). The bulk of the molluscan fauna (43%) consists of species with a wide distribution mainly to the north, while 14% reach their southern limits in Maputaland, 12% their nothern limits and 25% are endemic to the area (the rest are alien, or poorly known, species).

The importance of Maputaland as a tropical subtraction zone is further emphasized by the fact that six coastal lake/euryhaline fishes reach their southern limit of distribution in Lakes Sibaya and Mgobezeleni (*Sarotherodon placidus, Serranochromis meridianus, Croilia mossambica, Platygobius aenofuscus, Silhouettea sibayi, Eleotris melanosoma*) and eighteen primary freshwater species extend no further south than the Pongola River (Bruton

& Cooper, 1979). Another five species extend only slightly further south than Maputaland *i.e. Ctenopoma ctenotis* to Lake Bangazi–south (Blaber, 1978b), *Mugiligobius pongolensis* to lake St Lucia (Kok & Blaber, 1977), *Barbus toppini* to a southern tributary of the Mkuzi river, and *Engraulicypris brevianalis* and *Clarias theodorae* to the Umfolozi river (Crass, 1964). *Tilapia rendalli swierstrae* has been introduced into Natal but the natural southern limit of its distribution is considered to be the Pongola River (Jubb, 1967). None of the 13 characid, mormyrid or siluriform species in Maputaland (except *Clarias gariepinus*) extend further south than Zululand. The number of fish species south of Zululand decreases rapidly and between the Mtamvuna and Keiskamma rivers only one primary freshwater species is found (*Barbus anoplus*; Bowmaker *et al.*, 1978).

At the latitude of Lake St Lucia (28°S) the Mozambique plain narrows sharply and this large brackish lagoon is likely to have been a major barrier to the further southward distribution of primary freshwater fishes by a coastal route. Furthermore, the nearshore topography of Natal and Transkei is characterized by a narrow coastal plain and deeply-cut river valleys which would further hinder the distribution of primary fishes down the coast. Those species which have reached the Cape Province are thought to have done so via the Orange River drainage basin (Jubb, 1964).

Sarotherodon mossambicus, which is euryhaline and eurytopic, is well-adapted to disperse via coastal lakes, estuaries and the sea down the south-east African coast. It is the only east-coast species which extends further south than Natal along the coast. At the southernmost limit of their range they are more common in estuaries than in freshwater (Jubb, 1967), possibly because a brackish external milieu increases their tolerance of low temperatures (Allanson *et al.*, 1971).

Roberts (1975) includes all cichlids in his secondary division. *Pseudocrenilabrus philander* and *Tilapia rendalli swierstrae* occur in Lagoa Poelela at 8‰, and the former species inhabits the upper reaches of Sodwana estuary at 6‰, but *Tilapia sparrmanii* has not been caught in brackish water although it occurs in lakes connected to the sea e.g. St Lucia, Amanzimnyama and Mgobezeleni.

The inclusion of clariids and cyprinodontids in the secondary division is related more to their dispersion abilities than to proven euryhalinity. *Clarias lazera*, a close relative of *C. gariepinus*, has been spawned in a salinity of 2.2‰, and tolerates 9.5‰ under some physiological stress (Clay, 1977). *C. gariepinus* inhabits parts of Lagoa Poelela (pers. observation, 1972) at 8‰, but saline conditions are a barrier to their distribution as they are not found in the lower reaches of estuaries in Mozambique and Natal. The other clariids and cyprinodontids in Maputaland are apparently stenohaline and do not inhabit brackish water.

The peripheral fishes of Lake Sibaya are species which typically inhabit shallow sheltered sandy areas on the southern and eastern African marine coast. *Gilchristella aestuarius* is a small shoaling species which is distributed from the southern Cape coast to Madagascar and regularly enters Cape, Zululand and Mozambique coastal lagoons. *Hepsetia breviceps* is apparently confined to the South African coast (Smith 1965) but related atherinids inhabit freshwater and brackish lakes in Mozambique (*e.g. H. pingius* in Quissico lake, Hill *et al.*, 1975) and further north. *Glossogobius giurus* is widespread in the tropical Indo-Pacific (Smith, 1965) and penetrates lagoons and rivers as far inland as 400 km from the sea (Jubb, 1967). *Croilia mossambica* is confined to the placid parts of coastal lakes on the Mozambique plain (Blaber & Whitfield, 1977) but may inhabit sheltered sandy areas in the sea. *Silhouettea sibayi* is unique to Lake Sibaya but a close relative, *S. insinuans*, is known from sheltered marine localities on the Mozambique coast (Smith, 1959).

The relict estuarine component which at present survives in Lake Sibaya is that part of an earlier, probably more diverse estuarine/marine population which adapted to feeding and breeding in a closed freshwater lake. Other species which are likely to have been common in the early history of Lake Sibaya, but which have not survived the transition to freshwater, include *Ambassis commersoni, Megalops cyprinoides, Eleotris fusca, Mugil cephalus, Acanthopagrus berda* and *Mugil euronotus.* The secondary and primary freshwater fish component in the population is likely to increase in future as the lake continues to stabilize as a freshwater system. Some fish species probably only inhabit Lake Sibaya after a period of sustained high water levels, and have not as yet been caught there *e.g. Anguilla bicolor bicolor* which inhabits Vasa pan on the northern edge of the Lake Sibaya catchment.

Habitat Preferences and Distribution

Compared with other coastal lakes on the Mozambique plain, such as lakes St Lucia, Nhlange, Piti and Poelela, Lake Sibaya has a large deep-water body with a relatively narrow littoral zone. The littoral zone is nevertheless important in the ecology of the lake as most fish species are concentrated there, mainly as a consequence of the presence of rich food resources in the form of aquatic macrophytes, their epiphytic biota, and benthic bacteria and diatoms. The plant beds also provide shelter from predators for adults and juveniles of some species.

Due to the flat nature of the surrounding terrain, and the proximity of the Indian Ocean, Lake Sibaya is often subject to strong wind action, especially in summer. As a result the shallow terraces which fringe some shores are too wave-washed for the establishment of dense macrophyte beds, and most plants are confined to a band from two to seven metres below the effect of wave action. In sheltered bays, and particularly along the convoluted shores of the western and northern arms, dense macrophyte beds do occur close inshore in shallow water. The distribution and density of several fish species closely follows that of the submerged macrophytes.

The distribution of some fishes varies diurnally and seasonally, mainly in response to water temperature fluctuations. Lake level changes may modify the diel and seasonal movements of the fishes. At high lake level, when terrestrial zones are inundated, a sheltered inshore zone is created which is rapidly inhabited by those species which were previously confined to densely-vegetated slopes, sheltered bays and streams.

Seven main habitats can be distinguished in Lake Sibaya according to bottom profiles, substratum types, depth and macrophyte cover: marginal, terrace, gradual slopes, steep slopes, profundal, sheltered bays, limnetic zone (Fig. 7). A checklist with notes on the fauna and flora of these zones is given by Allanson *et al.* (1974). The distribution of fishes has been investigated using diving apparatus, especially SCUBA, as well as specially designed seine nets, trawl nets, fishtraps, longlines and conventional gillnets (described by Bruton & Boltt, 1975; Bruton, 1978). A recording echosounder was most useful for determining the distribution of larger fishes. In most cases collecting and recording gear were used in a standardized way at a series of pre-selected sites so that the catches were comparable between different seasons and years.

Marginal Habitats

The nature of marginal habitats changes at different lake levels. At receding or low lake level the shore is fringed with wide sandy beaches, and barren shallow pools (up to 400 m^2 in area) are found along gently sloping shores. These pools, which may dry up in winter, are inhabited in summer almost exclusively by the fry and juveniles of *Sarotherodon mossambicus* which constituted over 99% of the catch ($N = 8453$). Occasional specimens of *Tilapia sparrmanii* and *Pseudocrenilabrus philander* were found in the *S. mossambicus* shoals. Marginal pools represented a harsh environment with wide diel temperature fluctuations of 10° to 15°C in summer (Fig. 8) and maximum temperatures in excess of 35°C (occasionally 41°C). Juvenile *S. mossambicus* inhabit this zone as they have high temperature preferences (about 35°C, Badenhuizen 1967; Josman, 1971) and high upper lethal temperatures (41,7°C for fishes 25 to 35 mm TL, Kemp, 1966). Juvenile *T. sparrmanii* and *P. philander* are probably excluded from marginal pools by their lower temperature preferenda (<33°C, Donnelly, 1969a). At night *S. mossambicus* fry and juveniles migrate back into water deeper than 100 cm when shallow water temperatures fall below those of the main water body. Throughout the year the young of *S. mossambicus* select shallow water at temperatures from 19°C to 35°C, but had a tolerance range from 16.5°C (Minshull, 1969) to 39°C.

At increasing and high lake levels marginal beaches, pools and forest fringes are flooded and well-vegetated pools are formed in the littoral zone. These pools persist from year to year, and may link adjacent swamps and pans to the main lake. Many semi-aquatic and aquatic plants become established along flooded shores and provide shelter for a diversity of fishes and invertebrates. Flooded marginal habitats are inhabited by all the fish species known from Lake Sibaya except *Croilia mossambica, Silhouettea sibayi* (offshore benthic psammophils), *Labeo molybdinus* (a rare pelagic species) and *Marcusenius macrolepidotus* (which is confined to dense offshore macrophyte beds).

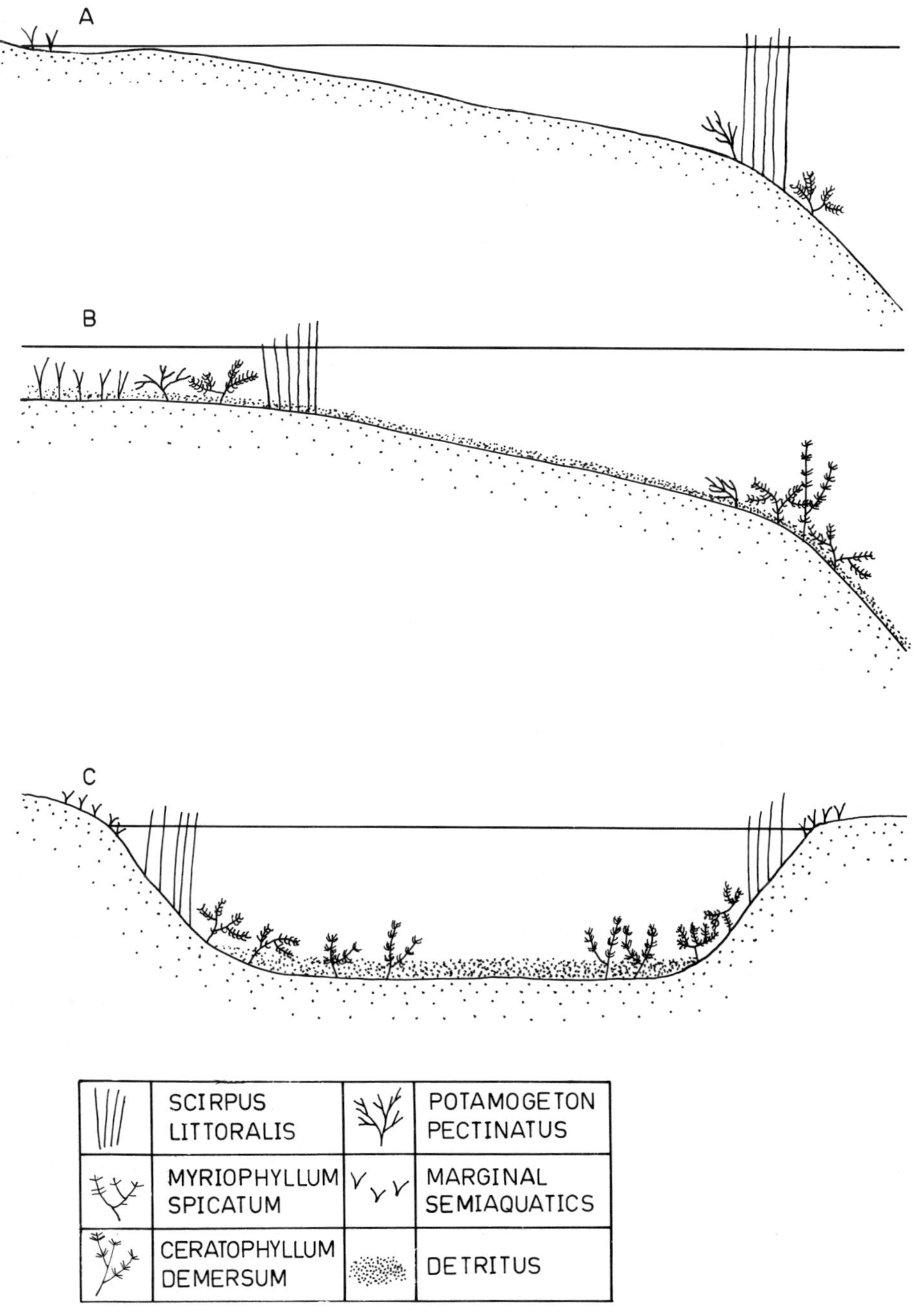

Fig. 7. Diagrammatic profiles of different inshore habitats in Lake Sibaya. A. Terrace and slope at low lake level. B. Terrace and slope at high lake level. C. Sheltered bay.

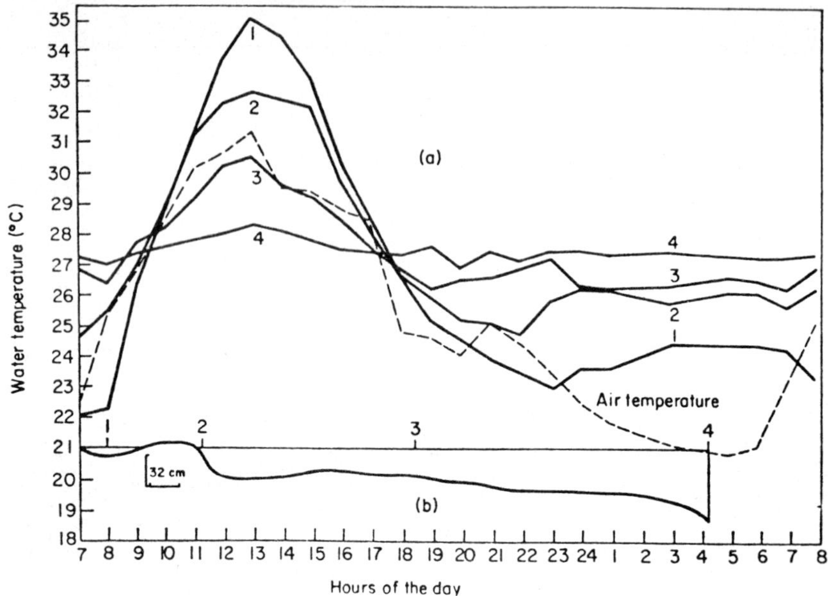

Fig. 8. Water temperatures and bottom profiles across a marginal pool and terrace on the eastern shore of the south basin of Lake Sibaya, 10–11 March 1971. The numbers 1 to 4 on the temperature profiles (a) indicate the position of temperature measurement points on the bottom profile (b) (from Bruton & Boltt, 1975).

The relative density of different fish species in shallow (<30 cm) and deeper (50–150 cm) flooded marginal pools during daylight hours in the summer of 1975/76 is given in Table 3. *Apocheilichthys katangae Sarotherodon mossambicus* and *Barbus paludinosus* were the most abundant species in shallow pools, and *P. philander, T. sparrmanii* and *A. katangae* in deeper pools. Samples obtained using seine nets and fishtraps confirmed these trends.

Night catches in marginal pools indicated that the populations were the same as those during the day. During the cool season the numbers of all species in flooded marginal pools decreased, although the population composition did not change.

The sheltered, well-vegetated lakelets and swamps which surround Lake Sibaya, have a similar fish fauna (Table 4) to deep marginal pools (Table 3), the habitat which they most closely resemble *i.e.* dominated by *P. philander, A. katangae* and *T. sparrmanii,* with lesser numbers of *Tilapia rendalli, Barbus paludinosus* and *S. mossambicus.* The collection methods used in adjacent lakelets – fishtraps and rotenone – are likely to have caught a representative sample of the total population although the latter method tends to underestimate the proportion of air-breathing fishes *e.g. Clarias gariepinus, C. theodorae* and *Ctenopoma multispinis.*

Table 3. Composition of fish populations in shallow marginal pools (<30 cm), and adjacent deeper marginal pools (50–150 cm) in Lake Sibaya in the Summer of 1975/76.

Fish species	Number of rotenone treatments in netted 10 m × 10 m area			
	3		6	
	Shallow marginal pools		Deeper marginal pools	
	No.	%	No.	%
P. philander	2	0.8	718	38.3
T. sparrmanii	5	2.1	422	22.5
A. katangae	176	73.3	331	17.6
T. rendalli swierstrae	6	2.5	199	10.6
B. paludinosus	19	7.9	121	6.4
S. mossambicus	26	10.8	42	2.2
C. gariepinus	0	0	13	0.7
B. viviparus	3	1.3	13	0.7
G. giurus	3	1.3	12	0.7
C. theodorae	0	0	2	0.2
C. multispinis	0	0	1	0.1
Total	240	100	1877	100

Terrace Habitats

Terrace habitats are gradually-sloping littoral shelves between the lake shore and the steep offshore slope. The substratum is sandy with variable amounts of overlying detritus which may be rich in diatoms. Terrace habitats increased in depth from 0.8 to 4.4 m at the deep edge during the study period. Diel extremes in temperatures were more marked at low than at high lake level, with mid-terrace water temperatures ranging between 26° and 31.3°C over 24 h at low lake level in March 1971, and between 26.3° and 28.8°C at high lake level in March 1976.

Table 4. Composition of fish catches using fishtraps and rotenone in lakelets and swamps (max. depth about 4 m) within the catchment of Lake Sibaya.

Fish species	Total	% of total
P. philander	231	35.6
A. katangae	175	27.0
T. sparrmanii	125	19.3
T. rendalli swierstrae	31	4.8
S. mossambicus	26	4.0
B. paludinosus	26	4.0
C. gariepinus	21	3.2
G. giurus	4	0.6
Ctenopoma multispinis	4	0.6
B. viviparus	3	0.5
Clarias theodorae	3	0.5

Table 5. Habitat preferences of four cichlid species (4–8 cm SL) in the south basin in the period September 1970 to May 1971 as revealed by seine net catches (from Bruton & Boltt, 1975).

			Catch												
			S. mossambicus			*T. sparrmanii*			*T. rendalli*			*P. philander*			*Total*
Habitat	Depth (m)	Pulls of slope seine	No.	*%	CPUE	No.	*%	CPUE	No.	*%	CPUE	No.	*%	CPUE	CPUE
Terrace	0–2	56	3727	98.1	66.5	25	0.6	0.4	3	0.1	0.1	47	1.2	0.8	67.8
Barren slope	2–7	40	843	61.5	21.1	179	13.1	4.4	7	0.5	0.1	341	24.9	8.5	34.1
Sparsely vegetated slope	2–7	115	1738	64.5	15.1	508	18.9	4.4	7	0.3	0.1	439	16.3	3.8	23.4
Densely vegetated slope	1–7	42	172	19.4	4.1	175	19.8	4.2	0	0	0	539	60.8	12.8	21.1
Total		253	6480	74.0	25.6	887	10.1	3.5	17	1.9	0.1	1366	14.0	5.4	

* Percentage of total catch for the four cichlids in one habitat type.

At low lake level *S. mossambicus* constituted over 98% of the catch on the terrace during the day (Table 5), with small numbers of the other cichlids, as well as *Glossogobius giurus*, *Croilia mossambica* and *Clarias gariepinus*. The latter three species were the only fishes regularly encountered on the terrace at night at low lake level.

S. mossambicus and *C. gariepinus* both perform diel movements into and out of shallow water. At low lake level their movements were complementary with the cichlid occupying the terrace during the day and the clariid at night.

In the period May to July the catch per unit effort (CPUE) from seine netting falls to zero for adult *S. mossambicus* (>8 cm SL) and less than 12 for juveniles (Fig. 9). In the cool to warm transition period juveniles move into terrace waters in greater numbers in August, followed in September by adults. The CPUE for juveniles increased from 7 in July to 16 in August and 29 in September, and for adults from 0 in July to 4 in September and 16 in October. Adults moving into shallow water in September were mainly breeding individuals (Fig. 10). During the peak of the breeding season in Summer (late September to January) the number of adult fishes on the terrace increases. The number of juveniles in the catch decreased from November to January, as a result of recruitment into the adult population but increased again in February, March and April when the new cohort from the recent breeding season entered the catchable population (>4 cm SL).

The proportion of the three cichlid species on the terrace changed markedly at intermediate and high lake level. At intermediate lake level (1974–75) terrace areas were deeper and less exposed to wave action at

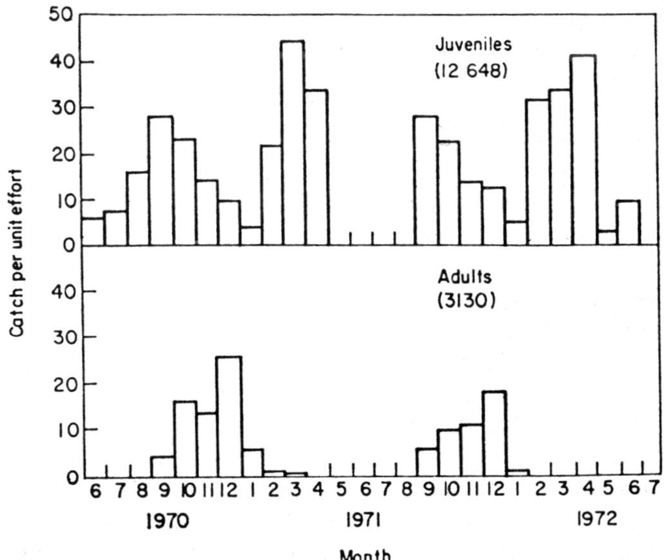

Fig. 9. The catch per unit effort of juvenile and adult *S. mossambicus* on the terrace and slope in the south basin of Lake Sibaya by means of a seine net. No netting was performed in the period May–August 1971. The sample size is given in brackets (from Bruton & Boltt, 1975).

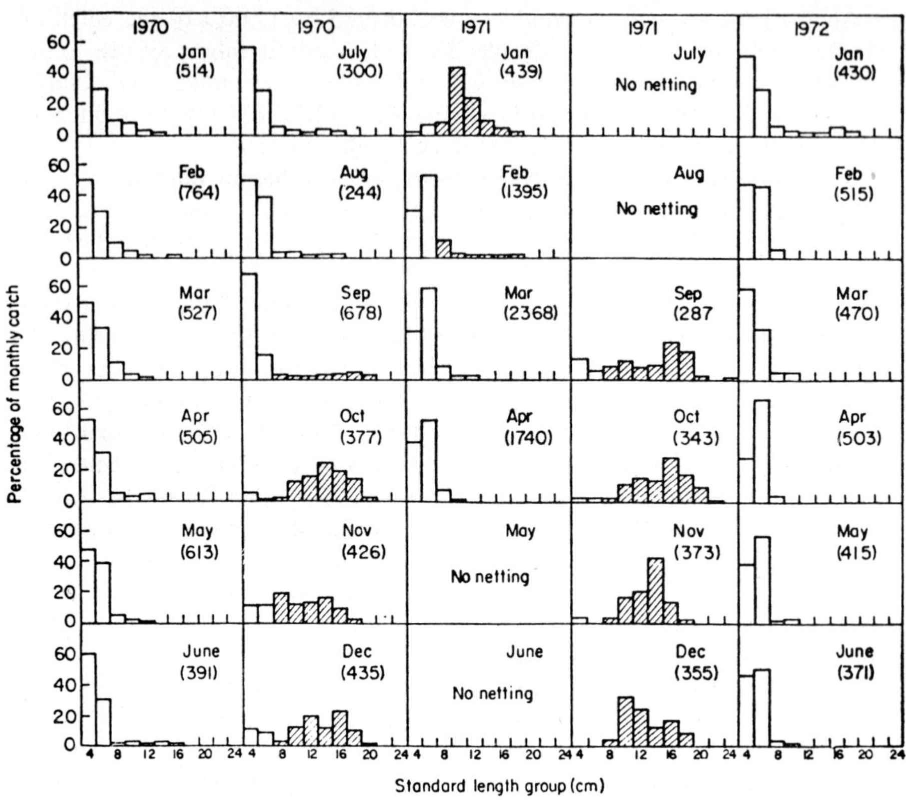

Fig. 10. The length frequency of *S. mossambicus* caught on the terrace and slope in the south basin by means of seine nets. The sample size is given in brackets. Shaded histograms: breeding fish; open histograms: non-breeding fish (from Bruton & Boltt, 1975).

substrate level. Higher numbers of *T. sparrmanii* and *Pseudocrenilabrus philander* inhabited the terrace during the day, especially in areas where macrophytes had become established. *S. mossambicus* were uncommon on the terrace during the day and occurred only in inshore or deep terrace areas but not in the open mid-terrace where they had been so numerous at low lake levels. An invasion of *S. mossambicus* onto the terrace took place at night between about 20h00 and 01h00, but thereafter they were uncommon again (Bowen, 1976a). The dominance of *S. mossambicus* in the terrace population was reduced to 59.1% over *P. philander* (27.7%) and lower numbers of *T. sparrmanii* (13.2%).

At high lake levels the cichlids altered their behaviour as well as their relative proportions. *P. philander* was dominant (62.2%) on the terrace followed by *T. sparrmanii* (23%) with *S. mossambicus* sparse (14.8%). *T. sparrmanii* and *S. mossambicus* mainly occupied branch tangles and flooded vegetation along the inshore margin, whereas *P. philander* was abundant in these areas as well as in mid and deep terrace. *P. philander* and *T. sparrmanii* remained on the terrace during the day and night, whereas *S.*

mossambicus mainly occupied the terrace during the night. *G. giurus, C. gariepinus, A. myaposae* and *G. aestuarius* were common on the terrace as well.

At all lake levels the number of cichlids on the terrace decreased during heavy winds (Bruton, 1973; Bruton & Boltt, 1975; Bowen, 1976a) particularly at low lake level when wave action reached to substratum level.

A pattern therefore emerges in which *S. mossambicus* occupied the terrace during the day almost to the exclusion of other cichlids when the water was very shallow and environmental conditions, especially temperature, had wide diel fluctuations. As the lake level increased, terrace water temperatures were moderated, and deeper more sheltered habitats became available. These changes resulted in increasing occupation of the terrace by fishes which normally inhabit sheltered habitats. The significance of these fish movements is discussed later.

Evidence from gillnetting, longlining, underwater observations using SCUBA and above water observations from a 7 m tower indicate that *C. gariepinus* more often enters terrace waters at night than during the day (Table 6, Bruton 1978). This pattern was most marked at low lake level. As the lake level rose, increasing numbers were seen on the terrace during the day compared with the night. Gillnet catches in sheltered bay and offshore habitats such as the profundal zone were, however, nearly equal day and night.

The relative density of *C. gariepinus* in different habitats, as deduced from longline catches, is given in Table 7. The highest catches were made in terrace habitats, and the lowest in the profundal zone, with moderate catches in intermediate habitats. Marginal habitats were occupied mainly by juvenile *C. gariepinus*, and offshore habitats by adults.

The burrowing goby *Croilia mossambica* is present on the terrace from a depth of about 1 m (>3 m on wave-washed shores) to the edge of the slope down which it extends with decreasing abundance into the profundal zone. The density of this species can be estimated by counting the number of

Table 6. The proportion of *C. gariepinus* entering shallow terrace waters during the day and night in Lake Sibaya at low lake level (1970–1973) and high lake level (1974–1976).

Method of assessment	Lake level: Terrace <2m deep	Lake level: Terrace >2 m deep	Number of paired settings or observations
	day : night	day : night	
Catches			
Longlines and fixed lines	1 : 3	1 : 12	23
Gillnets	1 : 5	1 : 14	48
Observations			
Tower	1 : 5		17
SCUBA	1 : 5	1 : 7	18
		(gillnets: bay 1 : 1.5)	

Table 7. The relative density of *C. gariepinus* (day and night catches combined) in offshore habitats in Lake Sibaya, as deduced from longline catches made during 1974 and 1975 (terrace depth >2 m).

Habitat	Catch per unit effort (one 120 unit set laid for 6 hours)	No. of longline settings (120 unit sets)
Terrace	14.4	33
Sheltered bay	8.3	16
Steep slope	5.1	20
Gradual slope	4.4	14
Profundal	1.3	23

burrows, as each fish excavates one burrow at a time (Blaber & Whitfield 1977). Their density was highest on the eastern terrace and gradually sloping northern shores of the main basin (one burrow per 10 m^2) and least on the steep offshore slope along the eastern shore (one burrow per 70 m^2). *C. mossambica* was most common in shallow sheltered sandy areas without macrophyte cover and avoided substrata which were disturbed by water movement, or contained mud or sand grains with a diameter greater than 0.5 mm.

Gradual Sloping Habitats

These replace the shallow terraces on windward shores and occur mainly on the southern and northern perimeters of the main and south basins. These areas have an inshore curtain of emergent macrophytes and sparse beds of submerged *M. spicatum*, *P. pectinatus* and *C. demersum* which are situated several hundred metres offshore. This habitat gradually merges with the profundal zone at 10 to 15 m although macrophytes rarely extend deeper than 8 m.

The substratum consists of compacted sand with a thin layer of flocculent detritus and macrophyte debris. The fish fauna is the same as that of the terrace with the addition of *Marcusenius macrolepidotus* and, rarely, *Labeo molybdinus*. In common with terraces and steep slopes, these are important areas for cichlid breeding (see below).

Steep Slopes

These are found on the deep edge of the terrace and represent a contracted transition zone between terrace and profundal habitats. The slope extends from a depth of one to four metres to 7 to 15 m where the gradient lessens. The sandy substratum is usually unstable but may support dense stands of submerged macrophytes. The composition of the fish population is determined by the nature of the macrophyte cover, and shows little variation in different seasons or in response to lake level fluctuations which is in contrast to the situation on the adjacent terrace.

On steep slopes which have no macrophyte cover, *S. mossambicus* is the most abundant cichlid (61.5%) followed by *P. philander* (25.4%) and *T. sparrmanii* (13.1% Table 5). Lesser numbers of *C. gariepinus*, *T. rendalli*, as well as the gobies, *Croilia mossambica*, *Glossogobius giurus* and *Silhouettea sibayi*, are found.

On sparsely-vegetated steep slopes, which have some *Scirpus littoralis* and *Potamogeton pectinatus* cover, the same species are found in addition to small numbers of *Aplocheilichthys katangae S. mossambicus* again dominates the catch (64.5%) but *T. sparrmanii* is slightly more abundant (19.9%) than *P. philander* (16.3%).

Densely-vegetated slopes have a diverse fish population consisting of all the above species as well as *Barbus paludinosus*, *B. viviparus* and *Gilchristella aestuarius*. Of the cichlids, *P. philander* is most abundant (60.8%) with approximately equal numbers of *T. sparrmanii* and *S. mossambicus* (19.8%, 19.4%). The ichthyofauna of these areas resembles that of sheltered marginal zones at high lake level.

The Profundal Zone

This extends from about 15 m to the deepest point in the lake (40 m). There are no macrophytes and substrata consist mainly of stabilized sand with a varying layer of detritus, or fine silt which reaches a depth of over 1 m in some areas. *G. giurus*, *C. mossambicus*, *C. gariepinus* and *P. philander* commonly inhabit sandy parts of this zone.

The vertical distribution of the cichlids in offshore areas was determined by means of a recording echosounder, fishtraps, gillnets, seine nets and underwater observations using SCUBA (Bruton & Boltt, 1975). Discrete dots recorded on the echosounder charts were produced by fish, almost certainly *S. mossambicus* >12 cm SL as this species formed 60–90% of the catch in the areas traversed.

S. mossambicus juveniles were common to a depth of 7 m, uncommon from 7–12 m and rare deeper than 15 m. More than half the adults caught, recorded or seen occurred shallower than 7 m while 90% were shallower than 11 m (Figs. 11, 12 and 13). No adults were found deeper than 18 m. In contrast *P. philander* and *T. sparrmanii*, although they were most common in water shallower than 10 m, occurred to a depth of at least 30 m (Table 8). Large areas of the profundal zone would thus appear to be beyond the normal range of adult *S. mossambicus*.

Subsequent experimental findings by Caulton (1975) and Caulton & Hill (1973, 1975) on the depth tolerances of *S. mossambicus* have indicated that juveniles are very tolerant of pressure changes, and compensate to depths of at least 33 m in a few minutes, whereas adults are less able to compensate to deep water. Furthermore, the maximum compensation depths of adults are temperature dependent and would be greater in summer than in winter (Fig. 14).

At cool season water temperatures adult *S. mossambicus* are restricted to water less than 7–10 m deep (Caulton & Hill, 1973, 1975), and since they

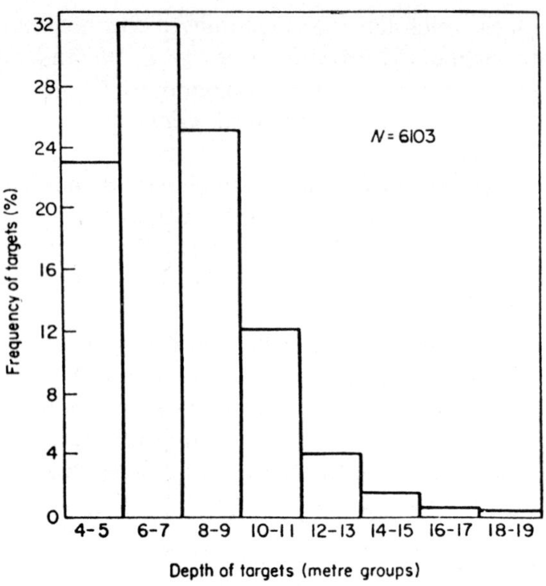

Fig. 11. The percentage frequency in different depth groups of echosounding targets thought to be *S. mossambicus* (from Bruton & Boltt, 1975).

are not found on the littoral terraces (Bruton & Boltt, 1975), their bottom distribution is restricted to an area of the slope from about 2 to 10 m. This finding was confirmed by underwater observations. The area of substrate available to the summer population is, however, greater as they can occupy, in addition to the terraces, bottom areas down to 20 m. The distribution of juvenile *S. mossambicus*, as well as *T. sparrmanii* and *P. philander* does not show the same seasonal variations since they can inhabit deep water throughout the year. Adult *T. sparrmanii* and *P. philander* are apparently depth-limited at lower temperatures (*i.e.* 15 and 16 m respectively at 22°C, Caulton, 1975) although field observations indicate that they do penetrate to a greater depth (30 m) than *S. mossambicus* in summer.

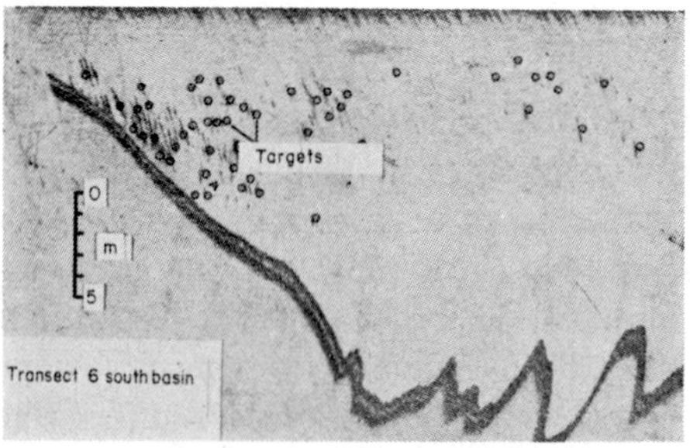

Fig. 12. Echotrace showing a typical pattern of target (? *S. mossambicus*) distribution with increasing depth, at 13h32 on 7 September 1970 (from Bruton & Boltt, 1975).

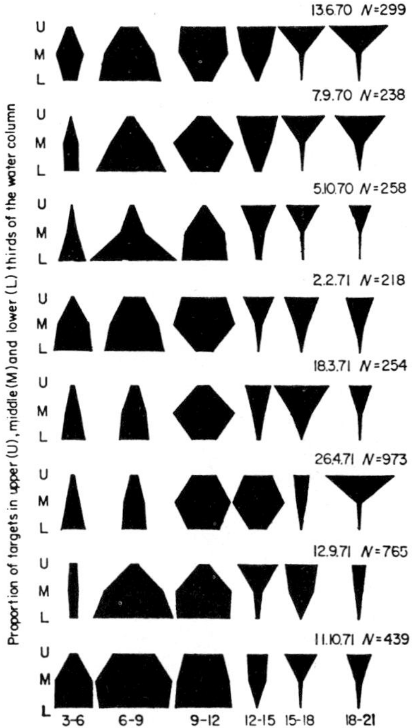

Fig. 13. The proportion of targets in the upper, middle and lower thirds of the water column in different depth zones showing the changing relationship of target position to the bottom and surface. The data are derived from discrete dots on the echotraces which were plotted on a scatter diagram of water depth versus target depth. The sample size (*N*) is given. (from Bruton & Boltt, 1975).

The maximum depth reached by various fish species in Lake Sibaya, based on SCUBA diving observations, is given in Table 8. Nine of the 18 species occur in areas shallower than 10 m.

Sheltered Bays

Sheltered bays and inlets are characterised by steep slopes close to shore and flat bottom profiles covered with a layer of organic ooze. Submerged and aquatic macrophytes are common. The most abundant cichlid is *P. philander* (51%) followed by *T. sparrmanii* (27%), *T. rendalli* (12%) and *S. mossambicus* (10%). Other species include *Glossogobius giurus*, *Clarias gariepinus* and *Aplocheilichthys katangae* with lesser numbers of *Barbus paludinosus*, *B. viviparus*, *Aplocheilichthys myaposae* and *Clarias theodorae*. *Marcusenius macrolepidotus* is the only persistent inhabitant of the central muddy zone although cichlids, clariids and cyprinodonts are found there in association with plants such as *Ceratophyllum demersum* and *Nymphaea capensis*. *Barbus viviparus* occasionally occurs in dense shoals in sheltered bays – Minshull (1970) caught 490 *B. viviparus* in a single baited fish trap on one occasion.

Table 8. The maximum depth recorded for various fish species in Lake Sibaya based on SCUBA diving observations and fishtrap catches.

		Max. depth.
Gilchristella aestuarius		5 m
Marcusenius macrolepidotus		6 m
Barbus viviparus		8 m
Barbus paludinosus		8 m
Labeo molybdinus		?
Clarias gariepinus		40 m
Clarias theodorae		2.5 m
Aplocheilichthys katangae		6 m
Aplocheilichthys myaposae		6 m
Hepsetia breviceps		5 m
Pseudocrenilabrus philander		30 m
Sarotherodon mossambicus	*adults*	18 m
	juveniles	25 m
Tilapia rendalli swierstrae		15 m
Tilapia sparrmanii		30 m
Croilia mossambica		27 m
Glossogobius giurus		40 m
Silhouettea sibayi		15 m
Ctenopoma multispinis		1.5 m

The Limnetic Zone

Lake Sibaya's limnetic zone is sparsely populated with fish, and the only permanent pelagics are the small planktivores *Gilchristella aestuarius* and *Hepsetia breviceps. S. mossambicus* and *C. gariepinus* have occasionally been seen and caught in open water, and echotrace targets of the former species were frequently recorded in surface waters within a few hundred metres of the shore.

In summary, the greatest diversity of fishes is found among dense vegetation in marginal and offshore habitats, and the least in barren terrace and profundal habitats. The number of fish species on the terrace increases at high lake level. These results parallel those of Allanson *et al* (1974) who found that the greatest diversity of fauna (mainly invertebrate) in Lake Sibaya occurred in sheltered marginal vegetation (83 taxa) and permanently submerged plant beds (55) with the least on the terrace (29) and in profundal and limnetic zones (24 and 16).

The density of those fish species which have been studied does not however, relate directly to faunal diversity. As shown in Table 5, the catch per unit effort of four cichlid species on the terrace (total 67.8) was far higher than on barren or sparsely vegetated slopes (34.1 and 23.4) or in sheltered bays (21.1). Furthermore, the relative density of *C. gariepinus* was over 40% higher on the terrace compared with any other habitat during the period 1974 to 1976 (Table 7). These trends can probably be explained in terms of food quality and availability, as discussed later.

Breeding

The fishes of Lake Sibaya mainly breed during the warm rainy season (October to February, Table 9) but ripe fishes may be found during all months of the year except June and July (midwinter). The presence of ripe females during the cool-to-warm transition months (August to October) does not necessarily imply that breeding takes place then, as some species, especially the non-guarders such as *Clarias gariepinus*, may delay spawning until suitable conditions occur.

Lake level fluctuations have a marked effect on the choice of spawning sites. At low lake level the terraces are wave-washed and barren and most breeding occurs among offshore stands of submerged macrophytes. At increasing and high lake levels, when marginal grassland is inundated, dense aquatic and semi-aquatic plant beds develop inshore and many species (especially cichlids and gobiids) breed in these areas as well as in the offshore plant beds. In contrast, *Clarias gariepinus* spawns inshore at all lake levels under widely differing conditions.

Glossogobius giurus deposits its eggs on logs, *Potamogeton* and *Myriophyllum* stems, and other plants. Small juveniles of this species, as well as *Croilia mossambica* and *Silhouettea sibayi*, have been seen by divers in dense plant

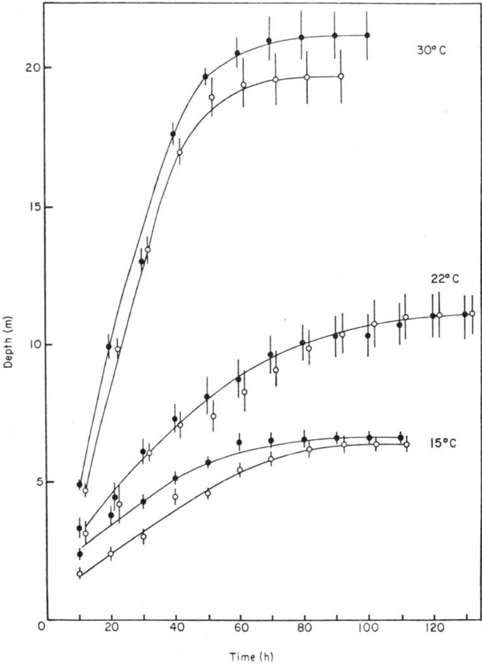

Fig. 14. Depth equilibration curves for mature adult *Sarotherodon mossambicus* at 15°, 22° and 30°C. Each point represents the mean of the equilibrations of ten fish and the vertical lines indicate two standard errors about the mean. Only one curve is given for fish at 22°C since males and females show no significant difference, ●, males; ○, females (from Caulton & Hill, 1975).

Table 9. The breeding seasons of the fishes of Lake Sibaya.

Species	Breeding season
Marcusenius macrolepidotus	November–February
Barbus paludinosus	October–February
Barbus viviparus	(Sept*)October–February
Labeo molybdinus	(December)(January)
Clarias gariepinus	(September)November–February (April)
Clarias theodorae	(November)December–February
Aplocheilichthys myaposae	October–February (March)
A. katangae	October–February
Sarotherodon mossambicus	♂ September–January (February) ♀ September–February (March)
Tilapia sparrmanii	September–March
T. rendalli swierstrae	October–March (April, May)
Pseudocrenilabrus philander	November–March
Glossogobius giurus	November–February
Croilia mossambica	♂ September–February ♀ October–March
Silhouettea sibayi	?
Gilchristella aestuarius	November–February
Hepsetia breviceps	October–February
Ctenopoma multispinis	November–February

* Months in brackets record rare or occasional breeding events.

beds. *C. mossambica* lays about 50 bright yellow adhesive eggs, with a diameter of 0.7 to 0.8 mm (Blaber & Whitfield, 1977).

C. multispinis leaves the lake in large groups (>50) during a spawning run, and migrates over low gradient swampy areas adjacent to streams. These migrations take place after and during heavy rain, mainly at night. The spawning fishes shuffle over damp ground using their serrated opercula as levers, and an arborescent suprabranchial organ for aerial breathing. Mating and egg deposition take place in shallow, temporary water.

The four cichlid species in Lake Sibaya exhibit different modes of parental care – two are monogamous, guarding substratum-spawners (*Tilapia rendalli swierstrae, Tilapia sparrmanii*) and two are maternal mouthbrooders with polygamous males (*Sarotherodon mossambicus, Pseudocrenilabrus philander*).

In *T. rendalli* the nest consists of a hollowed irregularly-shaped area of clean sand (diameter 50 to 90 cm) in which is interspersed a series of small pits (about 76 mm in diameter, 59 mm deep) and a larger central pit (180 × 113 mm). The eggs are deposited in the cleared area and diligently guarded and fanned by both parents. The newly hatched larvae are transferred by the female to small pits (personal observation using SCUBA). At low lake level breeding occurs in offshore plant beds, usually on the deep edge of sheltered bays among *Myriophyllum spicatum* and *Ceratophyllum demersum*, and not adjacent to exposed terraces where *Tilapia sparrmanii* and

Sarotherodon mossambicus mainly breed. At high lake level *T. rendalli* may breed in close proximity to the other cichlids in protected, well-vegetated inshore areas.

In *Tilapia sparrmanii* courtship and mating take place close to a simple, saucer-shaped nest (about 15 cm in diameter), which is constructed by the male. The eggs are guarded and fanned by both parents, and may be transferred by mouth to another cleared area, or to small pits about 2 cm in diameter and 2.5 cm deep, also excavated by the male. The larvae attach themselves to the sides of the pit by means of adhesive glands on the head, and execute progressively more active tail-flapping movements from their attached position. After the yolk has been absorbed the young become free-swimming and form a dense shoal which is guarded by both parents for seven to eight weeks (the latter figures are based on fish observed in a 10 × 12 m plastic-lined pool adjacent to Lake Sibaya, whereas the other observations were made on free-living fishes in the main lake). The adults assume an olive-green colouration with dark vertical bars when aggressive, and chase away potential predators such as *Pseudocrenilabrus philander* and *Glossogobius giurus*, which readily eat the fry if the parents are removed.

Male *Pseudocrenilabrus philander* in Lake Sibaya have a spectacular golden-yellow breeding colouration with a black ventrum, whereas the female is cryptically coloured. The breeding males (average SL 75 mm) dig one to four nests about 100 mm in diameter (max. 125 mm) and 5 to 6 mm deep in water depths ranging from 5 (rarely 3.5) to 24 m (Ribbink, 1975). The nests are arranged in groups or arenas, and usually occur in deeper water than those of *T. sparrmanii* and *S. mossambicus* (Fig. 15). *P. philander* males defend territories (diameter 2 to 3 m) around their nests and do not habituate to the close proximity of other breeding conspecifics (in contrast to *S. mossambicus*).

After an elaborate and vigorous courtship (described by Ribbink, 1971), the female deposits its eggs in the nest, then gathers the eggs, as well as milt released by the male in her mouth. Milt is also gathered from the male's anal fin which is bent to form a trough in the nest. The mouth-brooding females form schools, together with supernumerary males and juveniles, and inhabit submerged plant beds and the deeper areas of the profundal zone.

Oral incubation at 25°C takes 11 to 14 days after which the fry become increasingly independent and are eventually abandoned after five to six days (in aquaria, Ribbink, 1975). A single 40 to 50 mm female may produce as many as 90 9 mm fry. The juveniles form shoals in offshore zones from three to at least 30 m deep and exhibit appetitive sexual behaviour from a length of 25 mm. They usually remain in shoals for up to two years until a length of about 75 mm is reached before acquiring a territory.

The breeding season of *Sarotherodon mossambicus* may be divided into a nest-establishment phase by males (September, October), a period of intensive courtship (November, December), and a period of decreased nest-guarding by males, and mouth-brooding by females (January to March).

In early September, when water temperatures first exceed about 20°C, males move onto the terrace and commence aggressive conflicts prior to

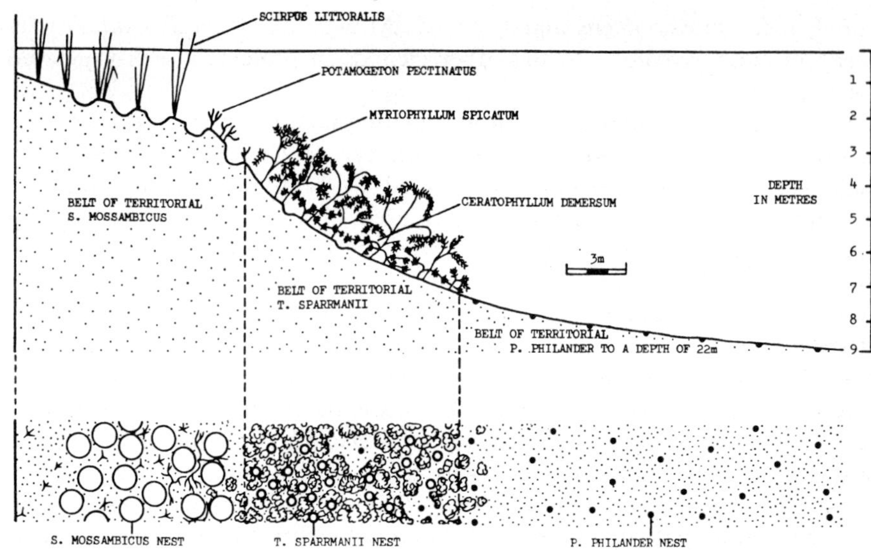

Fig. 15. Horizontal and vertical distribution of territories of *S. mossambicus. T. sparrmanii* and *P. philander*, in relation to each other and to macrophytes. The upper diagram represents a vertical section through the lake, and the lower diagram a plan view (from Ribbink, 1975).

excavating their nests. These conflicts, which involve broadside displays, mouth-locking, circling and chasing, probably serve to divide suitable nesting areas amongst the most aggressive individuals. Although the 'fights' involve only token aggression, some injury results to the fins, flanks and mouths of the aggressors. Small males are repulsed and apparently return to deeper water. This would explain the scarcity of 10–13 cm SL males in the breeding population, at least at low lake level (Bruton & Boltt, 1975). The modal length frequency of breeding males is 17 cm SL (Bruton & Allanson, 1974). The nests are simple hollows in the sand which are excavated by the males by means of their mouths.

Table 10. Number and density of *S. mossambicus* nests in the south basin of Lake Sibaya.

	Category			
	1	2	3	4
Number of nests	176	30	27	3
Nests/1000 m^2	13	5	7	1

The nests are divided into four distribution categories: (1) Associated with *S. littoralis* and *P. pectinatus* on the deep edge of the terrace; (2) on the deep edge of a barren terrace but closely associated with the *S. littoralis* fringes to the north and south; (3) in shallow protected water shoreward of a *S. littoralis* fringe; (4) on barren terraces shoreward of category 2, and not associated with aquatic macrophytes (from Bruton & Boltt, 1975).

The least favoured areas for nestbuilding are wave-washed shallow terraces with no macrophyte cover, whereas most nests occurred (at low lake level) on the deep edge of the terrace among *Scirpus littoralis* and *Potamogeton pectinatus* (Table 10). Larger nests were recorded from deeper water. The nests ranged in diameter from 20 to 185 cm, and in water depth from 41 cm to 8.5 m. When the lake level rises the area suitable for nesting increases and additional nests are found among flooded semi-aquatic plants and recently-established hydrophytes on the lake shore. These additional nests are apparently made by the small supernumerary males (10–13 cm SL) which form part of the breeding population at high lake level, but not at low lake level (Fig. 16).

After a brief, active courtship the female leaves the nest and mouthbroods the eggs and fry for 14 to 22 days in macrophyte beds. The advanced fry (>9 mm SL) are released into very shallow water along barren sandy shores or in marginal pools. On at least three occasions at low lake level mouthbrooding females (plus small numbers of other *S. mossambicus*) formed shoals of several hundred individuals which swam rapidly along the shallow terrace edge. Small groups left the main shoal and entered shallow (<4 cm deep) marginal pools where the fry were released. The large shoals consisted mainly of females with advanced fry, and assembled on all three occasions after heavy rain had caused extensive pools to form in the marginal zone. These pools were particularly suitable for occupation by the thermophilic juveniles of *S. mossambicus*. Shoaling by adult *S. mossambicus* was not

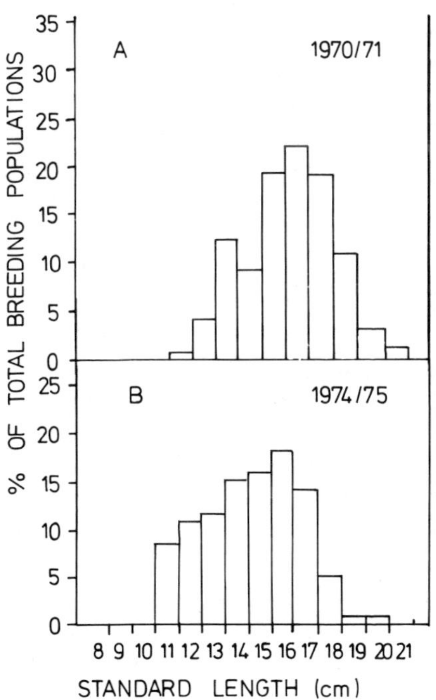

Fig. 16. The length frequency of breeding male *S. mossambicus* on the terrace in Lake Sibaya. A. Low lake level. $N = 550$. B. Intermediate lake level. $N = 300$.

observed under other circumstances in Lake Sibaya and may have some protective significance. Brock & Riffenburgh (1960) Eibl-Eibesfeldt (1962), Redakov (1973) and Neill & Cullen (1974) have shown that shoals of prey may confuse and reduce the effectiveness of predators.

The rate of recruitment of fry into the juvenile population on the northern, southern and eastern shores of the south basin of Lake Sibaya was estimated using mark/recapture techniques and the method of Chapman (1952). The figure obtained was 2400 fishes per month for an area of terrace of 2.1 km^2 for the period February to March 1971. In the period 17–19 March 1971 the standing crop of *S. mossambicus* on the terrace reached a peak, and juveniles (4–8 cm SL) were estimated to be 200 times more abundant than adults (Bruton, 1973).

In *C. gariepinus*, gonadal maturation is associated with increasing water temperatures and photoperiod from July to September (Fig. 17), and spawning is probably triggered by some factor(s) associated with the inundation of marginal grassland following heavy rain (Bruton 1979a). Sexual maturity is reached at lengths of 330 mm TL (♀) and 340 mm TL (♂) *i.e.* towards the end of the second year (Bruton & Allanson, 1979). The modal length of breeding females and males is 540 mm and 580 mm TL respectively. Individuals spawning in a stream which enters Lake Sibaya had a smaller modal length (380 mm TL) than a typical large group which spawned on the main lake edge (540 mm TL, Bruton, 1979a).

The fecundity of *C. gariepinus* in Lake Sibaya is related exponentially to total length ($y = 0.000009\,TL^{3.563}$, $R^2 = 0.8$) as is typical in fishes. Small (40 cm), modal (54 cm) and large (60 cm) females produce about 17,000, 50,000, and 70,000 eggs respectively. The largest Sibaya female actually examined (89.4 cm TL) produced about 163,000 eggs which is less than the figure given for other populations of this species where the largest females may produce over one million eggs (Gaigher, 1977, in Hardap dam, Namibia).

The main spawning season of *C. gariepinus* in Lake Sibaya extended from November to February although single observations were made in September and March. Twenty-one of the 22 spawning runs observed occurred at night with peak activity between 20h00 and 02h30, although preparatory behaviour started during the afternoon. One run took place during the afternoon (16h10 to 17h45). Rainfall was variable in relation to spawning but an average of 140 mm fell during the previous five days (range 0 to 414 mm). Spawning runs occurred most often on dark nights, *i.e.* within two or three days of new moon or last quarter and/or when the sky was overcast, possibly to reduce the risk of predation in very shallow water.

Spawning took place on exposed, gently sloping shores among inundated semi-aquatic and terrestrial plants, usually in water shallower than 30 cm. The catfish congregated in large groups at the lake edge at dusk and were noted to perform several intraspecific aggressive actions prior to courtship (Bruton, 1979a).

After a skirmish, the victor approaches a female and butts her lightly on the body. Receptive females swim inshore, followed by the male, and enter

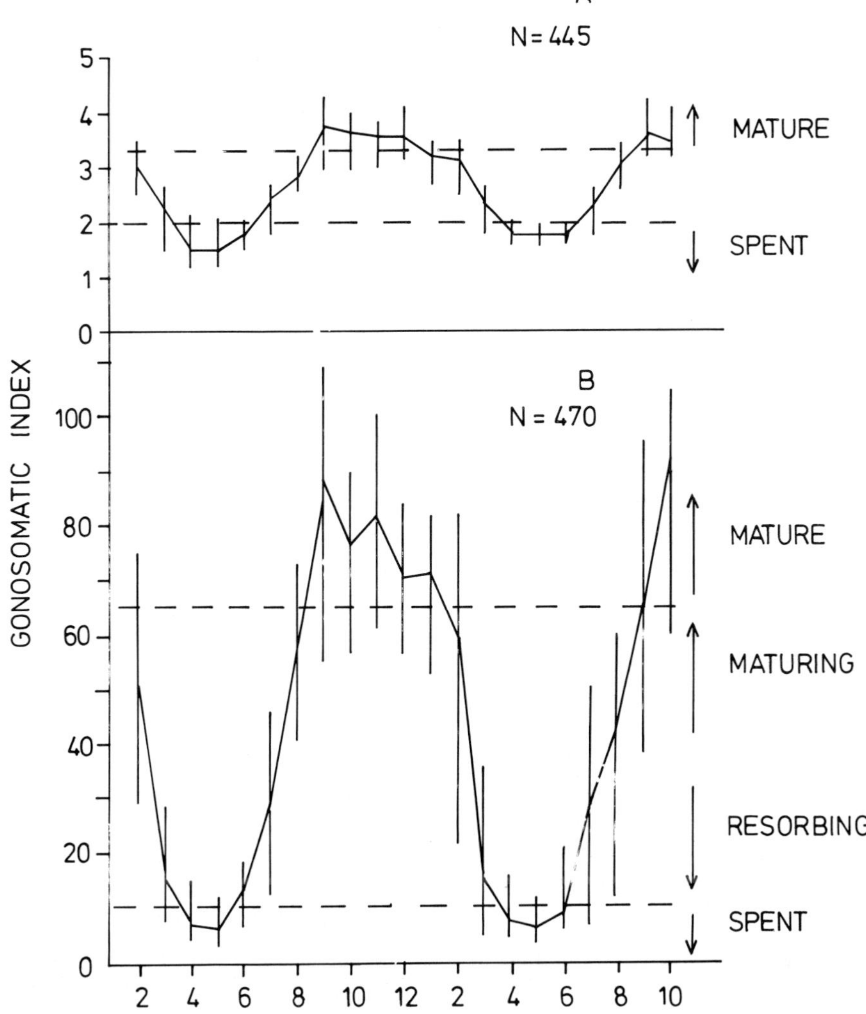

Fig. 17. Monthly mean gonosomatic indices and range (vertical bars) for male (A) and female (B) *C. gariepinus* during 21 consecutive months in Lake Sibaya. Minimum number in each month per sex: 20. (from Bruton, 1979a).

an area of inundated grassland. The pair wind their way through the vegetation for one or two minutes and may negotiate shallow sand bars, using the locked pectoral spines as levers to 'walk' (this species is also equipped with a suprabranchial organ for aerial respiration). Towards the end of following the male and female begin to quiver, and eventually stop swimming. The male then adopts a U-shaped position around the female's head. This mating posture is held for 17–18 secs (42 timings by stopwatch, range 12–20 secs) before the gonadal products are released. The eggs are released in a cloud of bubbles which may serve to transport them over a wide area. In addition, the female swishes its tail back and forth soon after egg release. The pair usually rest after mating and then resume courtship, which may be disrupted by an intruding male. Carefully watched pairs usually mated once or twice, rarely up to five times, before they were disturbed and separated.

There is no parental protection of the young in *C. gariepinus* as found in at least nine other siluroid families and two Asian species of *Clarias* (Breder & Rosen, 1966; Sidthimunka, 1972; Jhingran, 1977). The eggs, which are adhesive, are deposited on macrophyte leaves and stems, and abandoned.

Balon (1975) divided fishes into 32 eco-ethological guilds based on their breeding biology. Three main groups were recognised; non-guarders, guarders and bearers. These groups are represented in Lake Sibaya by *C. gariepinus*, *Tilapia sparrmanii* and *Sarotherodon mossambicus*, which exhibit increasing independance from external conditions in their breeding behaviour.

C. gariepinus, a non-guarding phytophilic open-substrate spawner, awaits suitable environmental conditions for spawning. In the Summer of 1968/9 when the rains were very late and occurred in March/April instead of September/October, this species delayed spawning in Lake Sibaya for seven months (Minshull, 1970). Likewise, females had mature (stage V) gonads in September 1975 but the first major spawning run only took place after heavy rains on 31 January 1976, although minor runs occurred in September and December 1975. *C. gariepinus* in Rhodesia (Holl, 1968) and Namibia (Gaigher, 1977) also delayed their spawning runs for several months until the first heavy rains.

In *Tilapia sparrmanii* the time of initiation of spawning in Lake Sibaya is influenced to a lesser extent by shallow water environmental conditions. At low lake level (1970–2), when the littoral terraces were shallow, wave-washed and experienced wide temperature fluctuations, *T. sparrmanii* commenced breeding in late November and December. At high lake level, when the terraces were three metres deeper and less subject to environmental extremes, breeding commenced in September.

Sarotherodon mossambicus creates a relatively stable environment for embryonic development by brooding the young in its mouth. This species started breeding in Lake Sibaya at almost exactly the same time of the year (the last three weeks of September) every year from 1970 to 1976 (the observation period) irrespective of prevailing, and widely differing, shallow water conditions.

Juvenile Biology

Well-vegetated offshore habitats and sheltered bays provide abundant cover and a diversity of food items for juvenile fishes, and are used as nurseries by many fish species in the lake at low lake level *e.g.* gobiids, cyprinids, cyprinodontids, most cichlids. At high lake level many fishes also utilise flooded inshore habitats as nurseries.

S. mossambicus is the only species which selects the extreme environment of the shallow terraces at low lake level as a habitat for its young.

Mouth-brooding females release the fry into water less than 10 cm deep in marginal pools and on the terrace edge. The fry occupy progressively deeper water during the day as they grow larger, but all move into deeper warmer terrace waters at night.

Minshull (1970) determined the size structure of the *S. mossambicus* population at various depths using baited fish traps. He found that areas shallower than 15 cm were occupied by fry (15–19 mm) and small juveniles (30–35 mm) with occasional larger juveniles (50–55 mm). The population at a depth of 45 cm was unimodal at 55–65 mm, whereas in deeper habitats (1 m and 5 m) a wider range of larger fishes was found (60–85 mm; 70–95 mm respectively). Subsequent sampling using fishtraps and seine nets confirmed these trends and emphasized that smaller fishes occupied shallower habitats (Fig. 18) although there was considerable mixing of shoals at times.

Bowen (1976a) found that *S. mossambicus* formed pockmarks when feeding. The diameter of these marks was directly proportional to fish size and could be used to identify the feeding areas of different size fishes. Small juveniles (<40 mm TL) fed in dense concentrations close inshore whereas large juveniles and adults fed in mid or deep terrace at lower densities, with occasional areas of concentrated feeding.

Juvenile *S. mossambicus* (<80 mm TL) showed a definite preference for open terraces (57.5% of a catch numbering 6480) and sparsely-vegetated slopes (26.8%) and were less common on barren slopes (13%) and in sheltered bays (2.7%).

A marked diel water temperature fluctuation occurred in shallow water (Fig. 8). Juveniles entered terrace waters in the morning when temperatures there first exceeded those of the main lake. They migrated back into water deeper than 100 cm at sunset when shallow water temperatures dropped below those of the main water body. At night juveniles lay motionless on

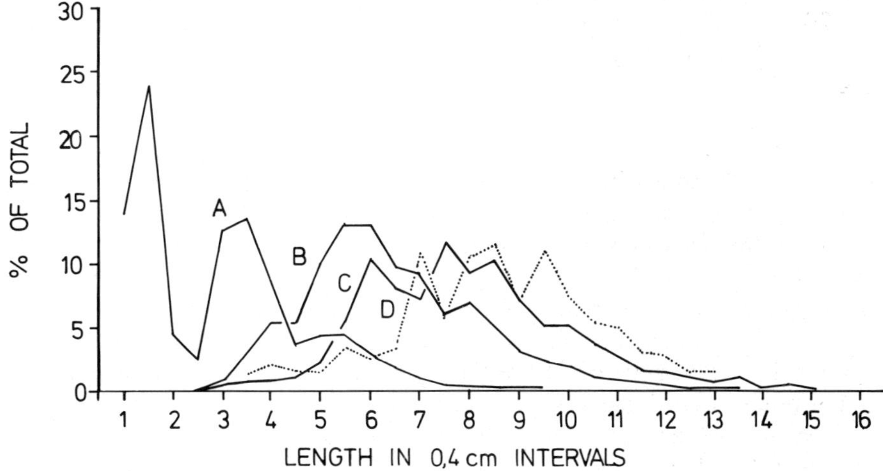

Fig. 18. Numbers of *S. mossambicus* caught in baited traps at sites A, 15 cm water depth, B, 45 cm, C, 1 m and D, 5 m expressed as a percentage of the total catch (from Minshull, 1970).

the slope at 1–5 m, or were dispersed in midwater; no shoaling was observed.

Juvenile *S. mossambicus* formed shoals on the terrace during the day which were largest in March and April immediately after breeding season, and smallest in the cool season (May to July). The shoals were made up of a limited range of sizes with the modal 2 cm length group comprising 40–60% of the shoal. During summer juveniles formed small schools (<20 fishes), when feeding, but large shoals (>300) when alarmed or moving rapidly along the shore (Bruton & Boltt, 1975; Bowen, 1976a). The composition and size of the feeding groups changed constantly. During winter shoals usually consisted of less than 100 fishes. The recovery of finclipped juveniles revealed that shoals may move rapidly across the terrace. Of 254 finclipped fish which were recaught, most had moved less than 400 m in two days, but two recaptures had moved 400 m in 2 h and 3100 m in four days respectively.

Clarias gariepinus also utilizes the shallow littoral zone as a nursery. The eggs are released in inundated well-vegetated areas and adhere to the leaves by means of an adhesive disk (Bruton 1979a). Egg and larval development is rapid. The ova hatch after 24 to 25 h (at 19°–24°C) and remain attached to the plant for a further 10 h while vigorously exercising their tails. At 50 h (about 52 mm TL) the larvae swim rapidly and at 80 h (62 mm TL) begin feeding. At this stage, and for the next five months they are thigmotactic and secretive and inhabit dense vegetation and flotsam. The larvae are able to utilize atmospheric oxygen at an age of eight days although the suprabranchial organ is not fully developed. This ability, as well as their rapid developmental rates are adaptations to ensure that the larvae can survive in shallow pools and move into deeper water before the pools dry up.

Small juvenile *C. gariepinus* (<50 mm TL) inhabit temporary marginal pools and the edges of sheltered bays at low lake level, and flooded marginal pools (<30 cm deep) at high lake level. Juveniles in the length range 50–200 mm also inhabited marginal pools but occasionally ventured onto the terrace edge at night. Catfish from 200–400 mm TL inhabit deep marginal pools and the shallow terrace edge.

The low number of *Tilapia rendalli* juveniles in shallow water habitats in Lake Sibaya is surprising as this species has wide temperature tolerances (11°–37°C; Whitfield & Blaber 1976). *T. rendalli* is known to reach such high population densities in some waterbodies that considerable damage results to the macrophyte community on which it feeds (Junor, 1969) and in many freshwater or brackish ecosystems they are known to be numerically dominant to *S. mossambicus* (*e.g.* Lake McIlwaine, Rhodesia, Caulton, 1976; Lake Poelela, Mozambique, Hill *et al.*, 1975). In Lake Sibaya this species is mainly confined to sheltered bays and offshore macrophyte beds, but juveniles occasionally venture onto the terrace.

T. sparrmanii and *P. philander* protect their young in offshore plant beds and sheltered bays at low lake levels, and also utilize inundated inshore areas at high lake levels. The gobiids, cyprinids and cyprinodontids also inhabit both offshore and inshore plant beds as juveniles.

Gilchristella aestuarius and *Hepsetia breviceps* are the only true pelagic species with planktonic larvae.

Growth and Population Structure

Growth studies have been performed on only two species – *Sarotherodon mossambicus* (Bruton & Allanson, 1974) and *Clarias gariepinus* (Bruton & Allanson, 1979), which were chosen as they are respectively the most prominent herbivore and predator in the lake.

Scales, otoliths and opercula were examined for age determination in *S. mossambicus*. The otoliths and opercula have regular but indistinct rings whereas the scales show clear rings formed by broken, widely-spaced circuli in the anterior field (Fig. 19). The time and rate of ring formation was determined by counting the number of circuli in the marginal increment *i.e.* the distance from the anterior edge of the scale to the most anterior scale ring. In fishes sampled soon after a ring has been formed, the ring is on the anterior margin of the scale and there are no, or few, circuli in the marginal increment. Five scales from the pectoral region of 2223 *S. mossambicus* caught between September 1970 and March 1971, and September 1971 and August 1972, were examined. The number of marginal circuli decreased suddenly in October 1970, January and September 1971, and January 1972, indicating that rings were formed in these months. No rings were formed during the cool season. Pre- and postbreeding feeding migrations, and increases in condition factor, were associated with the twice-yearly scale ring formation. Annual length increments were calculated using a modification of Lea's (1910) formula for 450 *S. mossambicus* collected at all times of the year in Lake Sibaya (Bruton & Allanson, 1974), and growth curves were plotted.

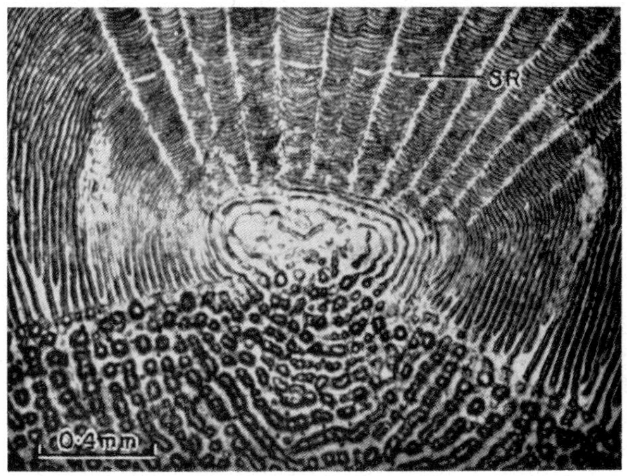

Fig. 19. Scale of *S. mossambicus* from Lake Sibaya showing scale ring (SR) formed by widely-spaced circuli (from Bruton & Allanson, 1974).

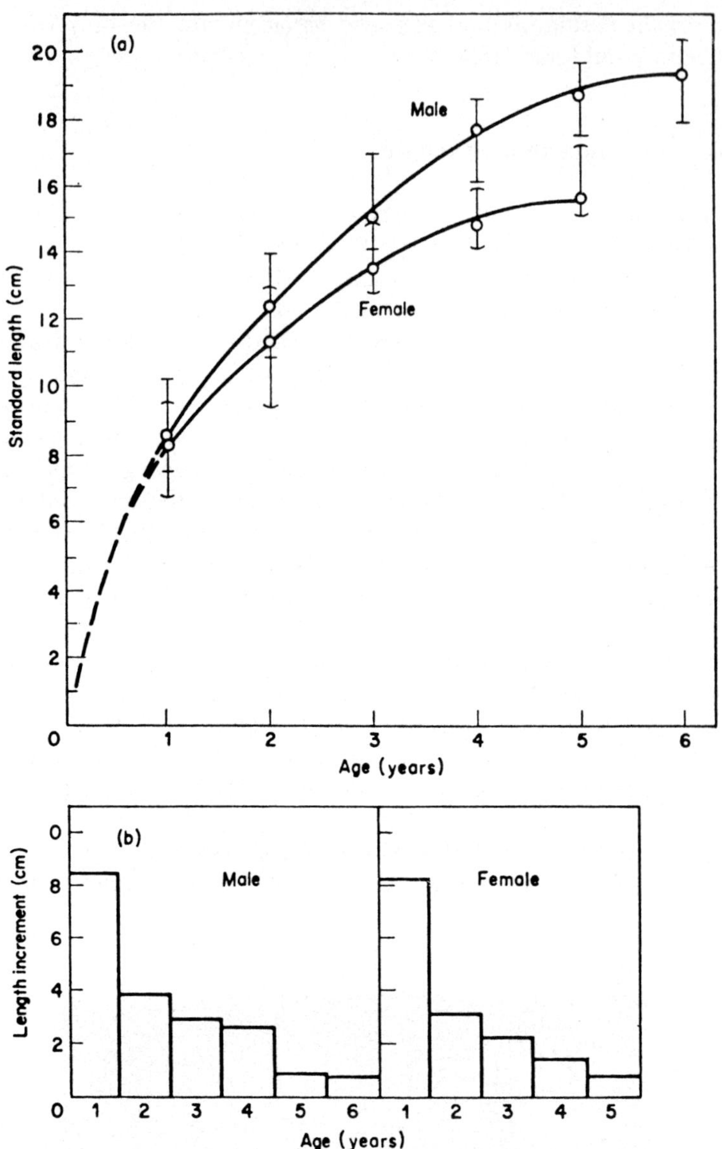

Fig. 20. Linear growth rates in standard length (a) and standard length increments (b) of male and female *S. mossambicus* from Lake Sibaya. Square brackets indicate the range for males, round brackets for females (from Bruton & Allanson, 1974).

Male *S. mossambicus* grow faster than females after the first year (Fig. 20a). In both sexes the growth rate is highest in the first year but decreases progressively thereafter (Fig. 20b). Somatic growth rates calculated from a length:weight curve, were also higher in males (Fig. 21a). Annual weight increments were highest in the fourth year (Fig. 21b) *i.e.* at a weight corresponding to the modal size of adult fishes. The maximum age of *S. mossambicus* in Lake Sibaya is about 7–8 years.

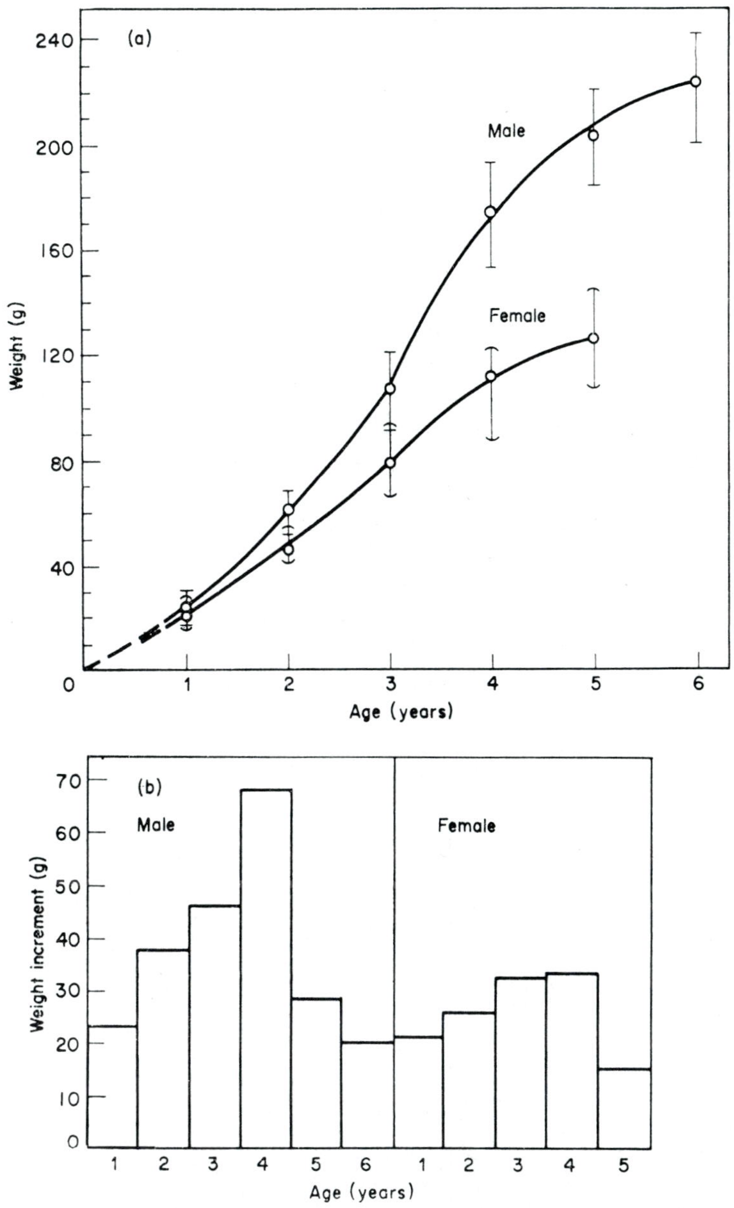

Fig. 21. Somatic growth rates (a) and weight increments (b) of male and female *S. mossambicus* from Lake Sibaya. Square brackets indicate the range for males, round brackets for females (from Bruton & Allanson, 1974).

The length composition of the breeding population of *S. mossambicus* (Fig. 22) reveals that the modal size of males is 17 cm SL and that of females 11–12 cm SL. The size at first maturity was 12 cm SL in males and 8–10 cm SL in females. Females first bred after one year, and males after one or two years. The largest male caught measured 23.6 cm SL (29.2 cm TL, mass

320.6 g) and the largest female 18.5 cm SL (22.8 cm TL, mass 215 g). Over 95% of the fishes examined measured less than 21 cm SL (26 cm TL).

The growth rate of *S. mossambicus* in Lake Sibaya follows the pattern shown by tilapiines in other systems, but not the dimension. As a result of breeding precociously, growth deceleration after maturity occurs earlier in the life cycle than in 'normal' populations and the final size of individuals is small. Populations of *S. mossambicus* in the Transvaal and Cape Province grew at the same rate, or slower than the Sibaya population in the first and/or second year, but grew faster in subsequent years (Le Roux, 1961; van Rensburg, 1966; du Toit, 1971).

The maximum size of *S. mossambicus* in Lake Sibaya is also smaller than in other natural systems. *S. mossambicus* reaches over 1500 g in Inyamithi pan, 60 km north-west of Lake Sibaya (Coke, pers. comm., 1973) and 2953 g in southern Africa (Jubb, 1967). In brackish water culture, this species may reach 36 cm TL in 8 months (Le Mare, in Hickling, 1970).

Dwarf populations of tilapiine species are known from other natural systems (Fryer & Iles, 1972; Iles, 1973). It is notable that many of these populations, *e.g.* those recorded by Hickling (1961), Whitehead (1962) and Coe (1966), were confined to hot springs or drought-stricken lakes, and subject to extreme environmental conditions. Others, such as the populations studied by Welcomme (1964, 1970), inhabited the hot, shallow margins of lakes, where mortalities were high due to intense predation and the

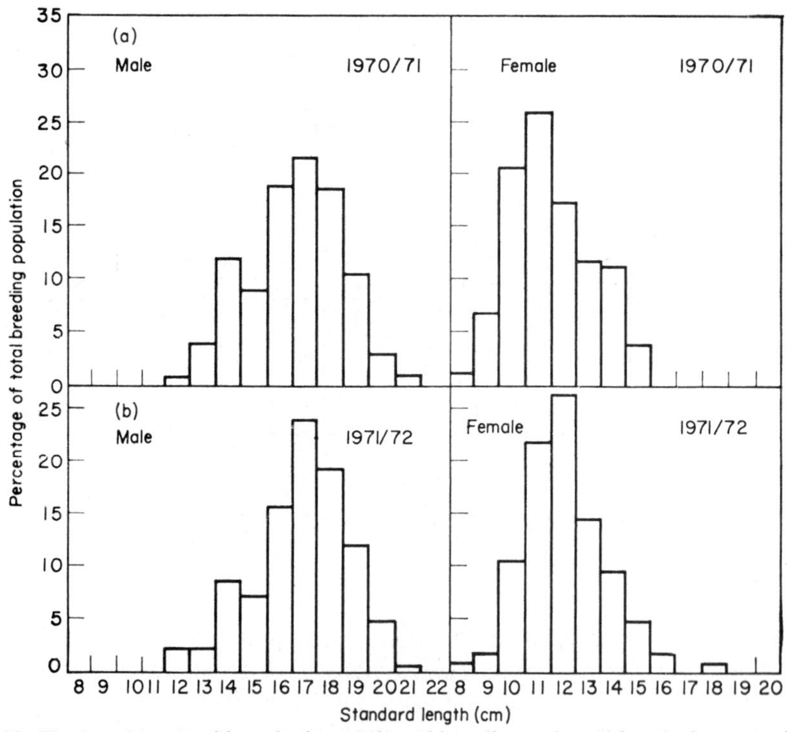

Fig. 22. The length composition of seine catches of breeding male and female *S. mossambicus* from Lake Sibaya in the summers of 1970/71 and 1971/72 (from Bruton & Allanson, 1974).

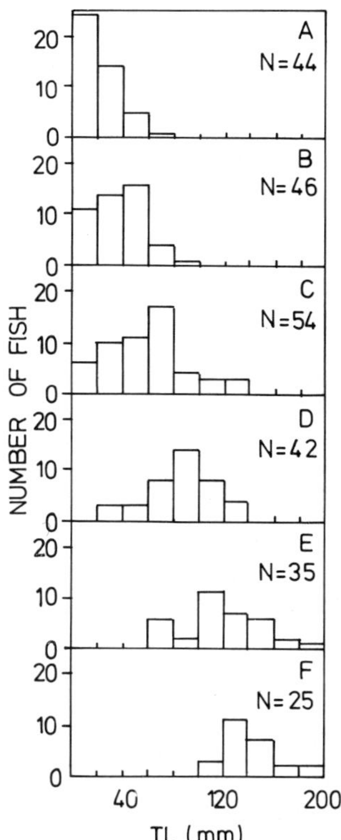

Fig. 23. The length distribution of fingerling *C. gariepinus* caught in the marginal shallows of Lake Sibaya, showing their growth rate by movement of the mode. The juveniles for this sample were all collected at the same site monthly over six months, and were released after capture. A: December 1975: mean length 18 mm B: January 1976: 43 mm C: February 1976: 81 mm D: March 1976: 102 mm E: April 1976: 126 mm F: May 1976: 144 mm.

drying up of marginal pools. The Sibaya population fits the latter category. Iles (1973) has suggested that dwarfing represents an adaptive mechanism involving reproductive and growth characteristics which enable tilapiine populations to withstand high mortality rates. This interesting hypothesis will be discussed later once the trophic ecology of these species has been outlined. The growth of juvenile *C. gariepinus* in Lake Sibaya was determined from the movement of length frequency modes of fishes caught in flooded marginal pools. Monthly length increments ranged between 18 and 38 mm (mean 24 mm, Fig. 23). The modal size after 6 months was 144 mm and after one year, 250 mm (Fig. 24, Bruton & Allanson, 1979).

Pectoral spines, lapillar otoliths, opercula and vertebrae were examined for possible use in age determination of adult *C. gariepinus* from Lake Sibaya, but only pectoral spines showed clear rings. Spine rings take the form of clearly-defined lines of denser tissue in the bone matrix (Fig. 25).

Total lengths were back-calculated from spine ring measurements using Fraser's formula as recommended by Ricker (1968). These results corresponded well with the lengths obtained for *C. gariepinus* whose spine rings

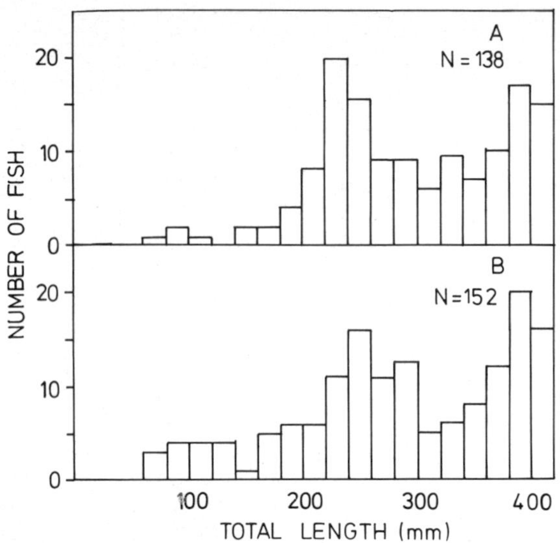

Fig. 24. The length frequency of *C. gariepinus* caught in shallow terrace and marginal habitats in Lake Sibaya, showing length modes of one year old fish. A: 240 mm in December 1974/January 1973. B: 260 mm in December 1975/January 1976.

were counted directly. The first spine ring is formed at a modal length of 400 to 410 mm TL, and no ring is formed at 240 to 260 mm TL *i.e.* at an age of about one year (Fig. 24), in the bulk of the catfish examined. If the first spine is assumed to be formed towards the end of the second year, a smooth curve is obtained which produces a good fit of von Bertalanffy's growth equation (von Bertalanffy, 1957). This assumption was tested by calculating von Bertalanffy's parameter, t_0 (the theoretical time when length

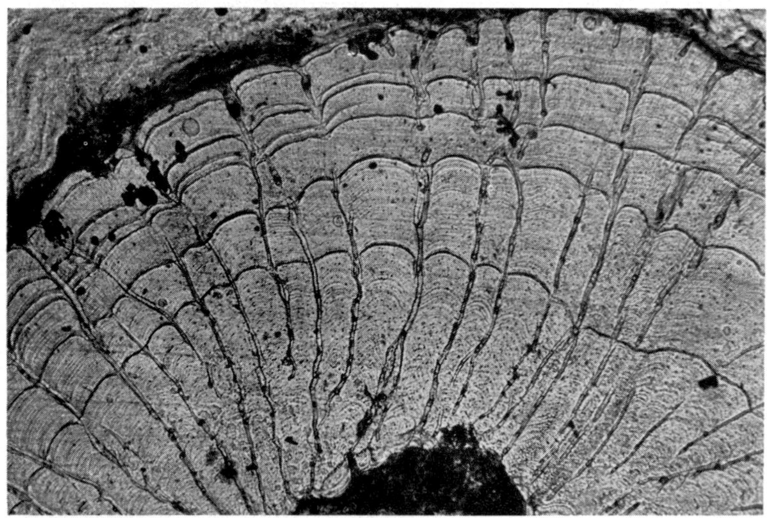

Fig. 25. Anterior section of a pectoral spine of *C. gariepinus* showing clear spine rings and the central lumen.

Table 11. Mean observed total lengths in mm ($\overline{TL}$), standard error of TL (SE), annual length increment (ΔTL), mean weight ($\bar{W}$) and number of catfish in sample (N) for different year classes of male and female *C. gariepinus* in Lake Sibaya. Mean length in the 1+ year class from Fig. 24.

Year	$\overline{TL}$	SE	ΔTL	$\bar{W}$	N
Males					
1+	240		240	97	
2+	399	0.49	159	419	81
3+	517	0.41	118	843	135
4+	575	0.53	58	1123	67
5+	629	0.67	54	1431	40
6+	659	1.06	30	1623	24
7+	695	1.21	36	1873	11
8+	726	1.47	31	2107	10
Females					
1+	240		240	97	
2+	406	0.48	166	455	91
3+	512	0.45	106	852	110
4+	564	0.52	52	1107	76
5+	608	0.41	44	1357	50
6+	639	0.98	31	1552	23
7+	648	2.46	9	1612	12

is zero) using Ricker's (1975) method. These calculations supported the assumption that no spine ring is laid down in the first year. The adequacy of the von Bertalanffy growth equation was also tested in this particular case using Ford-Walford plots, and found to be acceptable (Bruton & Allanson, 1979). The growth statistics are summarised in Table 11.

Growth in length for males and females was similar until 3+ years, after which males grew faster (Fig. 26). In both sexes the annual length increment was highest in the first year then decreased progressively. Large *C. gariepinus* (>700 mm TL) are occasionally encountered in Lake Sibaya. Back calculations of growth rates for four of these large catfish revealed that they attained a relatively high growth rate early in life and maintained this higher rate throughout life (Fig. 26). This phenomenon was also noted in aquaria where, as early as the second week, a small percentage of same-age catfish attain a larger size than the majority, and retained or increased this headstart in subsequent months. The largest male *C. gariepinus* caught in Lake Sibaya measured 1088 mm TL (8.79 kg) and the largest female 1036 mm (7.76 kg). This species exceeds 1300 mm TL and 20 kg in some southern African rivers (Bruton, 1976).

Growth in weight was nearly linear in males but decelerated after four years in females (Fig. 27). Annual weight increments were highest in the third year in both sexes. The modal age was 5+ years in males and 4+ years in females; few fishes exceeded 8+ years.

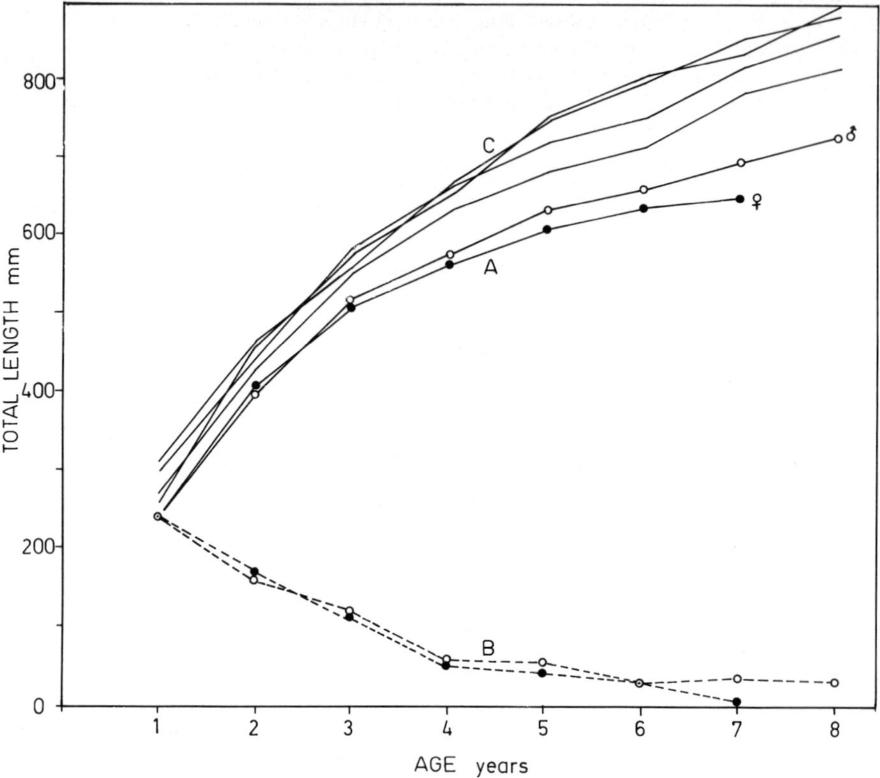

Fig. 26. Growth in length of *C. gariepinus* in Lake Sibaya. $N = 730$. A. Growth rate of average individuals in the population. B. Annual length increments of A. C. Growth rates of four large fast-growing fish.

The length frequency of juvenile and adult *C. gariepinus* caught in all habitats is given in Fig. 28. The length mode in males was 580–590 mm TL and in females 540–550 mm TL. Over one third of the population measured between 500 and 600 mm TL. This marked clumping is explained by the sharp decleration of growth rate between these lengths (Fig. 26). The overall sex ratio of 1980 *C. gariepinus* was 1.30 : 1 in favour of males.

The length frequency of four *C. gariepinus* populations is compared in Fig. 29. Sample B was taken from the flood plain pans of the Pongola river in Zululand by Kok (pers. comm., 1976). The Elands river collection (C) was made by van der Waal (1972) in the south-eastern Transvaal whereas sample D came from Hardap dam on a northern tributary of the Orange River in Namibia (Gaigher, pers. comm., 1976). The Sibaya collection is the only one which adequately sampled juvenile *C. gariepinus* <200 mm TL.

The length mode of the Sibaya population (500–600 mm TL) is 100–120 mm higher than that of the Pongola and Elands river populations, but 200 mm lower than the Hardap dam population. Length frequencies may, however, overemphasize the size of fish in a population if the fish are in poor condition. The populations were therefore compared in terms of weight structure and length-weight relationship as well (Figs. 30 & 31).

Although the Pongola catfish population had a smaller length mode, their weight structure is similar to that of the Sibaya population, with a greater proportion of fishes >2 kg. The Hardap dam population has a different weight structure with a very high component of large fish. This trend is emphasized by the following set of figures:

	% of *C. gariepinus* >2 kg
Lake Sibaya	4.1
Pongola river pans	17.4
Hardap dam	62.8

The length – weight relationship of juvenile *C. gariepinus* in the different populations was the same (Fig. 30a) but there were distinct differences in larger size groups of adults. *C. gariepinus* in Lake Sibaya are in relatively poor condition at lengths greater than 600 mm, whereas Pongola catfish are in good condition. Hardap dam and Elands river catfish are in an intermediate position. The good condition of the Pongola catfish is further emphasized by the following set of figures which are derived from the length: weight regressions given in Fig. 31.

	C. gariepinus population	Average weight at 1000 mm TL	% of D
A	Lake Sibaya	5539 g	57
B	Elands river	6971 g	72
C	Hardap dam	7078 g	73
D	Pongola floodplain pans	9681 g	100

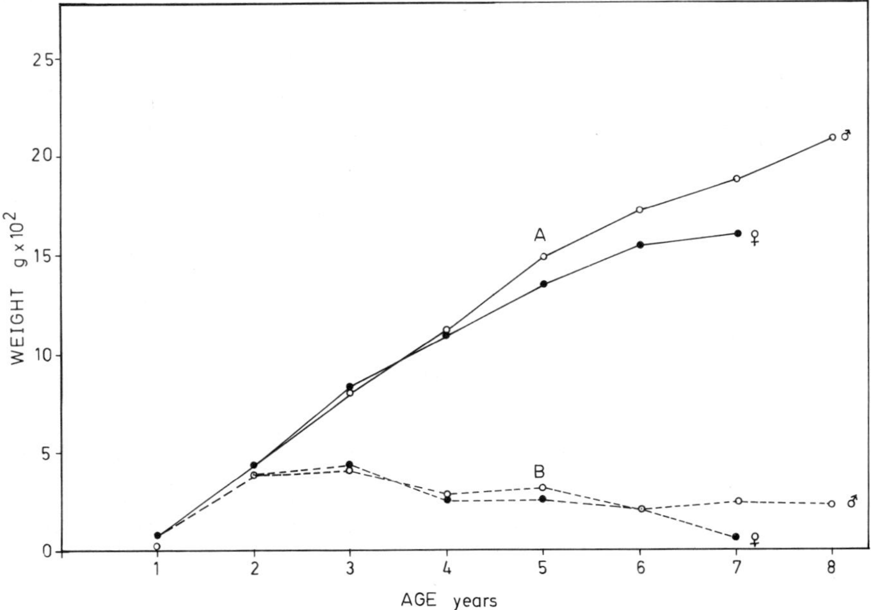

Fig. 27. Growth in weight of *C. gariepinus* in Lake Sibaya. $N = 730$. A. Weight increase. B. Annual weight increments.

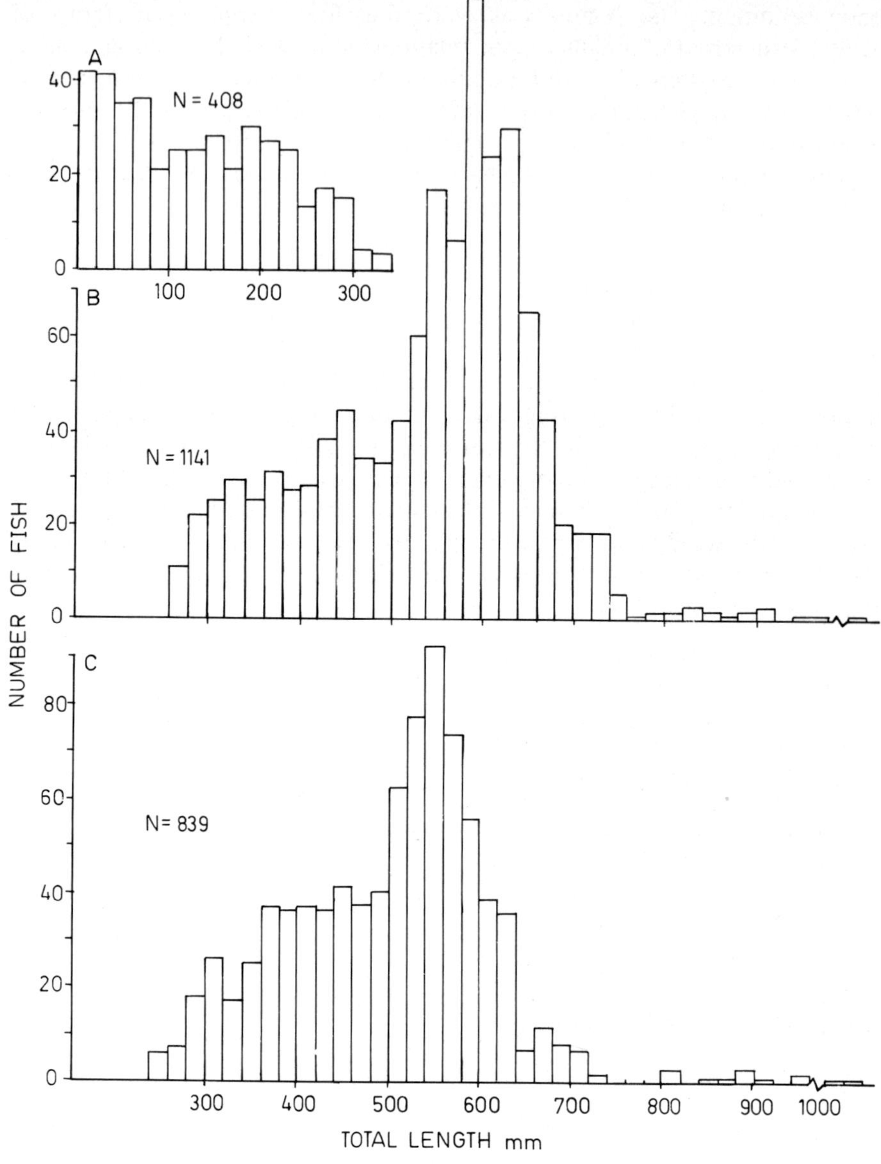

Fig. 28. Length structure of a sample of 2388 *C. gariepinus* from Lake Sibaya caught throughout the study period (1973–1976). A. Juveniles. B. Adult males. C. Adult females.

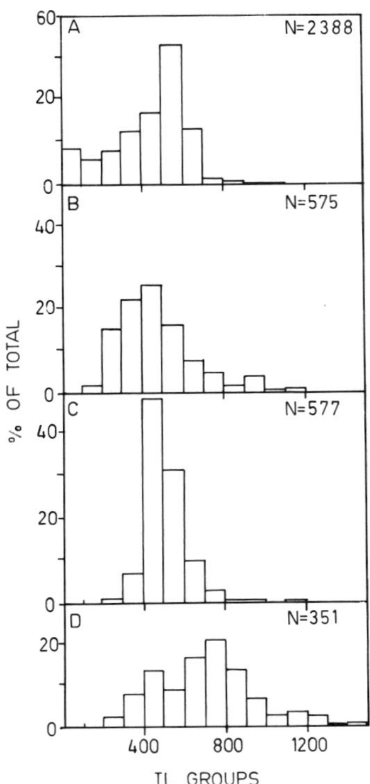

Fig. 29. The length frequency of three *C. gariepinus* populations. A. Lake Sibaya. B. Pongola floodplain pans (Kok, pers. comm.) C. Elands river (van der Wall, 1972) D. Hardap dam (Gaigher, pers comm.)

In summary, the condition of *C. gariepinus* in Lake Sibaya is similar to that in three other populations to a length of about 500 mm TL. In larger size groups Sibaya catfish are in poor condition. This trend is closely paralleled by *Sarotherodon mossambicus* in Lake Sibaya.

Food and Feeding Behaviour

The food and feeding behaviour of three species has been studied in detail: *Sarotherodon mossambicus*, a primary consumer, which feeds mainly on benthic diatoms and bacteria, *Croilia mossambica*, a small secondary consumer of epibenthic invertebrates, and *Clarias gariepinus*, a large tertiary consumer which feeds on fishes as well as many other organisms.

The feeding habits of two species (*Silhouettea sibayi* and *Labeo molybdinus*) have not been studied due to their apparent scarcity in the lake. The remaining thirteen species have only been investigated in sufficient detail to reveal their broad dietary habits.

The most striking feature of the trophic ecology of the fishes of Lake Sibaya is the wide range of food taxa taken by most species (Table 12) which makes it difficult to assign the fishes to particular feeding categories.

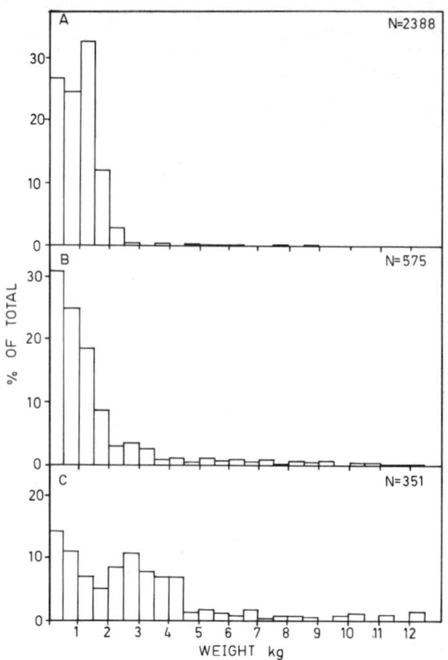

Fig. 30. The weight frequency of three *C. gariepinus* populations. A Lake Sibaya. B. Pongola floodplain pans (Kok, pers. comm.). C. Hardap dam (Gaigher, pers. comm.).

Nevertheless, three broad groups can be recognized: phytophages, which feed mainly on macrophytes or microphytes (two species); omnivores, which regularly ingest both plant and animal matter (one), and predators, the bulk of the population (13 species), which feed on insects, crustaceans, other invertebrates and fish. There are no specialist fish piscivores in the lake (*cf. Hydrocynus vittatus* in Lake Kariba) probably due to the lack of a well-developed pelagic fish community. Juvenile cichlids form a potential food source for a specialist piscivore, but they are mainly found on the shallow, well-lit terraces during the day (an inhospitable habitat for a large predator) or in the shelter of dense macrophyte stands. The largest piscivorous fish, *Clarias gariepinus*, is only able to catch fish prey under certain conditions (see below) and relies to a considerable extent on invertebrates as a source of food.

The low numbers of primary consumers is a reflection of the low primary productivity level of Lake Sibaya. No species is known to feed extensively on the phytoplankton, which has a low standing crop ($<5\,\mu g$ chlorophyll per litre; Allanson & Hart, 1975) although the planktonic diatoms *Melosira granulata* and *Synedra* sp. occur prominently in the diet of adult *Sarotherodon mossambicus* (Bowen, 1976a). The pelagic fish *Gilchristella aestuarius* and juvenile *Pseudocrenilabrus philander* feed to a limited extent on planktonic diatoms.

Only one fish species, *S. mossambicus*, relies predominantly on the benthic diatom community (and its associated microflora). This community is, however, only sufficient to maintain good growth and condition in that part of the population which can feed in the well-lit, warm parts of the terrace which are less than one metre deep *i.e.* the juveniles (Bowen, 1976a). The benthic algal population available to adult *S. mossambicus* in deeper water has a lower nutritive value, as described below, and they therefore also feed to a small extent on other algae, small fishes, aquatic insects, *Caridina nilotica*, gastropods and aquatic macrophytes (Minshull, 1968, 1969; Bruton & Boltt, 1975, Bowen, 1976a).

The steep slope around some shores, and the strong wave-action on gentle sloping shores, results in a restriction of the area suitable for growth of macrophytes. As a result, the only dense stands are found in a narrow band on certain slopes, and especially in sheltered bays, which only cover about 25% of the bottom area of Lake Sibaya. Only one specialist macrophytophage, *Tilapia rendalli swierstrae*, inhabits these areas. The mormyrid, *Marcusenius macrolepidotus* which feeds mainly on aquatic insects,

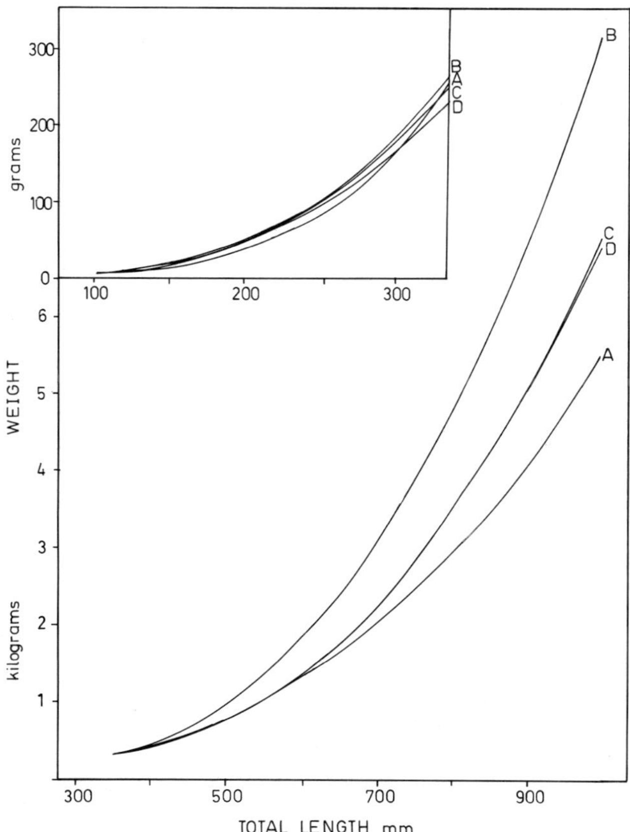

Fig. 31. The length: weight relationships of juvenile (upper diagram) and adult (lower diagram) *C. gariepinus* from four populations. Length: weight regressions for adult catfish: A. Lake Sibaya: $W = 0.00004\ TL^{2.699}$($R^2 = 0.92$, $N = 355$) B. Pongola floodplain pans: $W = 0.000003\ TL^{3.169}$($R^2 = 0.98$, $N = 165$. Kok, pers. comm). C. Hardap dam: $W = 0.000004\ TL^{3.071}$($R^2 = 0.95$, $N = 139$, Gaigher, pers. comm.). D. Elands river (van der Waal, 1972).

Table 12. Food preferences of the fishes of Lake Sibaya. Data from Minshull (1968, 1969), Ribbink (1971), Bruton & Boltt (1975), Bowen (1976a), Blaber & Whitfield (1977), Bruton (1979b) and unpublished observations of the author.

Feeding category	Fish Species	Main Food	Other Food	*N*
Macrophytophage	*Tilapia rendalli swierstrae*	*Potamogeton pectinatus* *Typha latifolia*	Terrestrial insects, *Caridina nilotica*, amphipods, *Potamon sidneyi*, juvenile cichlids, *Bulinus* spp., diatoms, filamentous algae	43
Microphytophage	*Sarotherodon mossambicus* juveniles	*Mastogloia elliptica*	Other diatoms, amphipods, ostracods	
	adults	*Melosira granulata*	Other diatoms, green algae, terrestrial insects, small cichlids, *Caridina nilotica*, gastropods, macrophytes.	1516
Omnivore	*Tilapia sparrmanii*	Algae (especially diatoms) *Potamogeton pectinatus*, amphipods	Aquatic and terrestrial insects, including chironomidae, other small crustaceans (mainly epiphytic)	42
Predator/ plankton	*Gilchristella aestuarius*	*Bosmina longirostris* (Cladocera)	Zoeae of *Hymenosoma orbiculare* copepods, amphipods, ostracods, diatoms, other algae copepods, flying insects (caught in the surface film)	33
	Hepsetia breviceps	*Bosmina longirostris*		35
Predator/ insects	*Marcusenius macrolepidotus*	Aquatic insects (especially Chironomidae, Odonata, Hemiptera)	Terrestrial insects, isopods, periphytic algae	20
	Clarias theodorae	Aquatic insects (especially Chironomidae, Trichoptera, Odonata, Ephemeroptera)	Culicid pupae, *Caridina nilotica*, *Melanoides tuberculatus*, cichlid fry.	28

Predator/small invertebrates	*Croilia mossambica*	chironomid larvae and small *Melanoides tuberculatus*	*Corophium triaenonyx*, ostracods, *Grandidierella lignorum*	132
	Barbus paludinosus	chironomid larvae, copepods, amphipods	other small aquatic insects and crustaceans, diatoms	18
	Barbus viviparus	chironomid larvae, copepods, amphipods	other small aquatic insects, diatoms	22
	Aplocheilichthys myaposae	*Bosmina longirostris* and aquatic insect larvae	small terrestrial insects, ostracods, diatoms	22
	A. katangae	*Bosmina longirostris*	filamentous algae, diatoms, small terrestrial and aquatic insects	21
Predator/ omnivore/ small piscivore	*Pseudocrenilabrus philander* juveniles	copepods, amphipods, chironomid larvae	ostracods, isopods, diatoms, macrophytes, *Botryococcus braunii*	44
	adults	cichlid, cyprinodontid and gobiid fry and small juveniles, copepods, amphipods	other fish fry, ostracods, isopods, aquatic and terrestrial insects, diatoms	40
	Glossogobius giurus	*Grandidierella lignorum*, *Apseudes digitalis*, *Caridina nilotica*, cichlid fry and small juveniles	*Aplocheilichthys* spp., amphipods, ostracods, *Melanoides tuberculatus*, *Bellamya capillata*	46
	Ctenopoma multispinis	*Aplocheilichthys* spp., juvenile cichlids, *Caridina nilotica*	Aquatic and terrestrial insects, *Bulinus* spp, isopods, amphipods, chironomids	18

Table 12 (*Continued*)

Feeding category	Fish Species	Main Food	Other Food	*N*
Predator/ omnivore/ large piscivore	*Clarias gariepinus*			
	<20 mm TL	Larval chironomidae and *Caridina nilotica*		6
	20–50 mm TL	*Grandidierella lignorum*	Chironomid larvae, *Caridina nilotica*, *Cyathura carinata*	19
	50–100 mm TL	*Povilla adusta*, *G. lignorum*, odonatan nymphs	*Caridina nilotica*, cichlid fry, chironomid larvae	37
	100–300 mm TL	*Caridina nilotica*, *Grandidierella lignorum*, cichlid fry	Odonatan nymphs, *Povilla adusta*, *H. orbiculare*	93
	300–700 mm TL	*H. orbiculare*, *S. mossambicus*	*G. lignorum*, *C. nilotica*, *Potamon sidneyi*, *P. philander*	291
	>700 mm TL	*S. mossambicus*	*P. sidneyi*, *Bellamya capillata*, *H. orbiculare*	29

but also on periphyton, small crustaceans and terrestrial insects, is likewise largely confined to dense macrophyte beds. These areas are also the main habitat of an abundant omnivorous cichlid, *Tilapia sparrmanii* which is a browser on epiphytic and benthic algae especially diatoms, a grazer of macrophytes, especially *Potamogeton pectinatus*, and a predator of amphipods, insects and other invertebrates.

The four small cyprinodontids and cyprinids (*Aplocheilichthys myaposae*, *A. katangae*, *Barbus paludinosus*, *B. viviparus*) also frequent macrophyte beds where they feed opportunistically on a variety of small pelagic, epiphytic and benthic crustaceans and insects. The two *Aplocheilichthys* species show an apparent preference for the cladoceran *Bosmina longirostris* but this finding may be the result of bias introduced by the small sample. They are likely to capture whichever accessible invertebrate is most abundant in the diverse habitat of the macrophyte beds.

Three species of piscivore/omnivores also frequent sheltered bays and flooded marginal pools at high lake level. *Glossogobius giurus* is a ubiquitous predator on amphipods, tanaids, *Caridina nilotica*, small gastropods, cichlid fry and *Aplocheilichthys* spp. Juvenile *Pseudocrenilabrus philander* feed on small crustaceans and aquatic insects, while the adults predate on these organisms as well as on small fishes, especially cichlid fry. Juvenile *Clarias gariepinus* (<300 mm TL) mainly inhabit densely vegetated areas where they feed on the most common epibenthic crustaceans and insects, as well as small fishes. These small fishes also fall prey to the fisheating spider *Thalassius spenceri* (Plate 1).

The diversity of food in barren sandy areas away from macrophyte beds is less than in sheltered bays (Allanson *et al.*, 1974).

Plate 1. An unusual predator of small cyprinid and cyprinodontid fishes at Lake Sibaya is the fisheating spider *Thalassius spenceri*. Photo: R. A. Holliday.

Croilia mossambica, the burrowing goby, feeds mainly on slow-moving benthic invertebrates such as chironomid larvae, gastropods, bivalves and amphipods, which are captured on the sand surface. All hunting appears to take place during the day while the goby is outside its burrow. Their eyes are large and sight is probably the primary sense used in locating prey which may explain why this species does not generally penetrate deeper than 16 m (Blaber & Whitfield, 1975) where less than 1% of surface illumination is found (Allanson & Hart, 1975). As a result of its slow-moving habits *Croilia mossambica* is itself vulnerable to predation from *Clarias gariepinus*, which may explain why the former species constructs a burrow into which it retreats when danger threatens. The burrow is not used for ambushing prey. The food preferences of another small benthic goby, *Silhouettea sibayi*, are not known but Farquharson (1970) has noted that this species buries itself in the sand with the aid of the pectoral fins, leaving the eyes uncovered. The long dorsal fin is erected above the sand surface and may serve as a lure to attract prey.

Two small predators, *Clarias theodorae* and *Ctenopoma multispinis* are usually found in adjacent swamps and streams but at high lake levels they penetrate into flooded marginal areas where they predate on the diversity of prey which develops there. The clariid is mostly insectivorous and feeds on the aquatic stages of odonatans, chironomids and the ephemeropteran *Povilla adusta*, and also on *Caridina nilotica*, *Melanoides tuberculatus* and fish fry. The anabantid feeds predominantly on small fishes (*Aplocheilichthys* spp., juvenile cichlids, *Barbus* spp.) as well as on crustaceans (*Caridina nilotica*, *Potamon sidneyi*) and aquatic and terrestrial insects.

The only two pelagic fishes in the lake, *Gilchristella aestuarius* and *Hepsetia breviceps* feed on the planktonic cladoceran *Bosmina longirostris* as well as the zoeae of *Hymenosoma orbiculare*, copepods, amphipods, ostracods, diatoms and other algae. These fishes are occasionally encountered in large shoals on the flooded terrace at high lake level where they were observed to feed on adult Ephemeroptera which had fallen onto the water surface.

(a) *The feeding ecology of Sarotherodon mossambicus* has been studied in detail by Bowen (1976a,b & 1978) with the primary aim of determining the cause of dwarfing and poor condition in adults compared with juveniles. Adult *S. mossambicus* in Lake Sibaya never develop the rounded, deep-bodied shape which is characteristic of adults from other populations (*e.g.* the Pongola river) but instead show extensive marasmus from apparent malnutrition (Fig. 3B). Because *S. mossambicus* is the most numerous fish in Lake Sibaya, the poor condition and small size of adults means that this large lake has negligible fisheries potential (Pike, 1969). Bowen identified food *quality* as the factor most likely to explain the different growth patterns of adults and juveniles, as the other two potentially important factors, the quantity of food available and the ability of the fish to utilize their food resources, appeared to be unimportant in Lake Sibaya.

Juvenile *S. mossambicus* feed predominantly on shallow sandy terraces

Plate 2. Juvenile *Sarotherodon mossambicus* about to feed on benthic floc on the sand. Photographed in an aquarium by S. H. Bowen.

around the margin of the lake (Plate 2), while adults usually feed in offshore waters at depths of three metres or greater. Both juveniles and adults feed on a mixture of detritus (dead organic matter), bacteria and diatoms which occur as a flocculent layer (benthic floc) on sandy substrata. In addition, Bowen demonstrated that *S. mossambicus* is able to collect phytoplankton from suspension (in an aquarium) and found plankton in the gut contents of some adults in the cool season. Le Roux (1956) also reported that *S. mossambicus* may feed on phytoplankton, while Bruton & Boltt (1975) noted that some adult *S. mossambicus* stomachs from Lake Sibaya contained a green fluid which may have been derived from plankton. Adult *S. mossambicus* also supplement their diet with other food items, as discussed above.

S. mossambicus has broadened bicuspid or tricuspid jaw teeth which closely resemble the teeth of the periphyton grazers illustrated by Greenwood (1974), while their pharyngeal teeth are numerous and fine, resembling those of plankton-eating cichlids (Fryer & Iles, 1972). The gut length: total fish length ratio is 3.93 (N = 121) which is near the ratio typical of herbivores (4.21, Bowen 1976a, recalculated from Fryer & Iles, 1972). The digestive tract of *S. mossambicus* is comprised of a stomach and an intestine which are separated by a well-developed pyloric sphincter. Concentrated gastric acid in the stomach, commonly at pH values of 1.5 and lower, lyses diatoms and bacteria which are subsequently digested in the intestine (Fig. 32, Bowen 1976b). This is the first report of digestion of bacteria by a fish. Similarly Moriarty (1973) found that the cichlid *Tilapia nilotica* in Lake George, Uganda, utilizes gastric acids below pH 2.0 to lyse blue-green algae.

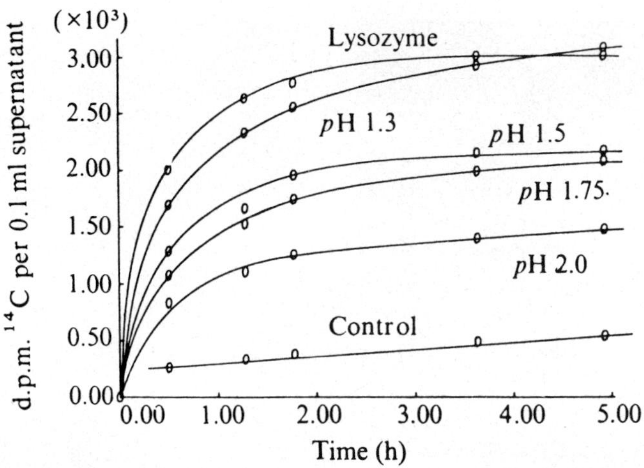

Fig. 32. Radioactivity released from bacteria during treatment with intestinal juice from *S. mossambicus* after acid treatment at pH values shown. Data given are experimental minus control with correction for volume reduction caused by sampling (from Bowen, 1976b).

Bowen found a daily cycle of diatom digestion in *S. mossambicus* which was related to feeding activity. Ingested diatoms passed through the stomach, where they were lysed by gastric acid, and into the duodenum where they were digested. The extent of lysis and subsequent digestion was inversely proportional to the gastric pH. As the stomach is filled at the beginning of a feeding period, gastric pH drops from 5 to as low as 1.25 (the lowest ever recorded from the stomach of an actively feeding fish). Gastric pH remains low while the fish feed but rises to pH 5 and above after feeding stops.

The proportions of different diatom species ingested by juvenile and adult *S. mossambicus* were different. Juveniles fed mainly on *Mastogloia elliptica*, *Stauroneis karstenii* and *Melosira granulata*, while adults ingested mainly *Melosira granulata*, *Synedra* sp. and *Closterium* sp. (Plate 3). Diatom concentrations in the guts of juveniles (6.29×10^4 per mg gut content) were over ten times the concentration in the guts of adults (0.59×10^4). Since Bowen (1976a) had shown that the diatoms which are consumed are largely digested, the results suggested that diatom concentration may play an important role in determining food quality in the diet of *S. mossambicus* in Lake Sibaya.

Further investigation by S. H. Bowen revealed that benthic floc from the feeding areas of juveniles and adults has similar concentrations of organic matter, total carbohydrate, soluble carbohydrate and calories but differed significantly with respect to diatom and protein concentrations. Diatom concentrations were in general higher in terrace floc (Fig. 33) and low in floc from deeper water. The protein concentration of benthic floc similarly decreased with increasing depth from 0 to 5 m (Fig. 34). Partial correlation analysis showed that protein and diatom concentrations had no correlation independant of their common relationship to depth, and further evidence revealed that detrital bacteria were the primary source of protein in benthic floc. Bowen suggested that the decreased availability of dissolved organic matter (DOM), on which bacteria are largely dependent for nutrition, (Jannasch & Pritchard, 1972; Paerl, 1974) at increasing depths may explain the distribution of bacteria implied by the distribution of protein found in benthic floc samples.

The protein content of benthic floc profoundly influences its nutritional value. The amount of digestible protein consumed by a fish depends directly on the ratio of digestible protein to digestible energy in the consumer's diet (Harper, 1967). As a result, the basic nutritional value of a given diet depends on its ratio of digestible protein to digestible energy (Boyd & Goodyear, 1971). For a given species under a particular set of environmental conditions, desirable protein to calorie ratios range from a minimum at

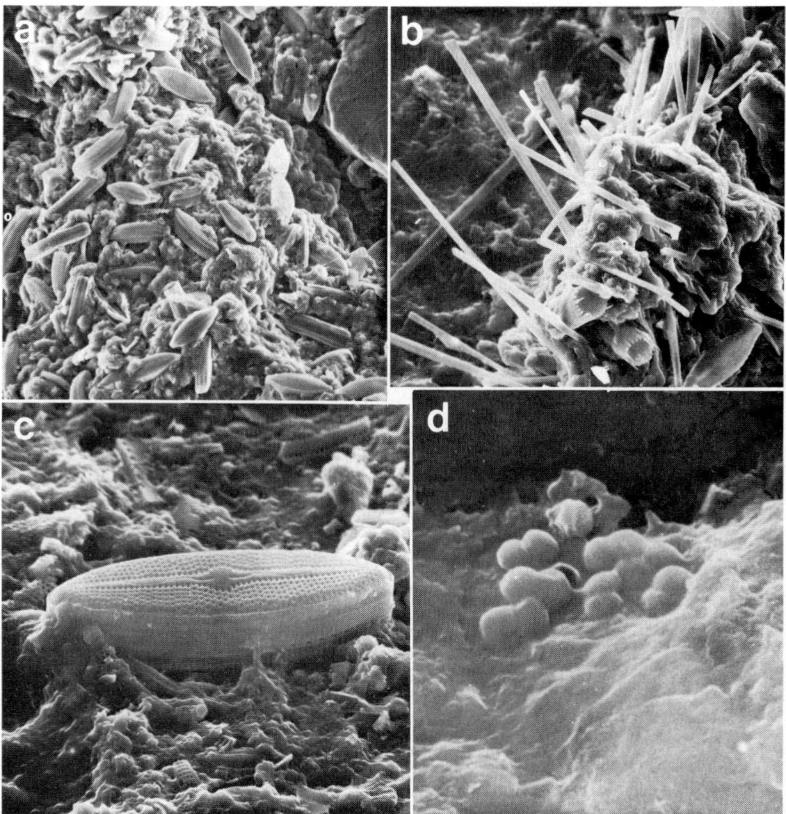

Plate 3. A. Typical gut content from juvenile *S. mossambicus* in Lake Sibaya. Scanning electron micrograph (SEM), 400 × magnification. B, as A, 800 × magnification. C. *Mastogloia elliptica*, the most common diatom in the alimentary tract of juvenile *S. mossambicus*. SEM, 2000 × magnification. D. Cells of bacterial dimensions occasionally found attached to detritus in the alimentary tract of juvenile and adult *S. mossambicus*. SEM, 10,000 × magnification (from Bowen, 1976a).

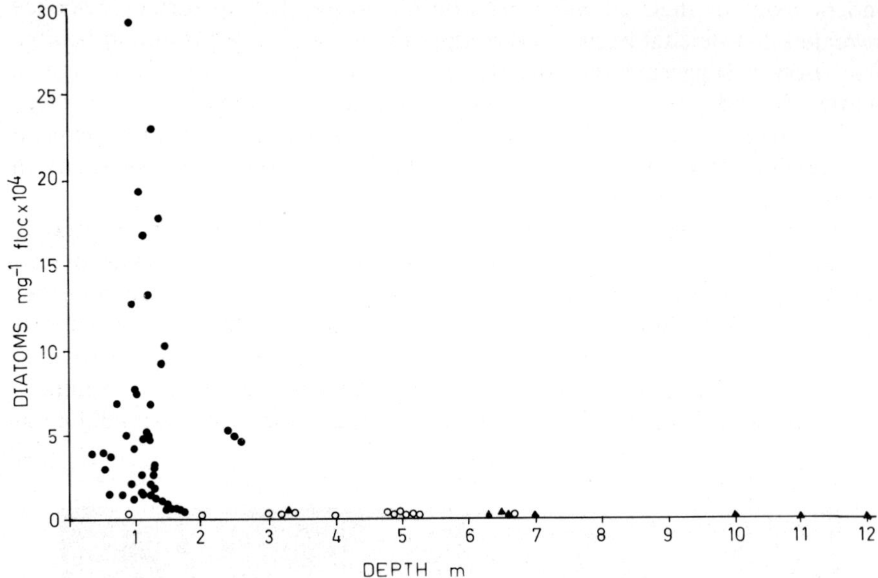

Fig. 33. Diatom concentration in benthic floc in terrace (●), gradual slope (○) and steep slope (▲) habitats (after Bowen, 1978).

which just enough protein is consumed for maintenance to a maximum at which growth is maximal. Below this range retarded growth, poor condition and other symptoms of protein deficiency became apparent (Bowen, 1976a). Above this range growth is reduced, presumably as a result of the high energetic cost associated with digestion and metabolism of protein (Boyd & Goodyear, 1971). Within this range growth is usually directly proportional to the protein content of the diet (Ogino & Saito, 1970; Russell-Hunter, 1970; Sabaut & Luquet, 1973).

Using the C:N ratio as a rough indication of the protein to calorie ratio, Russell-Hunter (1970) generalizes that most animals require diets with C:N ratios of 17:1 or less for maintenance. Diets with ratios greater than 20:1 result in conspicuous malnutrition. In diets comprised of only carbohydrate and protein, these ratios correspond to values of 22.0 and 17.9 mg of protein per kcal, respectively (Bowen, 1976a).

Bowen (1976a) used mean assimilation efficiencies for protein and calories to calculate mg of digestible protein per digestible kcal for 39 benthic floc samples at different depths in Lake Sibaya (Fig. 35). In the light of Bowen's results, it is not surprising that adult *S. mossambicus* in Lake Sibaya are dwarfed and in poor condition. (Fig. 3B).

By Russell-Hunter's criteria fish that feed consistently at depths greater than about 1.2 m would be expected to show conspicuous signs of malnutrition (Fig. 35). Fish feeding at depths between 1.2 and 0.75 should have enough dietary protein for maintenance and those that feed in water less than 0.75 m deep would be expected to have sufficient dietary protein for the good growth they exhibit (Bruton & Allanson, 1974). Dietary protein deficiency was therefore identified, clearly and brilliantly, as the cause of

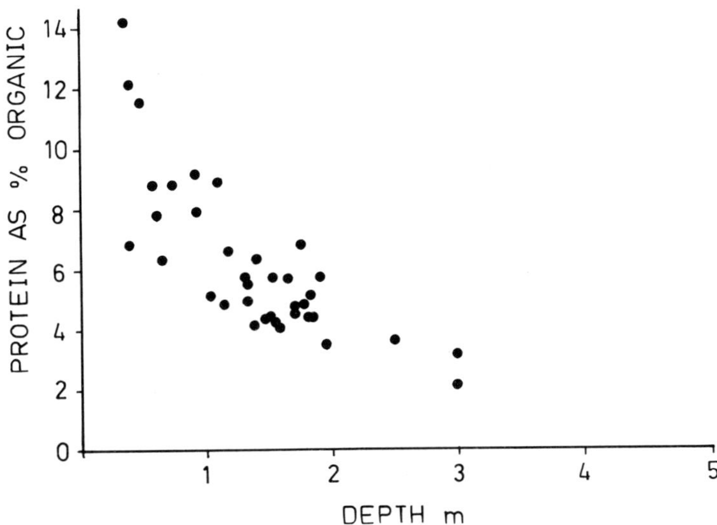

Fig. 34. The distribution of protein in benthic floc as a function of depth in Lake Sibaya (after Bowen, 1976a).

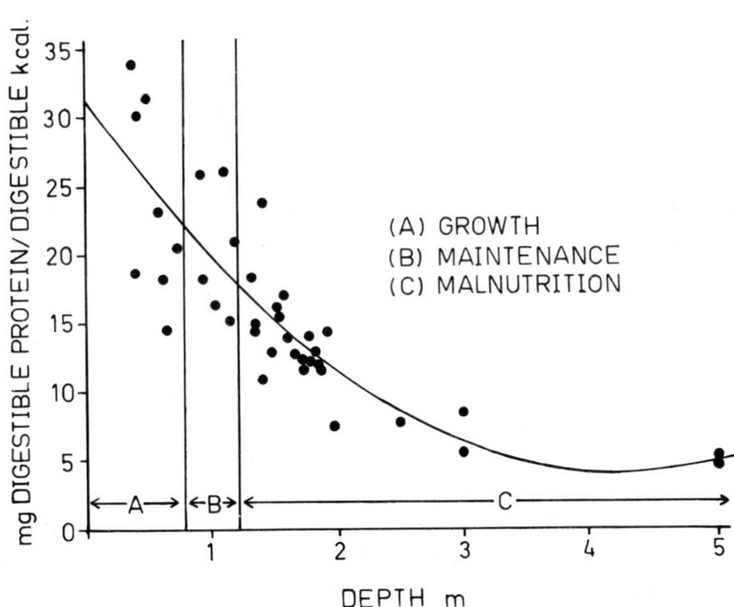

Fig. 35. The ratio of digestible protein to digestible kcal in benthic floc as a function of depth in Lake Sibaya. Curve fitted by parabolic regression. Expected nutritional significance of this ratio for fish feeding at different depths is indicated in the text according to Russell-Hunter's (1970) criteria (after Bowen, 1976a).

dwarfing and poor condition of adult *S. mossambicus* in Lake Sibaya by Bowen (1976a).

(b) *Feeding ecology of Clarias gariepinus.* In contrast to the oligophagous habits of *S. mossambicus, Clarias gariepinus* feeds on a variety of organisms including diatoms, macrophyte debris, molluscs, crustaceans, arachnids, insects, fish and Amphibia. This study served to illustrate the way in which a closed oligo-mesotrophic lake with a fluctuating shoreline and relatively narrow productive littoral zone supports an abundant population of large, euryphagous predators. The feeding strategy adopted by *Clarias gariepinus* is to switch from one prey to another as prey availability changes. To achieve this end, they are equipped with an impressive array of anatomical adaptations for feeding on a variety of prey in different habitats. *S. mossambicus* is an important prey of *Clarias gariepinus* in Lake Sibaya, and both Bruton (1973) and Bowen (1976a) recognized the possible effect of the predator on the cichlid's movements and feeding activity. A further reason for studying the feeding ecology of *C. gariepinus* was to assess the impact of the catfish on other vertebrate, as well as invertebrate prey, and contribute to an understanding of trophic pathways in the lake. Furthermore, no one has established the principles of predation by an omnivore, as Holling (1965, 1966) has done for an ambushing insect predator, and Beukema (1968) for a small carnivorous fish. Omnivores are interesting and widespread predators, and it is essential that their feeding ethology is known if we are to understand the predation process fully. The present study (Bruton, 1979b,c,d, this section) can be considered as the first step along this road.

C. gariepinus is predominantly a nocturnal, benthic feeder, which is reflected in its many anatomical adaptations for bottom feeding in low visibility: wide mouth, ability to depress the hyomandibular apparatus for 'sink-suck' feeding (see below), retention of the palatine-maxillary articulation as a hinge for probing movements of the maxillary barbel (Alexander, 1965), reduction in the importance of the eyes, and development of an abundant network of sensory organs on the body, head, lips and barbels (Angelopoulo, 1947).

C. gariepinus has four pairs of barbels of which the maxillary is the longest and most mobile. The nature of the sense organs on the barbels has not been investigated, but in *C. batrachus* they are both tactile and chemosensory in function (Srivastava & Sinha, 1961; Welsch & Starch, 1969). The barbels act as probes during foraging and as a forwardly directed 'seine' which may limit the chance of escape by prey. The eyes have little apparent role in prey fixation whereas olfactory organs and the lateral line system are probably important.

The mouth is wide and capable of considerable vertical displacement for engulfing large prey or large volumes of water. When opened, the volume of the buccopharyngeal chamber is increased which creates a strong negative pressure in the vicinity of the food, which is carried towards the mouth in the consequent current of water (Gosline, 1973; pers. obs.) The prey are prevented from escaping by broad bands of recurved teeth on the premaxillary and dentary bones. The vomerine (on the pre-vomer) and upper and lower pharyngeal bands of teeth also perform this function as well as that of incapacitating the prey, which is then swallowed whole. The teeth are conical and sharp on all tooth bands except the vomer where they are mainly molar-like.

C. gariepinus has long gillrakers on the anterior borders of the five branchial arches, and additional gillrakers on the posterior margin of the third and fourth arches, which interdigitate with those from the anterior row of the next arch. The number of gillrakers increases with length (Bruton, 1979b) although the scatter is wide, as found in *C. mossambicus* (Worthington, 1933), *C. lazera* (Mills, 1956) and *C. senegalensis* (Thomas, 1966). The mean width between developed gillrakers varies between <0.1 to 0.6 mm depending on fish size (Murray 1975).

The oesophagus is short and opens into a distinct, distensible stomach. The intestine is simple, thin-walled and short. In the stomach the food is crushed and tumbled by movements of the muscular stomach wall (Bruton, 1979b) and digested by means of enzymes secreted from single digestive glands (Khanna, 1966, in *C. batrachus*). Protease is present in the stomach and intestine of *C. mossambicus* with maximal activity at pH 4 and 6 in the respective tissues (Cockson & Bourne, 1972). Enzyme activity in the stomach is about four times that in the intestine.

Several modes of feeding were observed in *C. gariepinus* in Lake Sibaya (Fig. 36). Individual foraging is the normal method of feeding on the bottom. The catfish swims slowly with the barbels extended forward in a cone. When a prey is detected the catfish darts rapidly and accurately forward and sucks it in. When stationary, the catfish also erects the barbels in a cone; any prey which swims into the cone is immediately detected and attacked. Where deposits of macrophyte debris cover the substrate, *C. gariepinus* feeds by shovelling the head under the detritus, lifting the detritus up and taking any organisms which are exposed.

Small *C. gariepinus* (<200 mm TL) occasionally feed by positioning the body perpendicularly with the barbels spread across the water surface. A strong current of water is drawn into the mouth, and flotsam and neuston on the water surface, such as insects and plankton, are ingested. Larger *C. gariepinus* also surface feed in this way, especially after heavy rain when

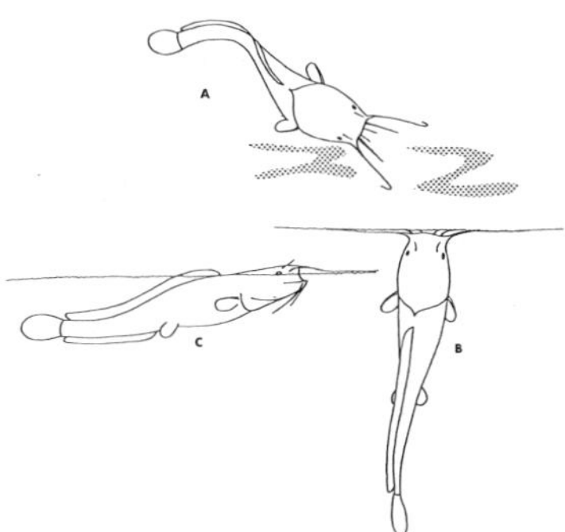

Fig. 36. Feeding behaviour of *C. gariepinus*. A. Individual benthic foraging. B. Perpendicular surface feeding. C. Horizontal surface feeding (sometimes performed in groups).

large groups of catfish assemble to feed on flotsam near the water edge. This mode of feeding was accompanied by loud sucking and intensive interaction and synchronization between the participating catfish (Bruton, 1979b).

Social hunting was also observed in *C. gariepinus* swimming near the water surface in a regular formation. This behaviour occurred at night over deep water, and the catfish appeared to be feeding on neuston, flotsam and plankton. Similar behaviour has been reported in *C. gariepinus* in Lake Kariba (Bowmaker, 1973), where the catfish had fed mainly on planktonic calanoid copepods, in *C. lazera* in East Africa (Worthington, 1933) and in *Heterobranchus longifilis* (another clariid) in Lake Kariba on zooplankton (Jubb, pers. comm., 1976). During this 'horizontal' surface feeding the catfish swim with the mouth open in a fixed position and water flows through the orobranchial chamber and across the gills.

Social hunting is also performed by *C. gariepinus* in shallow water on the edge of the terrace. A tightly-knit group of catfish in a rough sickle-shaped formation swim slowly inshore with their mouths open, herding shoals of small (20–80 mm TL) cichlids (especially *Sarotherodon mossambicus, Tilapia sparrmanii* and *Pseudocrenilabrus philander*). The catfish swim steadily inshore, opening and closing their mouths more or less in unison and eventually encircle the prey which form a dense, panic-stricken mass which is readily caught by the catfish. When all the prey have dispersed or been eaten, the catfish submerge and swim along shore where they reform as a pack and swim inshore again. This cycle may continue every few minutes for over an hour, but usually terminates after 20 to 30 mins. The groups of *C. gariepinus* observed in Lake Sibaya all numbered between 15 and 40 individuals, but larger groups have been observed feeding in this way in the Dobi river, Botswana (Donnelly, 1966, in *C. ngamensis*, about 400 individuals) and in Nyamithi Pan in Zululand (Pooley, 1972, in *C. gariepinus*, 75–100 individuals) *C. gariepinus* has also been observed to hunt in packs at the edge of river flood-plains when receding water levels force large numbers of small fishes to migrate back into the main river (Tait, 1965; Williams, 1971; Bell-Cross, 1974, 1976; van der Waal, pers. comm., 1976).

The advantages of social hunting in fishes are reviewed by Redakov (1973), Curio (1976), Shaw (1978) and Bruton in prep. In the particular case of *C. gariepinus* in Lake Sibaya, social hunting has several functions:

a) to locate dispersed prey, particularly when feeding on neuston, plankton and flotsam at the water surface,

b) to disorientate elusive prey *e.g. Sarotherodon mossambicus*, and cause them to lose their normal defensive shoaling synchronisation and become more vulnerable,

c) to restrict prey spatially by herding it to the water surface or into shallow water

d) to chase prey away from shelter.

The combined effect is for social hunting to increase the predation efficiency

Table 13. *C. gariepinus* caught for food preference analyses in Lake Sibaya.

Time period	Lake level range (m)	Collecting methods	No. of catfish caught (with stomach contents)	Total length range (mm)
1. 1970–1972	0.92–1.71	seine net, rod and line	92	300–700
2. January 1974–December 1975	1.83–3.03	long lines, gillnets, handnets	469	12–1088
3. March–April 1976	3.86–3.98	long lines, gillnets, handnets	123	246–836

of the individual predator and allow it to capture prey which is normally too elusive or dispersed.

As collections of *C. gariepinus* for food analysis were made at low, intermediate and high lake levels (Δ3.06 m, Table 13), the effect of the changing ecology of marginal areas on the diet of the catfish could be determined. The main collection was made at intermediate lake level.

The food class appearing in most stomachs was Crustacea (Table 14, Fig. 37, 77%) followed by fish, insects and molluscs, and the most important prey species or groups were *Hymenosoma orbiculare*, *Grandidierella lignorum*, *Caridina nilotica*, *Sarotherodon mossambicus*, odonatan nymphs, Chironomid larvae, *Povilla adusta* and *Glossogobius giurus*.

The class contributing the highest number of prey organisms was Crustacea (Table 15, 65%) followed by fish (mainly fry), insects and molluscs. The most numerous prey species were *Grandidierella lignorum*, *Hymenosoma orbiculare*, *Caridina nilotica*, *Povilla adusta*, cichlid fry, chironomid larvae and odonatan nymphs.

As a class, fishes contributed the greatest proportion of total prey dry weight (Table 15, 75%) followed by crustaceans, insects and molluscs. The most important food species in the dry weight analysis were *Sarotherodon mossambicus*, *Potamon sidneyi*, *H. orbiculare*, *Glossogobius giurus* and *Pseudocrenilabrus philander*.

Animals of marine or estuarine origin were well-represented in the diet of *C. gariepinus*, occurring in 87% of all stomachs examined (with contents). They were, in order of frequency of occurrence: *H. orbiculare*, *G. lignorum*, *G. giurus*, *Hepsetia breviceps*, *Gilchristella aestuarius*, *Croilia mossambica*, *Corophium triaenonyx*, *Apseudes digitalis*, *Cyathura carinata* and *Pontogeloides latipes*.

One incidence of cannibalism was recorded; a *C. gariepinus* measuring 1088 mm TL (weight 8790 g) contained a *C. gariepinus* measuring 280 mm TL (217 g).

C. gariepinus ingested 1.75 times more *Bellamya capillata* than *Melanoides tuberculatus* (Table 14) although the latter species was 17 times

Table 14. The food of *C. gariepinus* at intermediate lake levels (1974–75) in Lake Sibaya – frequency analysis. Number of catfish examined with stomach contents = 469. Total number of catfish examined = 683.

Food species or group	Frequency of occurrence in stomachs examined	% of total no. of stomachs examined
Fish: Total	239	50.9
Sarotherodon mossambicus	91	19.4
Pseudocrenilabrus philander	41	8.7
Cichlid fry	28	6.0
C. gariepinus eggs	5	1.1
Glossogobius giurus	39	8.3
Croilia mossambica	8	1.7
Aplocheilichthys spp.	41	8.7
Barbus spp.	10	2.1
Gilchristella aestuarius	6	1.3
Unidentified fish	59	12.6
Crustacea: Total	361	77.0
Hymenosoma orbiculare	231	49.3
Potamon sidneyi	30	6.4
Caridina nilotica	116	24.7
Grandidierella lignorum	122	26.0
Corophium triaenonyx	10	2.1
Afrochiltonia capensis	9	1.9
Apseudes digitalis	9	1.9
Cyathura carinata	44	9.4
Pontogeloides latipes	31	6.6
Unidentified Crustacea	35	7.5
Insecta: Total	177	37.7
Povilla adusta	67	14.3
Odonatan nymphs	68	14.5
Hemiptera	15	3.2
Trichoptera	6	1.3
Chironomid larvae	73	15.6
Terrestrial Coleoptera	42	9.0
Terrestrial Hymenoptera	8	1.7
Terrestrial Orthoptera	11	2.3
Other terrestrial insects	9	1.9
Unidentified Insecta	1	0.2
Mollusca: Total	103	22.0
Melanoides tuberculatus	54	11.5
Bellamya capillata	28	6.0
Bulinus spp.	35	7.5
Corbicula africana	7	1.5
Arachnida	10	2.1
Plant debris	42	9.0
Diatoms	12	2.6
Sand	46	9.8

more common in belt transects measured by SCUBA divers in Lake Sibaya (Boltt, 1969). This marked selection of the thin-walled globose snail rather than the heavy-walled spiralled one may be related to the relative ease with which *B. capillata* can be crushed.

The almost total lack of planktonic food in the diet of *C. gariepinus* in Lake Sibaya is in contrast to findings elsewhere (Groenewald, 1964; Munro, 1967; Schoonbee, 1969; Bowmaker, 1973; Murray 1975). The only planktonic organisms recorded were *Pseudodiaptomus hessei* (4 stomachs) and *Spirogyra* (1). *C. gariepinus* has been seen feeding on the water surface at night, as they do in Lake Kariba for plankton, but none of these surface feeders has been caught.

The food preferences of *C. gariepinus* in different length groups have been detailed in Table 12. Invertebrates, especially crustaceans, dominate the diet of small juveniles (<100 mm TL) but fish prey, initially as fry or fingerlings, become increasingly important in the diet of larger catfish. Insects are only important components of the diet of catfish smaller than 300 mm TL. Large adults (>700 mm TL) feed almost exclusively on *S. mossambicus*.

Besides the items mentioned above, *C. gariepinus* in Lake Sibaya has been observed to feed on frogs, figs, fish regurgitated by birds, *Sarotherodon mossambicus* fry, fledgling birds, rotting flesh and *Hippopotamus amphibius* dung (Bruton, 1979b).

There were few seasonal trends in the catfish diet except that the mean weight of *S. mossambicus* prey was higher in Autumn and Winter (15.8 and 23.0 g) than in Spring and Summer (2.7 and 3.2 g), indicating a preference for small cichlids when they are available in shallow water during the warm months.

An analysis of food preferences by habitat revealed that insects were taken most often in marginal and sheltered bay habitats, and molluscs in

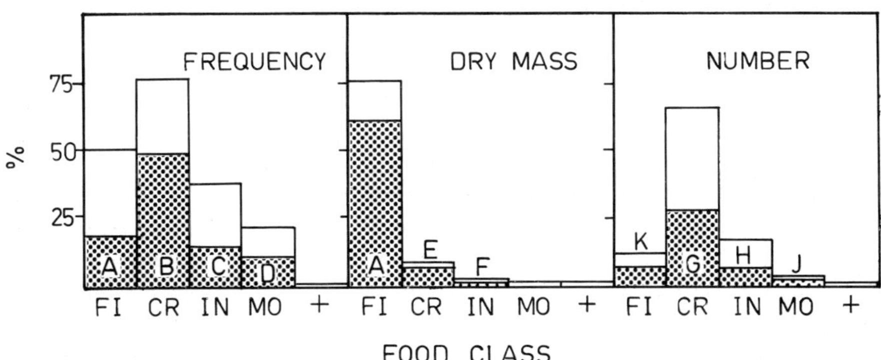

Fig. 37. The percentage contribution of different prey classes to the diet of 469 *C. gariepinus* caught at intermediate lake level in Lake Sibaya, analysed by frequency, dry weight and number, with the dominant prey species or group shaded. Food classes: FI: fish; CR: Crustacea; IN: insecta; MO: Mollusca; +: Other. Dominant prey: A: *S. mossambicus* B: *H. orbiculare* C: Chironomid larvae D: *M. tuberculatus* E: *P. sidneyi* F: Odonatan nymphs G: *G. lignorum* H: *P. adusta* J: *Bulinus* spp. K: Cichlid fry (from Bruton, 1979b).

Table 15. The food of *C. gariepinus* at intermediate lake levels (1974–75) in Lake Sibaya – numerical and dry weight analysis of most important food items. Number of catfish examined with stomach contents = 469. Total number of catfish examined = 683. Total number of prey found in 469 catfish = 5549.

	Numerical analysis		Dry weight analysis
Food species or group	No. individual prey	Percentage of total	Percentage of total prey dry weight
Fish: Total	695	12.5	75.0
S. mossambicus	126	2.3	60.6
P. philander	44	0.8	5.3
Cichlid fry	383	6.9	2.6
G. giurus	38	0.7	4.9
Aplocheilichthys spp.	104	1.9	1.6
Crustacea: Total	3615	65.1	18.6
H. orbiculare	1302	23.5	7.1
C. nilotica	606	10.9	0.6
P. sidneyi	27	0.5	10.7
G. lignorum	1548	27.9	0.1
P. latipes	44	0.8	0.04
C. carinata	88	1.6	0.07
Insecta: Total	1014	18.3	4.5
Odonatan nymphs	197	3.6	2.0
Chironomid larvae	350	6.3	0.1
P. adusta	388	7.0	0.7
Terrestrial insects	79	1.4	1.7
Mollusca: Total	225	4.1	1.9
M. tuberculatus	44	0.8	0.7
B. capillata	77	1.4	1.1
Bulinus spp.	104	1.9	0.1

sheltered bay and profundal habitats. Fish and crustaceans were captured with approximately equal frequency in all habitats.

The majority of prey taken (89.8% by weight, 79.7% by number) lived in close proximity to the substrate: water interface. Highly accessible prey was taken in preference to prey with moderate or low accessibility (for details, see Bruton, 1979b). The larvae of *Povilla adusta* are normally tubicolous but swim about actively in the dark (Hartland-Rowe, 1958; Petr, 1970; pers. obs.). They are presumably caught by the catfish in the free swimming stage as no silken cases were found in stomach contents. *Grandidierella lignorum*, an amphipod, builds burrows in the sand which would offer some protection against predation. The majority of those eaten by *C. gariepinus* were appetitive males (with an enlarged cheliped) which had probably left their burrows at night, as observed by Boltt (pers. comm., 1973).

C. gariepinus preyed on large benthic *Hymenosoma orbiculare* and not their planktonic zoeae. The mean size taken (8.3 mm) is about the mean size of *H. orbiculare* in the benthos of Lake Sibaya (9 mm, Forbes & Hill, 1969). The size of *S. mossambicus* prey increased linearly with predator length, although large catfish were observed to feed on fry and small juveniles. The largest *S. mossambicus* ingested measured 185 mm SL.

Feeding intensity in adult *C. gariepinus* was not high and, with the exception of fish caught in November, stomachs were on average less than half full. Small catfish (<200 m TL) had higher mean fullness indices (4.9) than medium (200–600 mm TL, 3.8) or large catfish (>600 mm TL, index 2.6; Bruton, 1979b).

The condition of juvenile *C. gariepinus* (<350 mm TL) remained good throughout the year whereas that of adults decreased markedly in winter and increased in summer (Fig. 38).

The diet of *C. gariepinus* at different lake levels is compared in Table 16 by frequency occurrence of different items. Three main trends are apparent:

a) The diet at all lake levels is dominated by *H. orbiculare* followed by a cichlid fish or *G. lignorum*.

b) The average number of food items per stomach was lower (1.8) at low lake level than at intermediate (4.8) or high lake level (6.9). Their diet was therefore more diverse when marginal areas had been flooded. These habitats support a richer biota than sandy wave-washed terraces.

c) Prey items varied markedly in importance at different lake levels. The increased abundance of *Povilla adusta* in the diet of *C. gariepinus* is a consequence of the population explosion of the ephemeropterans which took place after sedges and trees were drowned along the lake edge at high lake level. These plants provided suitable substrates for *P. adusta* nymphs to excavate their burrows.

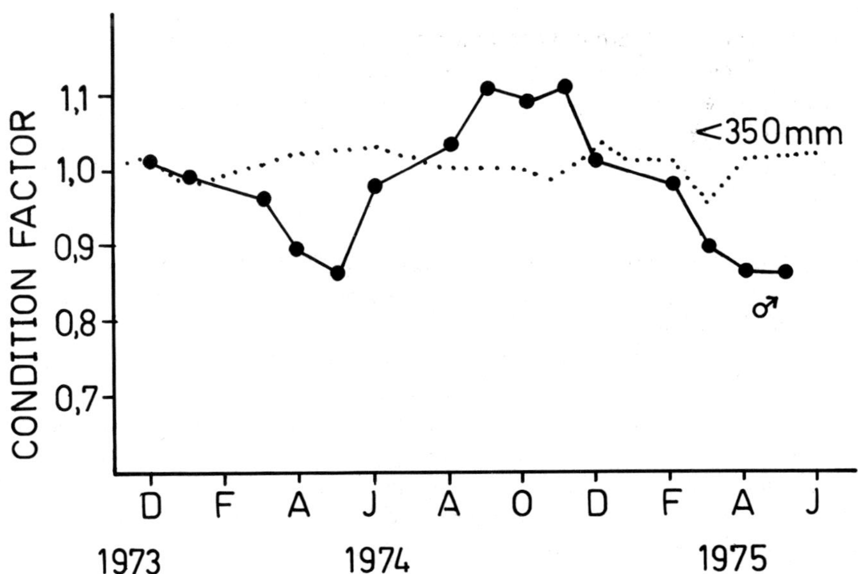

Fig. 38. Monthly condition factors of juvenile (<350 mm TL, dotted line) and adult male (solid line) *C. gariepinus* in Lake Sibaya (from Bruton, 1979b).

Table 16. Frequency of occurrence of major food items in *C. gariepinus* (TL 300–700 mm) at three different lake levels (lake levels defined in Table 13). N = number of stomachs examined with contents.

	Lake level		
	Low	Intermediate	High
	$N = 92$	$N = 469$	$N = 123$
Fish: Total	31.5	52.9	55.3
S. mossambicus	25.0	20.3	1.6
P. philander	4.3	6.5	36.6
T. sparrmanii	3.3	0.3	12.2
G. giurus	15.2	11.7	10.6
Crustacea: Total	64.2	85.2	80.5
H. orbiculare	47.8	62.9	75.6
G. lignorum	6.5	21.3	26.8
P. sidneyi	4.3	7.9	0
C. nilotica	10.9	17.5	8.9
Insecta: Total	15.4	25.4	14.6
Odonatan nymphs	8.7	5.5	0
Chironomid larvae	4.3	6.9	3.3
P. adusta nymphs	2.2	5.2	12.2
Mollusca: Total	8.7	22.0	23.3
B. capillata	6.5	5.8	8.9
M. tuberculatus	4.3	12.4	8.1
C. africana	0	2.4	11.4
Bulinus spp.	0	5.8	4.9
Mean no. food items per stomach	1.8	4.8	6.9

Some items which are only common in Summer, such as certain insects with aquatic larval stages, may be underestimated in the diet of high lake level catfish as these collections were only made in March and April.

The changing proportions of cichlid prey in the main catfish feeding area (littoral terraces) provided an opportunity to study the effect of relative prey density on catfish selectivity. If the proportions of each cichlid species in the diet of *C. gariepinus* and on the terrace at each lake level are about the same, the catfish would have taken prey in proportion to its abundance.

At all three lake levels the observed and expected differences were just significantly different (Table 17, and Bruton, 1979b). Nevertheless, as shown by the electivity indices, there was a trend towards predation on at least one or two species in proportion to their abundance.

This finding was tested further by observing predation by *C. gariepinus* on cichlid prey provided at different densities in a specially constructed observation pool. The general results confirmed the trend found in the field collections that *C. gariepinus* captures the three cichlid species more or less in proportion to their relative density (Bruton, 1979b). *T. sparrmanii* were negatively selected in all three sets of experiments, probably because of their more secretive, less active behaviour compared with other cichlids.

Table 17. Proportions of *S. mossambicus*, *T. sparrmanii* and *P. philander* on the terrace of Lake Sibaya and in the diet of *C. gariepinus* on the terrace at different lake levels from 1970–1976. The null hypothesis, that there was no significant difference between the number of cichlid prey eaten and the number expected to be eaten calculated from the proportions of prey in the terrace population, was tested using the G-test of Sokal & Rohlf (1973, p. 294) which compares observed and expected values. The G-statistic given in each case is for two degrees of freedom. The modified electivity index was calculated from observed and expected values as explained in Bruton (1979b). N = number of stomachs (with contents) examined from catfish caught on the terrace. The lake levels are defined in Table 13.

	Cichlid population sampled on the terrace using standarized seine net pulls		Cichlids in diet (number)		Modified electivity index
	number	% of total	Observed	Expected	
Low Lake Level (1970–1972)			$N=92$		
S. mossambicus	4570	88.3	27	31.8	−0.08
T. sparrmanii	204	3.9	3 } 9	4.2	+0.36
P. philander	388	7.5	6 }		+0.36
	5162		36 G = 4.880 P < 0.05 Significant		
Intermediate Lake Level (1974–1975)			$N=275$		
S. mossambicus	2009	59.1	126	101.1	+0.11
T. sparrmanii	449	13.2	2	22.6	−0.84
P. philander	942	27.7	44	47.4	−0.04
	3400		171 G = 39.23 P < 0.001 Significant		
High Lake Level (1976)			$N=123$		
S. mossambicus	273	14.8	2	9.2	−0.64
T. sparrmanii	424	23.0	15	14.3	+0.02
P. philander	1146	62.2	45	38.6	+0.08
	1843		62 G = 9.13 P < 0.01 Significant		

Further field and experimental observations were conducted to determine the efficiency of *C. gariepinus* as a predator of its main fish prey, *S. mossambicus*. Field observations revealed that the catfish mainly occupied the terraces at night whereas the cichlid inhabited these areas during the day (at least at low and intermediate lake levels). Furthermore, it was found that

C. gariepinus more readily catches the cichlid during its nocturnal inshore migrations. Three factors were identified which may increase the efficiency of *C. gariepinus* as a piscivore: (i) decreased water depth (ii) decreased light intensity (iii) increased predator numbers. The influence of these factors on the predation sequence: search → encounter → discover → grasp → eat was determined by counting the number of these appetitive events which were performed by catfish under standardized conditions, but exposed to different water depths, light regimes and predator densities. A greater number of positive appetitive events to the right *i.e.* towards the completion of the predation sequence, would indicate increased prey risk and/or predation efficiency under the prevailing conditions.

The observations (reported in detail by Bruton, 1979c) revealed that low light intensities, shallow water and increased predator density may increase the predation efficiency of *C. gariepinus* on *S. mossambicus*. These findings explain why *C. gariepinus* undertakes regular nocturnal forays into shallow water at night, where it feeds on cichlids.

Further experiments conducted on the main invertebrate prey of *C. gariepinus* revealed that the catfish is a more efficient predator of invertebrates than fish (Bruton, 1979b). The predation failure rate *i.e.* the percentage of discovered prey which was not successfully attacked was 79.3% for fish and 15.3% for invertebrates. If the tanaid *Apseudes digitalis*, which is not a preferred prey, is excluded from the calculation, the figure for invertebrate prey falls to 12.3%. Why then does *C. gariepinus* switch from an invertebrate diet to a largely fish diet (in terms of dry weight) when they are apparently more efficient at catching invertebrates? Food quality is the obvious answer. A predator is thought to allocate its resources of time and energy in the most profitable or 'optimal' manner *i.e.* energy yield from the prey is balanced against the energy costs of catching, ingesting and assimilating that prey (MacArthur, 1972). While this was not a study on energetics, it is relevant to show that fish prey provided far more energy than the other prey classes eaten by *C. gariepinus* in Lake Sibaya.

As shown in Table 18, fish prey only constituted 12.5% of all prey eaten yet their yield in terms of energy content represented 76.5% of the total. In addition, the average energy content per individual fish prey was 14 times higher than for other individual prey items. Clearly, fishes are the most valuable prey of *C. gariepinus* in Lake Sibaya for fuelling metabolic processes and the catfish can afford to expend considerable energy on their capture, including that used during their diel inshore movements.

Fish prey also provide more protein than invertebrate prey, as well as the correct amino acid balance to fish predators (Love, 1970; Dupree & Halver, 1970; Phillips, 1972). By increasing the efficiency of *C. gariepinus* as predators of other fishes, the nocturnal inshore migrations may thus play an important role in providing an adequate diet for the maintenance of good growth and condition in the predator.

Large adult *C. gariepinus* (>600 mm TL) feed on fish prey, but their feeding rate is low (judging from the proportion of empty stomachs, >45%), and their food source is apparently not adequate for the maintenance of

Table 18. Comparison of the contribution of different prey classes to the diet of 469 *Clarias gariepinus* caught at intermediate lake level in Lake Sibaya. Energy contents of prey organisms from Caulton (1976, fish); Thayer *et al* (1973, Crustacea and Mollusca) and Odum (1971, Insecta).

Prey class	% of total prey number	% of total prey dry weight	% of total prey kJ	Average kJ per prey item
Fish	12.5	75.0	76.5	22.88
Crustacea	65.1	18.6	16.5	0.95
Insecta	16.8	4.5	5.2	1.16
Mollusca	4.1	1.9	1.7	1.56

good growth and condition. As a result, these catfish are emaciated and have large heads and thin bodies. Their main fish prey consists of adult *S. mossambicus* which Bruton (1973) and Bowen (1976a) showed are in poorer condition than juveniles. The energy yield from adult *S. mossambicus* prey may be less per gram than for juveniles as Caulton (1976) has found that *Tilapia rendalli* in poor condition yield 8.1% less energy per gram than good condition fish. In addition, adult *S. mossambicus* may be more difficult to catch than juveniles on the terrace due to the greater water depth and apparent low density of predators in offshore habitats (Table 7).

S. mossambicus is largely confined to water shallower than 12 m in Lake Sibaya (Bruton & Boltt, 1975) and no other large prey are regularly taken by *C. gariepinus* in deeper water. *Glossogobius giurus, Pseudocrenilabrus philander* and *Croilia mossambica* are present in variable numbers in the profundal zone, but they rarely occur in the diet of catfish caught there. Only *H. orbiculare* among the profundal benthos contributes significantly to the diet of *C. gariepinus. Bellamya capillata, Melanoides tuberculatus, Caridina nilotica* and the various infaunal microcrustacea are too small, unpalatable and/or sparse to support a large predator. The standing crop of zooplankton is certainly inadequate. Thus poor food quality and low availability of large food items can be identified as the cause of decreased growth rates and poor condition of large *C. gariepinus* in Lake Sibaya.

Discussion

The most successful and abundant fishes in Lake Sibaya are all eurytopic species with a wide distribution on the southern Mozambique plain and elsewhere in southern Africa (Jubb, 1967). They consist of four 'secondary' freshwater species – *S. mossambicus, P. philander, T. sparrmanii* and *C. gariepinus*, and one 'peripheral' form, *Glossogobius giurus*. These adaptable species are able to make extensive use of the rich, but harsh and variable, littoral zone, which appears to be the key to their successful utilization of Lake Sibaya. Some other species frequent recently flooded littoral areas *e.g. Aplocheilichthys katangae, Ctenopoma multispinis* and *Clarias theodorae*, but their preference for vegetated areas precludes use of barren or sparsely-vegetated terrace zones. The more stable, offshore profundal and sheltered bay zones are mainly populated by more stenotopic species including the 'primary' freshwater fishes (mormyrid, cyprinids), two 'peripheral' species (small gobiids) and the other 'secondary' forms. The overall low nutrient status of Lake Sibaya, and the consequent low standing crop of plankton (Allanson & Hart, 1975) account for the poor development of a pelagic fish community which consists of only two small species (a clupeid and an atherinid).

The two fish species for which growth data are available (*S. mossambicus, C. gariepinus*) both grow rapidly and maintain good condition compared with other populations during their inshore feeding phase, but reach a small final size and have poor condition in later life due to the poorer quality of food which is utilized by larger, offshore feeding individuals. Clearly, occupation of the littoral zone for a part of their life cycle is an important prerequisite for good growth, good condition, and therefore high reproductive potential in these species. The obvious question is: why do large *S. mossambicus* and *C. gariepinus* not feed in areas where protein-rich food is readily accessible *i.e.* the inshore littoral zone? Both biotic and abiotic factors are likely to play a role.

The diel movements and penetration into shallow water of cichlids have been interpreted in terms of many factors including water temperatures and depth selection (Welcomme, 1964; Donnelly, 1969b; Bruton & Boltt, 1975; Caulton, 1975, 1977; Whitfield & Blaber, 1976), the influence of predators (Fryer, 1961; Jackson, 1961; Donnelly, 1969b; Bruton & Boltt, 1975), the role of metabolic energy budgets (Caulton, 1977, 1978a,b), and the importance of food quality (Bowen, 1976a). Bruton (1973), Caulton (1975) and Bowen (1976a) showed that wave action and cloud cover may decrease inshore feeding activity. Clearly, the rhythm and extent of inshore migrations by cichlids may be controlled by several factors, and the position of a fish in shallow water is determined by a compromise of several preferences, rather than one preference.

The higher temperature tolerances and preferences of juvenile cichlids compared with adults has been demonstrated by several workers (Allanson & Noble, 1964; Welcomme 1964; Badenhuizen, 1967; Donnelly, 1969a; Josman, 1971; Bruton & Boltt, 1975; Caulton, 1975, 1978a). This factor is undoubtedly important in allowing juvenile cichlids to penetrate further into warmer shallow well-lit areas during the day than adults. Caulton (1975) noted that the consequence of this thermophilia is that the high metabolic cost, resulting from increased routine energy requirements at elevated temperatures, would be expected to be distinctly disadvantageous to growth unless this cost could be offset by some intrinsic metabolic homeostasis or a superabundance of food in the shallow margins. Later Caulton (1977) showed for *Tilapia rendalli* that the pattern of routine metabolism does not, in fact, follow the general form expected of poikilotherms. Instead of the characteristic exponential increase in metabolic costs with increasing temperature, there was a plateau over the temperature range characteristic of the

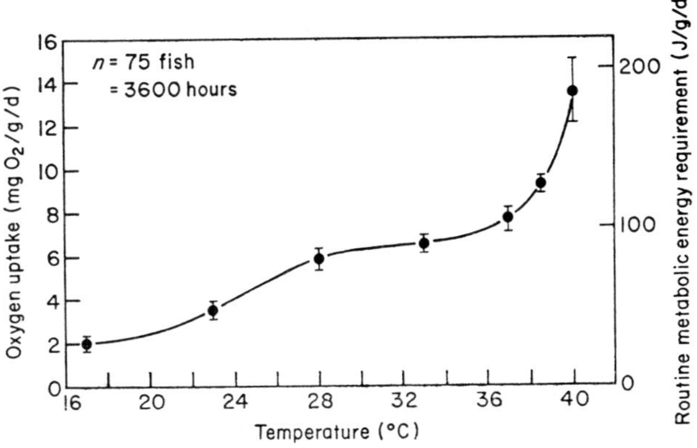

Fig. 39. The effect of temperature on oxygen uptake by young *Tilapia rendalli*. The vertical bars represent two standard errors about the mean (from Caulton, 1977).

daytime inshore habitat of the species (28° to 37°C) which resulted in a suppression of the expected high metabolic cost at these warm temperatures (Fig. 39). During the day maximal occupation of the preferred temperatures results in optimal energy storage. The maximum retention of this energy gain was facilitated by the fish moving into cool deeper water during the night, when no feeding took place and metabolic energy demands were low. A similar plateau function was reported to be characteristic of the oxygen demand of *Sarotherodon mossambicus* by Job (1969), but Caulton (1978b) found no evidence of this phenomenon in this species and suggested that Job's finding may be the result of an artifact due to experimental design. Nevertheless, the need to occupy high temperature areas for the ingestive phase of feeding, and relatively low temperature areas for the digestive stage, is likely to be important in many thermophilic fishes, especially the cichlids, and may explain in part why they perform regular diel movements into and out of shallow water. Caulton (1978a) has subsequently demonstrated with immature *Tilapia rendalli* that growth is enhanced by these diel movements and is predicted to be optimal if the fish occupy water at 30°C during the day but return to nocturnal habitats at 18°C. This conclusion was reached using a bio-energetic model incorporating measurements of feeding intensity, metabolic energy demands and assimilation efficiencies at temperatures ranging from 18° and 34°C. He concludes that the utilization of thermal fluctuations to ensure efficient growth rates may be a more important explanation for the daytime inshore concentrations of young cichlids than food abundance or predation pressures.

The high energy cost of feeding at elevated temperatures may also be offset by a superabundance of food in marginal areas, as these areas, with their high light intensities are likely to provide optimal conditions for the production of diatoms, the main food resource of *Sarotherodon mossambicus* in Lake Sibaya (Bowen, 1978). In deep lakes with a relatively narrow littoral zone, such as Sibaya, the area of shallows is restricted, and larger individuals in the population are forced to feed in poorly-illuminated areas due to their preference for cooler, slightly deeper water. In systems with extensive shallow areas and no deep basin, larger individuals in the population will feed in relatively shallower water, especially if the turbidity of the water is high (which offers some protection against bird predators) as is typical of many shallow water ecosystems *e.g.* floodplain pans, small eutrophic impoundments. In these systems fish of all sizes will have access to protein-rich benthic floc where its food-quality is likely to be adequate for growth. The *S. mossambicus* population in Nyamithi pan, a floodplain pan 60 km north-west of Lake Sibaya, may be cited as an example. In this shallow, eutrophic system *S. mossambicus* reaches a size of 43.2 cm TL and 1543 g. In deep water systems occupied by *S. mossambicus* which have an abundance of food in offshore habitats this species also reaches a large size. In Lake McIlwaine, Rhodesia, the phytoplankton has a permanently high standing crop (Marshall & Falconer, 1973) which is utilized by adults while the juveniles feed on benthic diatoms (Munro, 1967). The adults also feed on periphytic algae and reach a large maximum size of 47.3 cm (Munro, 1967).

In spite of its small size, *S. mossambicus* is one of the most abundant fishes in Lake Sibaya. Bruton (1973) and Bruton & Allanson (1974) suggested that precocious breeding and small maximum size are part of an adaptive strategy which enables this species to obtain maximum benefit from the rich food resources in shallow water. By breeding precociously, generation time is shortened and a greater intrinsic rate of population increase results. Thus high mortality rates, to which the juveniles may be subject as a result of the drying up of marginal pools and predation by birds, may be balanced by high recruitment rates. The maturity length ratio (length at first maturity/maximum length) for *S. mossambicus* in Lake Sibaya is 0.43 which is considerably lower than the characteristic value for cichlids (about 0.70, Iles, 1971). According to Iles, low values are typical of dwarfed tilapiine populations and imply that a larger proportion of the population is an actual or potential reproductive state.

Precocious breeding would also have the important effect of changing the size and age structure of the population such that small, young fish, which are best able to utilize rich inshore food resources, would form a larger proportion of the total population. I have argued that the small maximum size of *S. mossambicus* adults results from an unusually early shunting of resources from growth to reproduction which allows this species to best utilize the richest food resources in the lake.

In the light of his further investigations, Bowen (1976a) was able to clarify the role of precocious breeding in the success of *S. mossambicus* in Lake Sibaya. As reproductive potential is directly proportional to body weight in fishes (Gadgil & Bossert, 1970) energy expended for reproduction at one age will result in reduced growth and thus reduced reproductive potential at all subsequent ages. On the other hand, early reproduction contributes towards increased reproductive output by shortening both the generation time and the time during which individuals are exposed to mortality before they reproduce. Thus Gadgil and Bossert argue that in order to maximize reproductive output, natural selection will favour the onset of maturity and reproduction at an age that strikes the optimal balance between the advantages of shortened generation time and reduced mortality conferred by early reproduction, and the advantage of increased reproduction potential conferred by late reproduction. Bowen (1976a) has defined the stage at which food resources become limiting for *S. mossambicus* and thus deny adults the growth and increase in reproductive potential which is available to adults of this species in other systems. The poor condition and small final size of adult *S. mossambicus* in Lake Sibaya is therefore due to both the poor quality of their food resources (Bowen, 1976a) and to the early shunting of resources from growth to reproduction. Precocious breeding is identified as an adaptation to maximize reproductive output and to make best use of inshore feeding areas, given the limited resources available to adults, and doubtless makes an important contribution to the success of the population. In Lake Sibaya, *S. mossambicus* has utilized its phenotypic plasticity to such an extent that the juvenile stages are adapted to utilize the best available food resources efficiently, whereas the adults are mainly concerned with reproduction and maintenance. The dwarfing of individuals in this population is therefore not related to overtaxing the food supply, as in 'stunted' populations in overstocked ponds, resulting in slow growth and small final size, but to abundant food, fast growth (in immature fishes) and early maturity.

The possible effect of predation pressure on the diel movements of *S. mossambicus* and other cichlids must now be considered. Various birds feed on juvenile cichlids in Lake Sibaya, including cormorants, darters, herons, kingfishers and gulls (see next chapter). The only known bird predators of adult *S. mossambicus* ($>$8 cm SL) in the lake are the fish eagle *Haliaetus vocifer*, giant kingfisher *Megaceryle maxima*, goliath heron *Ardea goliath* and osprey *Pandion haliaetus*. Adult fisheagles are capable of catching the largest *S. mossambicus* in Lake Sibaya and feed mainly on large adults (average 15 cm TL) in the lake (next chapter). Their constant presence along some shores must inhibit the penetration of adult *S. mossambicus* into shallow water. Juvenile *S. mossambicus* react to any disturbance by forming dense shoals which move rapidly along the terrace, sometimes into deeper water. *S. mossambicus* is included in the prey of *Crocodylus niloticus* in Lake Sibaya, but their effect on cichlid distribution is not known.

Cichlids occur in the diet of *C. gariepinus* in many lakes and rivers including Barberspan, Transvaal (Groenewald, 1964), Lake Sibaya (Minshull, 1969; Bruton & Boltt, 1975; Bruton, 1979b), Elands river, Transvaal (van der Waal, 1972), lower Shire River, Malawi (Willoughby & Tweddle, 1976) and Kafue river (Bell-Cross, 1976). *S. mossambicus* is the main prey of *C. gariepinus* (in terms of weight) in Lake Sibaya, but predation pressures may differ according to lake level. At low lake level, few catfish venture onto the terrace during the day when juvenile *S. mossambicus* are present there. At night the cichlids move into offshore areas where they rest on the substratum or swim slowly in midwater. Occasional individuals 'perch' among the fronds of aquatic macrophytes, in common with *T. sparrmanii* and *G. giurus* (SCUBA observations). Dispersal and perching can be regarded as strategies which reduce contact with a potential predator, but predation risk is nevertheless higher than during the day as *C. gariepinus* is an active nocturnal predator.

As the lake level rises, the factors which inhibit catfish occupation of the terrace during the day (see below) apparently diminish in importance, and greater numbers venture onto the terrace, which may explain why the proportion of fishes in their diet increased by over 40% at intermediate and high lake level compared with low lake level (Table 16). The reversed diel inshore migration of *S. mossambicus* at high lake level (when this species most often ventures onto the terrace at night, Bowen, 1976a) may be a consequence of the increased density of *C. gariepinus* on the terrace during the day, as well as the slightly warmer water temperatures experienced on the deepened terrace at night compared with the day (see Chapter 4).

The mortality caused by *C. gariepinus* in Lake Sibaya is likely to be on weak, slow, inattentive or cornered fishes, as the catfish has been shown to be inefficient at catching free-ranging, healthy *S. mossambicus*, except at night or in groups in shallow water (Bruton, 1979b,c). Nevertheless, the number of prey taken is considerable and we cannot discount *C. gariepinus* as an influence on the diel movements of juvenile and small adult *S. mossambicus*. Adult *S. mossambicus* larger than 150 mm TL (the approximate length at which adults return to the terrace for breeding) are too large for *C. gariepinus* in Lake Sibaya where their prey rarely exceeds 110 mm TL (*i.e.* 20% of the total length of the predator; the modal length of *C. gariepinus* on the terrace being 560 mm TL; Bruton, 1978). Thus the catfish would appear to have little effect on the movements, or on the growth and condition, of adult *S. mossambicus* in Lake Sibaya.

In summary, temperature effects on feeding and growth efficiency, and the high quality food resources in shallow water, may be regarded as the overriding factors which induce cichlids, such as *S. mossambicus*, to perform diel inshore movements. The pattern of these movements may be influenced by other factors including predation pressure, water depth, and wave action.

The diel movements of *C. gariepinus* may also be regulated in part by temperature preferenda. In Lake Sibaya, *C. gariepinus* inhabits terrace areas at temperatures from 18° to 35°C, but avoids shallow water at temperatures above 35°C which excludes them from these areas during the day in summer at low lake level. At higher lake levels, when terrace water temperatures are moderated, *C. gariepinus* is common in shallow water.

C. gariepinus falls prey to both *Crocodylus niloticus* and *Haliaëtus vocifer* in Lake Sibaya and elsewhere (see next chapter). The threat of the fisheagle is particularly apparent in a clear lake such as Sibaya as these soaring raptors are able to stoop at great speed onto a catfish feeding in shallow water or rising to the surface for air.

The relative predation efficiency of *C. gariepinus* on fishes is most likely to be an important factor influencing the inshore movements of the catfish. Their greater efficiency at night, in groups and in shallow water must explain in part their tendency to feed on the terrace at night at low lake level. As the lake level rises, their risk of predation from fisheagles (which can only capture prey within a metre of the water surface) decreases, and they are able to venture onto the terraces in search of fish prey.

The dominant role of inshore food resources in the life history of certain fishes in an oligo/mesotrophic lake has been shown by these results. The extensive profundal and limnetic zone of Lake Sibaya is comparatively lifeless compared with the littoral fringe in so far as fish populations are concerned.

Fluctuations in water level, either as a result of summer rain or longterm high rainfall periods, have a marked effect on this important zone. At low lake levels the littoral terraces are barren and harsh, and are inhabited predominantly by only one species of fish. Rising lake levels establish a more moderate inshore habitat which is occupied by a more diverse biota and supports an abundant population of fish. Through these fluctuations the littoral zone persists as a crucial and dynamic 'trophic storeroom' for the lake.

Acknowledgements

My research on the fishes of Lake Sibaya was performed during a six and a half year stay at Lake Sibaya Research Station. I am grateful to Professor B. R. Allanson for this unique opportunity, and to Professor Allanson, Dr B. J. Hill and the late Dr R. E. Boltt for supervising my research. I am grateful to my wife, Carolynn, and to Tony and Nola Bruton for field assistance, and to Dr S. H. Bowen, Dr I. Gaigher, Dr B. van der Waal and Mr H. Kok for valuable discussions and permission to use unpublished information and published figures. Our fish research would not have been possible without the co-operation of the KwaZulu Government Service and Natal Parks Board. I am grateful to Mrs M. M. Smith of the J. L. B. Smith Institute of Ichthyology for allowing me to complete this work while in her Institute, and to Mrs H. Tomlinson for typing. The excellent fish drawings were kindly done by Ms E. M. Tarr.

References

Acocks, J. P. H. 1975. Veld types of South Africa. Mem. Bot. Surv. S. Afr. No. 40 (2nd ed.): 1–128.

Alexander, R. McN. 1965. Structure and function in the catfish. J. Zool. 148: 88–152.

Allanson, B. R., A. Bok & N. I. van Wyk. 1971. The influence of exposure to low temperature on *Tilapia mossambica* Peters (Cichlidae). II. Changes in plasma osmolarity, sodium and chloride ion concentrations. J. Fish Biol. 3: 181–185.

Allanson, B. R., M. N. Bruton & R. C. Hart. 1974. The plants and animals of Lake Sibaya, KwaZulu, South Africa: a checklist. Rev. Zool. afr. 88: 507–532.

Allanson, B. R. & R. C. Hart. 1975. The primary production of Lake Sibaya, KwaZulu, South Africa. Verh. int. Verein. Limnol. 19: 1426–1433.

Allanson, B. R., B. J. Hill, R. E. Boltt & V. Schultz. 1966. An estuarine fauna in a freshwater lake in South Africa. Nature, Lond. 209: 532–533.

Allanson, B. R. & R. G. Noble. 1964. The tolerance of *Tilapia mossambica* (Peters) to high temperature. Trans. Am. Fish. Soc. 93: 323–332.
Angelopoulo, V. 1947. The morphology of *Clarias gariepinus* (Burchell). Unpubl. M.Sc. thesis, University of Pretoria. 65 pp.
Badenhuizen, T. R. 1967. Temperatures selected by *Tilapia mossambica* Peters in a test tank with a horizontal temperature gradient. Hydrobiologia 30: 541–554.
Balon, E. K. 1975. Reproductive guilds of fishes: a proposal and definition. J. Fish. res. Bd Can. 32: 812–864.
Barnard, K. H. 1948. Report on a collection of fishes from the Okavango River with notes on Zambezi fishes. Ann. S. Afr. Mus. 36 (5): 407–458.
Bell-Cross, G. 1974. A fisheries survey of the Upper Zambezi River system. Occ. Pap. natn. Mus. Rhod. *B*5: 279–338.
Bell-Cross, G. 1976. *The Fishes of Rhodesia.* Trustees Natn. Mus. & Monuments of Rhod., Salisbury, Rhodesia. 268 pp.
Beukema, J. J. 1968. Predation by the three-spined stickleback (*Gasterosteus aculeatus* L.): The influence of hunger and experience. Behaviour 31: 1–126.
Blaber, S. J. M. 1978a. Fishes of the Kosi system. Lammergeyer 24: 28–41.
Blaber, S. J. M. 1978b. First record of *Ctenopoma ctenotis* from South Africa. Lammergeyer 26: 63.
Blaber, S. J. M. & A. K. Whitfield. 1977. The biology of the burrowing goby *Croilia mossambica* Smith (Teleostei, Gobiidae). Environ. Biol. Fishes 1 (2): 197–204.
Boltt, R. E. 1969. The benthos of some southern African lakes. Part II. The epifauna and infauna of the benthos of Lake Sibayi. Trans. R. Soc. S. Afr. 38: 249–269.
Boltt, R. E., B. J. Hill & A. T. Forbes. 1969. The benthos of some southern African lakes. Part I: Distribution of aquatic macrophytes and fish in Lake Sibayi. Trans. R. Soc. S. Afr. 38: 241–248.
Boulenger, G. A. 1915. Catalogue of the freshwater fishes of Africa in the British Museum (Nat. Hist) 3, Brit. Mus. (Nat. Hist.), London.
Bowen, S. H. 1976a. Feeding ecology of the cichlid fish *Sarotherodon mossambicus* in Lake Sibaya, KwaZulu. Unpubl. Ph.D thesis, Rhodes University, South Africa.
Bowen, S. H. 1976b. Mechanism for digestion of detrital bacteria by the cichlid fish *Sarotherodon mossambicus* (Peters). Nature, Lond. 260: 137–138.
Bowen, S. H. 1978. Benthic diatom distribution and grazing by *Sarotherodon mossambicus* in Lake Sibaya, South Africa. Freshwater Biol. 8: 449–453.
Bowmaker, A. P. 1973. Potamodromesis in the Mwenda River, Lake Kariba. Geophys. Monog. 17: 159–164.
Bowmaker, A. P., P. B. N. Jackson & R. A. Jubb. 1978. Freshwater fishes. In 'Biogeography and ecology of southern Africa' (Ed. M. J. A. Werger). W. Junk, The Hague.
Boyd, C. E. & C. P. Goodyear. 1971. Nutritive quality of food in ecological systems. Arch. Hydrobiol. 69: 265–270.
Breder, C. M. & D. E. Rosen. 1966. Modes of reproduction in fishes. TFH Publications, New Jersey.
Brock, V. E. & R. H. Riffenburgh. 1960. Fish schooling: A possible factor in reducing predation. J. Cons. perm. int. Explor. Mer. 25: 307–317.
Bruton, M. N. 1973. A contribution to the biology of *Tilapia mossambica* Peters in Lake Sibaya, South Africa. M.Sc thesis, Rhodes University, Grahamstown, South Africa.
Bruton, M. N. 1974. Two new fish records from Lake Sibaya. News Lett. Limnol. Soc. sth. Afr. no. 22: 47–48.
Bruton, M. N. 1975. First record of *Sarotherodon placidus* (Pisces: Cichlidae) from South Africa. Lammergeyer 22: 33–36.
Bruton, M. N. 1976. On the size reached by *Clarias gariepinus.* J. Limnol. Soc. sth. Afr. 2 (2): 57–58.
Bruton, M. N. 1977. The biology of *Clarias gariepinus* (Burchell, 1822) in Lake Sibaya, KwaZulu, with emphasis on its role as a predator. Unpubl. Ph. D. thesis, Rhodes University, South Africa.
Bruton, M. N. 1978. The habitats and habitat preferences of *Clarias gariepinus* (Pisces:

Clariidae) in a clear, coastal lake (Lake Sibaya, South Africa). J. Limnol. Soc. sth. Afr. 4 (2): 81–88.
Bruton, M. N. 1979a. The breeding biology and early development of *Clarias gariepinus* (Pisces: Clariidae) in Lake Sibaya, South Africa, with a review of breeding in species of the subgenus *Clarias* (*Clarias*). Trans. zool. Soc. Lond. 35: 1–45.
Bruton, M. N. 1979b. The food and feeding behaviour of *Clarias gariepinus* (Pisces: Clariidae) in Lake Sibaya, South Africa, with emphasis on its role as a predator of cichlids. Trans. zool. Soc. Lond. 35:
Bruton, M. N. 1979c. The role of diel inshore movements by *Clarias gariepinus* (Pisces: Clariidae) for the capture of fish prey. Trans. zool. Soc. Lond. 35: 115–138.
Bruton, M. N. (in prep.). Social hunting in fishes.
Bruton, M. N. & B. R. Allanson. 1974. The growth of *Tilapia mossambica* (Pisces: Cichlidae) in Lake Sibaya, South Africa. J. Fish Biol. 6: 701–715.
Bruton, M. N. & B. R. Allanson. (in press). The growth of *Clarias gariepinus* (Pisces: Clariidae) in Lake Sibaya, South Africa. S. Afr. J. Zool.
Bruton, M. N. & C. C. Appleton. 1975. Survey of Mgobezeleni lake-system in Zululand, with a note on the effect of a bridge on the mangrove swamp. Trans. R. Soc. S. Afr. 41 (3): 283–294.
Bruton, M. N. & R. E. Boltt. 1975. Aspects of the biology of *Tilapia mossambica* Peters (Pisces: Cichlidae) in a natural freshwater lake (Lake Sibaya, South Africa). J. Fish Biol. 7: 423–445.
Bruton, M. N. & K. H. Cooper. (ed.). 1979. Studies on the ecology of Maputaland. Rhodes University Press, Grahamstown.
Campbell, G. D. 1947. Fish life at Kosi Bay. Bull. Natal Soc. Preserv. wild Life nat. Resour. August mag., 3–4.
Campbell, G. D. & B. R. Allanson. 1952. The fishes of the 1947, 1948 and 1949 scientific investigations of the Kosi area. Bull. Natal Soc. Preserv. wild Life nat. Resour. April mag., 1–8.
Campbell, G. G. 1969. A review of scientific investigations in the Tongaland area of northern Natal. Trans. R. Soc. S. Afr. 38 (4): 305–316.
Caulton, M. S. 1975. Diurnal movement and temperature selection by juvenile and sub-adult *Tilapia rendalli* Boulenger (Cichlidae). Trans. Rhod. Sci. Assoc. 56: 51–56.
Caulton, M. S. 1976. The energetics of metabolism, feeding and growth of subadult *Tilapia rendalli* Boulenger. Unpublished D. Phil. thesis. University of Rhodesia, Salisbury.
Caulton, M. S. 1977. The effect of temperature on routine metabolism in *Tilapia rendalli* Boulenger. J. Fish Biol. 11: 549–553.
Caulton, M. S. 1978a. The importance of habitat temperatures for growth in the tropical cichlid *Tilapia rendalli* Boulenger. J. Fish Biol. 13: 99–112.
Caulton, M. S. 1978b. The effect of temperature and mass on routine metabolism in *Sarotherodon* (*Tilapia*) *mossambicus* (Peters). J. Fish Biol. 13: 195–201.
Caulton, M. S. & B. J. Hill. 1973. The ability of *Tilapia mossambica* Peters to enter deep water. J. Fish Biol. 5: 783–788.
Caulton, M. S. & B. J. Hill. 1975. The effect of temperature on the ability of *Tilapia mossambica* Peters to enter deep water. J. Fish Biol. 7: 221–226.
Chapman, D. G. 1952. Inverse, multiple and sequential sample censuses. Biometrics 8: 286–306.
Clay, D. 1977. Preliminary observations on salinity tolerance of *Clarias lazera* in Israel. Bamidgeh 29 (3): 102–109.
Cockson, A. & D. Bourne. 1972. Enzymes in the digestive tract of two species of euryhaline fish. Comp. Biochem. Physiol. 41A: 715–718.
Coe, M. J. 1966. The biology of *Tilapia grahami* Boulenger in Lake Magadi, Kenya. Acta trop. 23: 146–177.
Coe, M. J. 1967. Local migration of *Tilapia grahami* Boulenger in Lake Magadi, Kenya, in response to diurnal temperature changes in shallow water. E. Afr. Wildlife. J. 5: 171–174.
Crass, R. S. 1960. Notes on the freshwater fishes of Natal with descriptions of four new species. Ann. Natal Mus. 14: 405–458.

Crass, R. S. 1964. *Freshwater Fishes of Natal.* Shuter & Shooter, Pietermaritzburg.
Crass, R. S. 1966. Features of freshwater fish distribution in Natal and discussion of controlling factors. News Lett. Limnol. Soc. sth. Afr. 7: 31–35.
Curio, E. 1976. *The Ethology of Predation.* Springer-Verlag, Berlin.
Donnelly, B. G. 1966. Shoaling, communication and social hunting in the catfish *Clarias ngamensis.* Piscator 67: 54–55.
Donnelly, B. G. 1969a. Observations on the temperature preferenda of three cichlid fishes, *Tilapia mossambica* Peters, *Tilapia sparrmanii* A. Smith and *Hemihaplochromis philander* (M. Weber). Unpublished honours project, Rhodes University, Grahamstown.
Donnelly, B. G. 1969b. A preliminary survey of *Tilapia* nurseries on Lake Kariba during 1967/68. Hydrobiologia 34 (2): 195–206.
Dupree, H. K. & J. F. Halver. 1970. Amino acids essential for the growth of channel catfish, *Ictalurus punctatus.* Trans. Am. Fish. Soc. 99: 90–92.
Du Toit, P. 1971. Die verhaal tussen die ouderdom en lengte/massaverhouding van *Tilapia mossambica* Peters 1852, in Loskopdam, Oos-Transvaal. Unpubl. M.Sc. thesis, Rand Afrikaans University, South Africa.
Eibl-Eibesfeldt, I. 1962. Freiwasserbeobachtungen zur Dentung des Schwarmverhaltens verschiedener Fische. Z. Tierpsychol. 19: 165–182.
Farquharson, F. L. 1970. A new freshwater gobi (Pisces: Gobiidae) from Lake Sibayi, Zululand, South Africa. Ann. Cape Prov. Mus. (Nat. Hist.). 8 (10): 85–87.
Forbes, A. T. & B. J. Hill. 1969. The physiological ability of a marine crab, *Hymenosoma orbiculare* Desm., to live in a subtropical freshwater lake. Trans. R. Soc. S. Afr. 38: 281–284.
Fryer, G. 1961. Observations on the biology of the cichlid fish *Tilapia variabilis* Boulenger in the northern waters of Lake Victoria (East Africa). Revue Zool. Bot. afr. 64: 1–33.
Fryer, G. & T. D. Iles. 1972. *The Cichlid Fishes of the Great Lakes of Africa.* Oliver & Boyd, London.
Gabie, V. 1965. Problems associated with the distribution of freshwater fishes in Southern Africa. S. Afr. J. Sci. 61: 383–391.
Gadgil, M. & W. H. Bossert. 1970. Life historical consequences of natural selection. Am. Nat. 104: 1–24.
Gaigher, I. G. 1977. Reproduction of the catfish (*Clarias gariepinus*) in the Hardap dam, South-West Africa. Madoqua 10: 55–59.
Gaigher, I. G. & R. Mc. C. Pott. 1973. Distribution of fishes in Southern africa. S. Afr. J. Sci. 69: 25–27.
Gosline, W. A. 1973. Considerations regarding the phylogeny of Cypriniform fishes with special reference to structures associated with feeding. Copeia No. 4: 761–776.
Greenwood, P. H. 1974. The cichlid fishes of Lake Victoria, East Africa: the biology and evolution of a species flock. Bull. Br. Mus. nat. Hist. (Zool.) Suppl. 6: 1–134.
Groenewald, A. A. van J. 1964. Observations on the food habits of *Clarias gariepinus* Burchell, the South African freshwater barbel (Pisces: Clariidae) in Transvaal. Hydrobiologia 23: 287–291.
Harper, A. E. 1967. Effects of dietary protein content and amino acid pattern on food intake and preference, in 'Handbook of Physiology', Section 6, Vol. 1 (29: 399–410). American Physiological Society, Washington, D.C.
Harrison, A. C. 1936. Black bass in the Cape Province. Investl Rep. Fish. mar. biol. Surv. Div. Un. S. Afr. 7: 1–119.
Hart, R. C. & B. R. Allanson. 1975. Preliminary estimates of production by a calanoid copepod in subtropical Lake Sibaya. Verh. int. Verein. Limnol. 19: 1434–1441.
Hartland-Rowe, R. 1958. The biology of a tropical mayfly, *Povilla adusta* Navas, with special reference to the lunar rhythm of emergence. Rev. Zool. Bot. afr. 58: 185–195.
Hickling, C. F. 1961. *Tropical Inland Fisheries.* Trop. Agr. Ser. Longmans, London.
Hickling, C. F. 1970. Management of brackish water fish ponds. Advances in Marine Biology 8: 178–180.
Hill, B. J., S. J. M. Blaber & R. E. Boltt. 1975. The limnology of Lagoa Poelela. Trans. R. Soc. S. Afr. 41 (3): 263–271.
Holl, A. E. 1968. Notes on spawning behaviour of barbel *Clarias gariepinus* Burchell in

Rhodesia. Zool. Afr. 3: 185–188.
Holling, C. S. 1965. The functional response of predators to prey density and its role in mimicry and population regulation. Mem. entomol. Soc. Can. 45: 1–62.
Holling, C. S. 1966. The functional response of invertebrate predators to prey density. Mem. entomol. Soc. Can. 48: 1–86.
Iles, T. D. 1971. Ecological aspects of growth in African cichlid fishes. J. Cons. perm. Int. Explor. Mer 33: 363–385.
Iles, T. D. 1963. Dwarfing or stunting in the genus *Tilapia* (Cichlidae); a possibly unique recruitment mechanism. Rapp. P.v.-Réun. Cons. perm. int. Explor. Mer 164: 247–254.
Jackson, P. B. N. 1961. The impact of predation, especially by the tigerfish (*Hydrocynus vittatus* Cast.), on African Freshwater fishes. Proc. zool. Soc. Lond. 136: 603–662.
Jackson, P. B. N. 1975. Common and scientific names of the fishes of southern Africa. Part II. Special publication no. 14, J. L. B. Smith Institute of Ichthyology, Rhodes University, Grahamstown. pp. 179–213.
Jannasch, H. W. & P. H. Pitchard. 1972. The role of inert particulate matter in the activity of aquatic microorganisms. Mem. Ist. ital. Idrobiol. 29 (Suppl.): 289–308.
Jhingran, V. G. 1977. *Fish and fisheries of India.* Hindustan Publishing Corporation, Delhi.
Job, S. V. 1969. The respiratory metabolism of *Tilapia mossambica* (Teleostei) II. The effect of size, temperature, salinity and partial pressure of oxygen. J. mar. Biol. 3 (3): 222–226.
Josman, V. 1971. Some aspects of the effect of temperature on the respiratory and cardiac activities of the cichlid teleost *Tilapia mossambica.* Unpubl. M.Sc. thesis, Rhodes University, South Africa.
Jubb, R. A. 1952. A note on the distribution of *Hydrocyon vittatus* Cast. S. Afr. J. Sci. 49: 50–51.
Jubb, R. A. 1964. Freshwater fishes and drainage basins in Southern Africa. S. Afr. J. Sci. 60: 17–21.
Jubb, R. A. 1967. *Freshwater Fishes of Southern Africa.* Balkema, Cape Town.
Junor, F. J. R. 1969. *Tilapia melanopleura* Dum. in artificial lakes and dams in Rhodesia with special reference to its undesirable effects. Rhod. J. agric. Res. 7: 61–69.
Kemp, A. C. 1966. An investigation of the effect of rate of rise of temperature on the survival of *Tilapia mossambica* (Peters). Unpubl. B.Sc (Hons.) thesis, Rhodes University, South Africa.
Khanna, S. S. 1966. Histology of the digestive tract of a teleost, *Clarias batrachus* (Linn.). J. zool. Soc. India 16: 53–58.
Kok, H. M. & S. J. M. Blaber. 1977. A new freshwater goby (Teleostei: Gobiidae) from the Pongola floodplain, Zululand, South Africa. Zool. Afr. 12 (1): 163–168.
Lea, E. 1910. On the methods used in herring investigations. Publ Circonst. Cons. perm. int. Explor. Mer, no. 53.
Le Roux, P. J. 1956. Feeding habits of the young of four species of *Tilapia.* S. Afr. J. Sci. 53: 33.
Le Roux, P. J. 1961. Growth of *Tilapia mossambica* Peters in some Transvaal impoundments. Hydrobiologia 18: 165–176.
Love, R. M. 1970. *The chemical biology of fishes.* Academic press, London.
MacArthur, R. H. 1972. Geographical ecology. Harper & Row, New York.
Marshall, B. E. & A. C. Falconer. 1973. Physico-chemical aspects of Lake McIlwaine (Rhodesia), an eutrophic tropical impoundment. Hydrobiologia 42: 45–62.
Mills, H. D. 1956. The African mudfish, *Clarias lazera.* Ibadan University Press, Ibadan.
Minshull, J. L. 1968. A summary of the results obtained at the Lake Sibaya Research Station, for the period April to October 1968. Institute for Freshwater Studies, Annual Report, 1968, pp. 1–10.
Minshull, J. L. 1969. An introduction to the food web of Lake Sibaya, Northern Zululand. News Lett. Limnol. Soc. sth. Afr. 13 (suppl.): 20–25.
Minshull, J. L. 1970. Studies on the fish population: Lake Sibaya, 1968–1969. Unpubl. Report, Institute for Freshwater Studies, Rhodes University, South Africa. 24 pp.
Moriarty, D. J. W. 1973. The physiology of digestion of blue-green algae in the cichlid fish *Tilapia nilotica.* J. Zool., 171: 25–39.

Munro, S. L. 1967. The food of a community of East African freshwater fishes. J. Zool. 151: 389–415.
Murray, J. L. 1975. Selection of zooplanktion by *Clarias gariepinus* (Burchell) in Lake McIlwaine. An eutrophic Rhodesian reservoir. Unpubl. M.Sc. thesis, University of Rhodesia, Salisbury.
Neil, S. R. & J. M. Cullen. 1974. Experiments on whether schooling by their prey affects the hunting behaviour of cephalopod and fish predators. J. Zool. 172: 549–569.
Odum, E. P. 1971. *Fundamentals of Ecology*. W. B. Saunders Co., London.
Ogino, C. & K. Saito. 1970. Protein nutrition in fish. I. The utilization of dietary protein by young carp. Bull. Jap. Soc. scient. Fish. 36: 250–254.
Paerl, H. W. 1974. Bacterial uptake of dissolved organic matter in relation to detrital aggregation in marine and freshwater systems. Limnol. Oceanogr. 19: 966–972.
Petr, T. 1970. Macroinvertebrates of flooded trees in the man-made Volta lake (Ghana) with special reference to the burrowing mayfly *Povilla adusta*. Hydrobiologia 36: 373–398.
Phillips, A. M. 1972. Calorie and energy requirement. In 'Fish Nutrition.' (Ed. J. E. Halver). Academic Press, London. pp. 1–28.
Pike, T. 1969. Composition, relative abundance and size range of fish populations in Lake Sibaya. News Lett. Limnol. Soc. sth. Afr. 13 (Suppl.): 38–43.
Pooley, A. C. 1972. Feeding behaviour of barbel (catfish). Lammergeyer 17: 65.
Redakov, D. V. 1973. Schooling and the Ecology of Fish. (Transl. from Russian by H. Mills). John Wiley & Sons, New York.
Ribbink, A. J. 1971. The behaviour of *Hemihaplochromis philander*, a South African cichlid fish. Zool. Afr. 6 (2): 263–288.
Ribbink, A. J. 1975. A contribution to the understanding of the ethology of the cichlids of southern Africa. Unpubl. Ph.D. thesis, Rhodes University, South Africa.
Ricker, W. E. 1968. *Methods for Assessment of Fish Production in Fresh Waters*. IBP Handbook No. 3, Blackwell, Oxford.
Ricker, W. E. 1975. *Computation and Interpretation of Biological Statistics of Fish Populations*. Bull. Fish. Res. Bd Can. 191: 382 pp.
Roberts, T. R. 1975. Geographical distribution of African freshwater fishes. Zool. J. Linn. Soc. 57: 249–319.
Russell-Hunter, W. D. 1970. *Aquatic productivity: An introduction to some basic aspects of biological ocenography and limnology*. Collier-MacMillan, London.
Sabaut, J. J. & P. Luquet. 1973. Nutritional requirements of the gilt-head bream *Crysophys aurata*. Quantitative protein requirements. Mar. Biol. 18: 50–54.
Schoonbee, H. J. 1969. Notes on the food habits of fish in Lake Barberspan, Western Transvaal, South Africa. Verh. int. Verein. Limnol. 17: 689–701.
Shaw, E. 1978. Schooling fishes. Am. Scient. 66: 166–175.
Sidthimunka, A. 1972. The culture of Pla Duk (*Clarias* spp.). Report no. 12. Inland Fisheries Division, Dept. of Fisheries, Bangkok, Thailand.
Smith, J. L. B. 1951. Are there tigerfish at St. Lucia? S. Afr. Angler 5 (10): 6.
Smith, J. L. B. 1959. Gobioid fishes of the families Gobiidae, Periophthalmidae, Trypaucheridae, Taenioididae and Knaemeriidae of the western Indian Ocean. Ichthyol. Bull. Rhodes Univ. (13): 185–226.
Smith, J. L. B. 1965. The Sea Fishes of Southern Africa. Central News Agency, South Africa.
Sokal, R. R. & F. J. Rohlf. 1973. Introduction to biostatistics. W. H. Freeman, San Francisco.
Srivistava, C. M. & A. N. Sinha. 1961. Contribution to the study of barbels of fishes. Vest. csl. zool. Spol. 25: 12–15.
Stuckenberg, B. R. 1969. Effective temperature as an ecological factor in southern Africa. Zool. Afr. 4 (2): 145–197.
Tait, C. C. 1965. Observations on predation by the white pelican, *Pelicanus onocrotalus*, and by silurid fish of the genus *Clarias*, and a note on cormorant predation. Fish. Res. Bull. (Zambia) 3: 31–32.
Thayer, G. W., W. E. Schaaf, J. W. Angelovic & M. W. Lacroix. 1972. Caloric measurements of some estuarine organisms. Fish. Bull. natl. oceanic & atmos. adm. (US) 71: 289–296.
Thomas, J. D. 1966. On the biology of the catfish *Clarias senegalensis*, in a man-made lake in

the Ghanian savanna with particular reference to its feeding habits. J. Zool. 148: 476–514.
Tinley, K. L. 1958. A preliminary report on the ecology of Lake Sibayi, Tongaland. Natal Parks Board Report, unpublished.
Tinley, K. L. 1964. Fishing methods of the Thonga tribe in north-eastern Zululand and southern Mozambique. Lammergeyer, 3 (1): 9–39.
Tinley, K. L. 1976. The ecology of Tongaland. Wildlife Society of Southern Africa, Durban. 140 pp.
van Bruggen, A. C. & C. C. Appleton. 1977. Studies on the ecology and systematics of the terrestrial molluscs of the Lake Sibaya area of Zululand, South Africa. Zool. Verh., Leiden 154: 1–44. 4 pls.
Van der Elst, R. P., S. J. M. Blaber, J. H. Wallace & A. K. Whitfield. 1976. The fish fauna of Lake St Lucia under different salinity regimes. Unpubl. report, 22 pp.
Van der Waal, B. C. W. 1972. 'n Ondersoek na aspekte van die ekologie, teelt, en produksie van *Clarias gariepinus* (Burchell 1822). M.Sc. Thesis. Rand Afrikaans University, Johannesburg.
Van Rensburg, K. J. 1966. Growth of *Tilapia mossambica* Peters in De Hoop Vlei and Zeekoei Vlei. Dept. Nature Conserv. Invest. Report 9. South Africa.
Von Bertalanffy, L. 1957. Quantitative laws in metabolism and growth. Q. Rev. Biol. 32: 217–231.
Welcomme, R. L. 1964. The habits and habitat preferences of the young of Lake Victoria *Tilapia* (Pisces: Cichlidae). Rev. Zool. Bot. afr. 70: 1–28.
Welcomme, R. L. 1967. Observations on the biology of the introduced species of *Tilapia* in Lake Victoria. Rev. Zool. Bot. afr. 76: 249–279.
Welcomme, R. L. 1970. Studies on the effects of abnormally high water levels on the ecology of fish in certain shallow regions of Lake Victoria. J. Zool. 160: 405–436.
Welsch, V. & V. Storch. 1969. Die Feinstruktur der Geschmacksknospen von Welsen (*Clarias batrachus* (L) and *Kryptopterus bicirrhis* (C&V)). Z. Zellforsch. mikrosk. Anat. 100: 552–559.
Whitehead, P. J. 1962. The relationship between *Tilapia nigra* (Günther) and *T. mossambica* Peters in the eastern rivers of Kenya. Proc. zool. Soc. Lond. 138: 605–637.
Whitfield, A. K. & S. J. M. Blaber. 1976. The effects of temperature and salinity on *Tilapia rendalli* Boulenger 1896. J. Fish Biol. 9: 99–104.
Williams, R. 1971. Fish ecology of the Kafue River and flood plain environment. Fish. Res. Bull. (Zambia) 5: 305–330.
Willoughby, N. G. & D. Tweddle, 1976. The ecology of the commercially important species in the Shire Valley Fishery, Southern Malawi. Fisheries Report, Malawi Government Service.
Worthington, E. B. 1933. The fishes of Lake Nyasa (other than Cichlidae). Proc. zool. Soc. Lond. 1 & 2: 285–316.

9 The amphibians, reptiles, birds and mammals of Lake Sibaya

M. N. Bruton

Introduction

Prior to the establishment of a research station at Lake Sibaya in 1968, general collections of vertebrates had been made in eastern Maputaland (*e.g.* Roberts, 1936). The first ecological research was performed in 1958 by K. L. Tinley who surveyed the hippopotamus at Lake Sibaya as well as on the Mkuze and Pongola floodplains and at the Kosi lake system, and made lists of other vertebrates. These surveys, as well as the earlier expeditions of 1947–49, revealed the great diversity and abundance of vertebrates in Maputaland. In 1956, Natal Parks Board personnel began monitoring the abundance of the hippopotamus at Lake Sibaya on a regular basis.

During 1970–76, observations on the non-piscine vertebrates of Lake Sibaya were confined mainly to five important species – a large aquatic lizard, *Varanus niloticus*, the Nile crocodile, *Crocodylus niloticus*, two piscivorous birds (white-breasted cormorant, *Phalacrocorax carbo* and fisheagle *Haliaëtus vocifer*), and the hippopotamus, *Hippopotamus amphibius* (Plates 1–4) Although these species were dominant members of the ecosystem, they were not sufficiently common for us to carry out extensive culling programmes to determine food preferences, reproductive status *etc.* Our observations on food preferences are based on Amphibia and reptiles killed accidentally on roads or in fishtraps, a small sample of cormorants, and hippopotami culled by the conservation authorities. These direct observations, and the findings of others, are sufficient for a general statement on the role of these groups in the ecology of the lake. Further killing, as emphasized by Gans & Pooley (1976) with reference to crocodile research, would not advance our understanding significantly.

Zoogeography

The zoogeography of Maputaland vertebrates is particularly interesting as their affinities, on the whole, lie with faunae to the north. According to Köppens' (1931) climatic classification, the Lake Sibaya area of Maputaland belongs to the extreme southerly end of the Tropical Savanna type of climate which embraces much of the Mozambique coastal plain (Schulze, 1947), and is characterised by the coldest month (July) having a mean temperature above 18°C. The existence of this tropical peninsula on the south-east coast of Africa is due to the low-lying topography of the great Mozambique Plain, which stretches southwards to about 29°S, and the

Plate 1. Juvenile leguaan, *Varanus niloticus*. Photo: M. N. Bruton.

proximity of the warm Agulhas current (Poynton, 1964). Many elements of the tropical fauna and flora extend southwards down this plain to, at least, Lake St Lucia, at 28°S.

To the south and west of the plain the tropical association is progressively replaced by a complex of non-tropical elements, hence Maputaland acts as a replacement-transition zone. This trend has been shown in the Mollusca (van Bruggen, 1967, 1969, van Bruggen & Appleton, 1977); Amphibia (Poynton, 1964, 1967; Poynton & Wager, 1979) Reptilia (Stuckenberg, 1969; Bruton & Haacke, 1975) and Aves (Lawson, 1967). Checklists of all the vertebrates of Maputaland have been compiled by the author in collaboration with various workers, in order to make a clear statement on the zoogeography of this remarkable fauna (Bruton & Cooper, 1979).

Amphibia

Forty-five species and subspecies of frogs have been recorded from Maputaland (Poynton & Wager, 1979) of which at least 22 are found in Lake Sibaya and surrounding swamps (Table 1, Plates 5–7).

Poynton (1964) divided the southern African Amphibia into three zoogeographical groups: (a) tropical (b) transitional (c) Cape. Thirty-four of Maputaland's forty-five species, and all except *Xenopus l. laevis* and *Rana f. fasciata* among the Amphibia of Lake Sibaya, belong to the 'tropical' zoogeographical group *i.e.* they have a substantial part of their range to the

Plate 2. The Nile crocodile, *Crocodylus niloticus*. Photo: C. C. Appleton.

north in an area experiencing a tropical climate. The proportion of tropical species in the Amphibian fauna of Maputaland (76%) is far higher than that for the southern African Amphibian fauna as a whole (24.5%, Poynton & Broadley, 1978). There is a sharp subtraction of this tropical fauna on the coastal plain to the south of Maputaland and less than half the species extend as far south as Durban. The 18°C effective temperature isoline (Stuckenberg, 1969) coincides closely with their southern limit. Nevertheless, this precipitous subtraction is impossible to explain in terms of present climatological conditions such as temperature, which show a continuous gradient down the coastal plain. Rather, Poynton (1964, 1967) has suggested that the present faunal pattern may be due to past thermal conditions.

Plate 3. The white-breasted cormorant, *Phalacrocorax carbo*. Photo: C. Bruton.

Plate 4. The hippopotamus, *Hippopotamus amphibius*. Photo: R. C. Hart.

Table 1. Checklist of the Amphibia of Lake Sibaya and surrounding swamps, and their occurrence in three major habitat types.

	Adjacent swamps	Sheltered bays	Exposed shores
Anura			
Pipidae			
Xenopus laevis laevis (Daudin, 1802)	×	×	
Xenopus muelleri (Peters, 1844)	×		
Bufonidae			
Bufo gutturalis Power, 1927	×	×	×
Bufo maculatus Hallowell 1854	×		
Ranidae			
Pyxicephalus adspersus edulis Peters, 1854	×		
Rana angolensis Bocage, 1866	×	×	×
Rana fasciata fasciata Smith, 1849		×	×
Ptychadena oxyrhynchus (Smith, 1849)	×		
Ptychadena mascareniensis mascareniensis (Dumeril & Bibron, 1841)	×	×	
Ptychadena porosissima (Steindachner, 1867)	×		
Phrynobatrachus natalensis (Smith, 1849)	×		
Chiromantis xerampelina Peters, 1854	×	×	
Leptopelis cinnamomeus (Bocage, 1893)	×	×	
Kassina maculata (Dumeril, 1853)	×		
Kassina senegalensis (Dumeril & Bibron, 1841)	×		
Afrixalus brachycnemis brachycnemis (Boulenger, 1896)	×		
Afrixalus fornasinii fornasinii (Bianconi, 1850)	×		
Hyperolius argus Peters, 1854	×		
Hyperolius tuberilinguis Smith, 1849	×	×	×
Hyperolius pusillus (Cope, 1862)	×	×	
Hyperolius nasutus Günther, 1864.	×		
Hyperolius marmoratus taeniatus Peters, 1854	×	×	×
TOTAL:	21	10	5

Two Amphibian species from the 'Cape' fauna reach Maputaland and both are wide-ranging forms *i.e. Xenopus l. laevis* and *Cacostenum boettgeri.* Only *X. l. laevis,* the most ubiquitous, occurs in Lake Sibaya. The 'transitional' complex is represented in Maputaland, likewise, by wide-ranging species, *Bufo vertebralis fenoulheti, Breviceps adspersus, Tomopterna natalensis, Rana fasciata fasciata* and *Cacosternum nanum nanum* of which only *Rana f. fasciata* has been recorded from Lake Sibaya.

Several frogs in the catchment area of Lake Sibaya are not associated with permanent water (they are excluded from Table 1). These species inhabit

5

6

7

Plate 5. The painted reed frog *Hyperolius marmoratus*. Photo: R. A. Jubb.
Plate 6. Common platanna, *Xenopus laevis*. Photo: W. R. Branch.
Plate 7. Common rana, *Rana angolensis*. Photo: W. R. Branch.

open grassland and woodland, and shelter in termitaria, underground burrows and under leaf litter where conditions are sufficiently moist for reproduction e.g. *Breviceps adspersus adspersus*, *B. mossambicus*, *Phrynomerus bifasciatus bifasciatus* and *Hemisus marmoratum*. Other frogs are arboreal and occur mainly in the coastal dune forest e.g. *Arthroleptis stenodactylus*.

Those species which occur near permanent water are mainly associated with densely-vegetated swamps and sheltered bays of the main lake (Table 1). Exposed shores have only five common species. When the lake level rises, extensive marginal pools are created which form an ideal habitat for frogs. Many swamp and sheltered bay species inhabit these flooded areas. No frogs occur in open water habitats in Lake Sibaya. *Xenopus l. laevis* and *X. muelleri*, the most aquatic species, are confined to swamps and shallow sheltered bays of the main lake.

Rana angolensis is one of the few vertebrates at Lake Sibaya which breeds during the cool season. Most other frogs (fishes, reptiles, birds and mammals) breed between August and March. The method of protecting the eggs from predators varies considerably. *Bufo gutturalis* lays a large number of eggs in twin gelatinous tubes. These transparent strings are draped over submerged plants and are a common sight underwater in early summer. *Afrixalus b. brachycnemus* encloses the egg mass in a purse formed when the edges of a leaf are folded over and glued together, whereas *Hyperolius pusillus* lays its eggs in flat masses between two water-lily leaves (*Nymphaea capensis*). *Chiromantis xerampelina* encloses its eggs in a mass of froth which is deposited on vegetation overhanging permanent water.

Xenopus l. laevis is cannibalistic, and also feeds on small fish and invertebrates. The other species prey mainly on aquatic insects, especially Odonata, Hemiptera and Coleoptera, and small Crustacea (Poynton, 1964; Wager, 1965; personal observations).

Frogs, toads and tadpoles, in turn, fall prey to a large number of animals. Field observations at Lake Sibaya have shown that they are eaten by at least four species of fish and 15 species of reptiles (Table 2). Amphibia are not, however, an important item in the diet of the main fish predator in the lake, *Clarias gariepinus*, as only one frog, and no tadpoles, were found in 469 catfish stomachs (with contents). In addition, 37 species of terrestrial and aquatic birds which occur at Lake Sibaya are known to include Amphibia in their diet (McLachlan & Liversidge, 1970). The most numerous of these predaceous birds are the egrets, herons and storks.

Table 2. Fishes and reptiles which are known to eat frogs, toads and tadpoles at Lake Sibaya.

Pisces	
CICHLIDAE	*Pseudocrenilabrus philander* (M. Weber, 1897)
CLARIIDAE	*Clarias gariepinus* (Burchell, 1822) (only one frog in 469 stomachs)
	Clarias theodorae M. Weber, 1897 (tadpoles only)
ANABANTIDAE	*Ctenopoma multispinis* Peters, 1844 (tadpoles only)
Reptilia	
CROCODYLIDAE	*Crocodylus niloticus* Laurenti, 1768
VARANIDAE	*Varanus n. niloticus* (Linnaeus, 1762)
PYTHONIDAE	*Python sebae* (Gmelin, 1789)
COLUBRIDAE	*Lycodonomorphus r. rufulus* (Lichtenstein, 1823)
	Boaedon f. fuliginosus (Boie, 1827)
	Philothamnus i. irregularis (Leach, 1819)
	Philothamnus hoplogaster (Günther, 1863)
	Crotaphopeltis hotamboeia (Laurent, 1768)
	Dispholidus t. typus (A. Smith, 1829)
	Thelotornis c. capensis A. Smith, 1849
	Psammophis phillipsii (Hallowell, 1844)
ELAPIDAE	*Naja haje annulifera* (Linnaeus, 1758)
	Naja melanoleuca Hallowell, 1857
VIPERIDAE	*Causus rhombeatus* (Lichtenstein, 1823)
	Bitis a. arietans (Merrem, 1820)

Amphibia are an important link between aquatic and terrestrial biotopes. Tadpoles, and adults, feed on aquatic invertebrates, and many eventually fall prey to vertebrates, mainly semi-aquatic or terrestrial birds and reptiles. Their chosen habitat, the water:land interface, is a complex one, where food availability fluctuates seasonally and in response to changes in lake level. Their ability to feed on a variety of prey, both aquatic and terrestrial, is thus likely to be important for their survival. Under lake conditions which favour the rapid development of invertebrate populations, such as inundation of lowlands in summer, the Amphibia would play an important role by converting this increased lake productivity into food for higher organisms.

Reptilia

The reptile fauna of Maputaland is diverse and comprises at least 105 species of which 86 (81.9%) have tropical affinities (FitzSimons, 1943, 1962; Bruton & Haacke, 1975). The 'non-tropical' component is made up mainly of rupicolous lizards from arid and montane areas in western Maputaland. Most Maputaland reptiles are terrestrial (58 species) or arboreal (27) but 20 species are closely associated with water. Seven of the latter are marine (one snake, one lizard and five turtles) whereas four terrapins, a lizard, a crocodile and seven snakes live in or near freshwater habitats. Eight reptile species are considered here to be part of the fauna of Lake Sibaya (Table 3, Plates 8–11). Terrapins are common in the Mkuzi, Pongola and Usutu floodplains and rivers, and *Pelusios c. castaneus* occurs in Vasa pan 3 km north of Lake Sibaby. Only one specimen of the latter species, and no other terrapins, have been recorded from Lake Sibaya. Their absence is probably due to the unsuitability of the habitat as these species usually inhabit shallow muddy water.

The seven freshwater or semi-aquatic snake species in Maputaland include *Lycodonomorphus whytii*, known only from Ndumu Game Reserve in Natal (Pooley, 1970) and six species found in Lake Sibaya and elsewhere. *Python sebae* is common around the shores of the lake and in the coastal dune forest. This large species (a dead 5.3 m specimen was found) feeds mainly on warm-blooded prey (FitzSimons, 1962), which it may ambush from a submerged position at the water edge. Frogs, birds and, less commonly, fish, are also eaten. Two pythons were caught in fishtraps at a depth of 2 m off the northern shore of Crocodile Point. Both had gorged themselves on *Sarotherodon mossambicus*, *Pseudocrenilabrus philander* and *Tilapia sparrmanii* in the traps before they drowned. The brown water snake *Lycodonomorphus r. rufulus* and the olive marsh snake *Natriciteres variegata sylvatica* are rarely encountered. Both species feed on frogs, tadpoles and small fish which are expertly caught under water. The green water snakes *Philothamnus i. irregularis* and *P. hoplogaster* are common around the lake and adjacent swamps and feed mainly on frogs and tadpoles. *Naja melanoleuca* occurs along shores fringed with forest, and feeds on toads,

Table 3. The reptile fauna of Lake Sibaya, and their preference for three major habitat types.

	Adjacent swamps	Sheltered bays	Exposed shores
Squamata			
Sauria			
Varanidae			
Varanus niloticus niloticus (Linnaeus, 1762) Water leguaan	×	×	×
Serpentes			
Pythonidae			
Python sebae (Gmelin, 1789) African python	×	×	
Colubridae			
Lycodonomorphus rufulus rufulus (Lichtenstein, 1823) Brown water snake	×	×	
Natriciteres variegata sylvatica Broadley, 1966 olive marsh snake	×	×	
Philothamnus irregularis irregularis (Leach, 1819) northern green water snake	×	×	
Philothamnus hoplogaster (Günther, 1863) Common green water snake	×	×	
Elapidae			
Naja melanoleuca Hallowell, 1857 Forest cobra	×	×	
Crocodylidae			
Crocodylus niloticus Laurenti, 1768 Nile crocodile	×	×	×

frogs and fish (FitzSimons, 1962; Pitman, 1974) which may be caught under water (personal observations).

Besides the leguaan and crocodile mentioned below, another 59 species of reptiles occur within the catchment of Lake Sibaya but they are mainly associated with terrestrial and arboreal biotopes. *Philothamnus s. semivariegatus* is abundant in the coastal dune forest but commonly feeds along the lake shore as well. Certain burrowing forms, especially the limbless skinks *Scelotes arenicola* and *Typhlosaurus a. aurantiacus* are drowned in considerable numbers when sudden increases in lake level flood extensive areas of lowland.

The water leguaan, *Varanus n. niloticus*, is common around the shores of Lake Sibaya. These large lizards swim well with lateral sculling movements of the tail and may capture prey underwater. Their diet at Lake Sibaya

Plate 8. African python, *Python sebae*. Photo: W. R. Branch.

includes the crab *Potamon sidneyi*; insects, especially beetles, termites and ants; molluscs (the large terrestrial pulmonate *Metachatina kraussi* and the aquatic prosobranch *Melanoides tuberculatus*); stranded fish and carrion; *Clarias gariepinus* on a spawning run; *Ctenopoma multispinis*; *Xenopus l. laevis* and *Bufo regularis* when moving across damp ground; snakes, especially *Philothamnus* spp.; and white-breasted cormorant, *Phalacrocorax carbo*, eggs and fledglings. They are cunning and persistent robbers of crocodile nests, and also prey on the hatchlings. Leguaans also occur on the marine coast where they feed on carrion and on the eggs and hatchlings of the loggerhead turtle *Caretta caretta*.

Although no comprehensive research has been undertaken on the Nile crocodile, *Crocodylus niloticus*, in Lake Sibaya, a field project in the summer of 1970/71, continued field observations until 1976, and the research of Mr

Plate 9. Brown water snake, *Lycodonomorphus rufulus*. Photo: W. R. Branch.

Plate 10. Common green water snake, *Philothamnus hoplogaster*. Photo: W. R. Branch.

A. C. Pooley in adjacent areas of Zululand provide a fairly complete natural history of this species. Mr Pooley's experiments and observations were initiated at Mkuzi and Ndumu Game Reserves, and are currently being continued at Lake St Lucia, under the auspices of Natal Parks Board (Pooley, 1962; 1969a,b,c; 1974, 1977).

Crocodiles inhabit the main lake as well as surrounding swamps and Vasa pan. The number of crocodiles in the main lake is not known exactly. Night counts were performed from a slow boat in 1970 and 1973 along parts of the lake shore. The numbers obtained, 35 and 43, for crocodiles over 1 m in length are probably underestimates, whereas a figure of 60 (>1 m) for the whole lake is a reasonable approximation. An average of 17 crocodiles (range 12–21) of all sizes was counted around the shores of the south basin during regular day and night excursions in 1970 and 1971, but no hatchlings were seen. Using Attwell's (1963) correlation between hindfoot length and total length, the largest crocodile seen measured about 4.3 m. Adults usually measured between 2.5 and 3.5 m.

Few crocodiles were seen during the cool months (May to August). Increasing numbers of adults were encountered from September onwards, and the first sign of breeding activity was in November when large individuals were regularly seen along particular areas of shore. According to Modha (1967), the Nile crocodile is territorial and defends breeding territories by means of symbolic aggressive displays. Courtship is accompanied by a splash display and mating takes place in shallow water. Pooley (1977) found that the male and female form a pair bond which lasts for at least 4 months until the young become independent. The female digs a nest 30–50 cm deep with her forelimbs in sandy soil above high water mark in or close to shade. Grass is uprooted with the forelimbs and teeth to form a clearing around the nest. The clutch size varies considerably. Modha (1967) recorded an average of 33 (range 14 to 46, $n = 15$) for a Lake Rudolph population, and Pooley (1969a) an average of 45 (18 to 61, $n = 75$) for Zululand populations. The eggs are oval and measure 50×80 mm in diameter.

In 1958 Tinley observed crocodile nesting sites in the northern and western arms, Etsheni Bay and Mbibi Pans. The former two areas are still used but no nests were found (or reported by local inhabitants) in the latter two areas, probably due to increased human and cattle interference, during the period 1970 to 1976.

The location of 30 nesting sites at Lake Sibaya is shown in Fig. 1. These records are based on observations of nests which had been excavated by leguaans or Africans, or from which the young had emerged. In several instances the continued presence of adult crocodiles in a particular area revealed the approximate sites of nests, which could be confirmed later once hatching had taken place. A large arena of 18 nests (11 in one group with seven more widely spaced) occurred along the eastern shore of a small inlet on the western arm. This shore had a narrow *Scirpus littoralis* fringe along the water edge, and a grassy slope leading to closed woodland about 20 m from the lake. The soil was loose and sandy. Although there were indications of *Varanus* predation in 1970/71, hatching success during the summer was high. In contrast, no nests were successful during the following summer due to disturbance from cattle, and predation on the eggs by *Varanus niloticus* and poachers. The other twelve nests were used in 1970/71 but in 1971/72 ten of these nests had been abandoned due to disturbance by cattle. In 1975 the lake level rose to the highest level in 34 years (Pitman & Hutchison, 1975) and many crocodile nesting sites were flooded. Only three of the original 30 nests were used in the summer of 1975/76. Other nests undoubtedly occur along parts of the lake shore which could not be visited regularly. The remains of egg shells were found around small lakes on the

Plate 11. Hadedah *Hagedashia hagedash* and Nile crocodile *Crocodylus niloticus*. Photo: D. Wolff © Wildlife Society of Southern Africa.

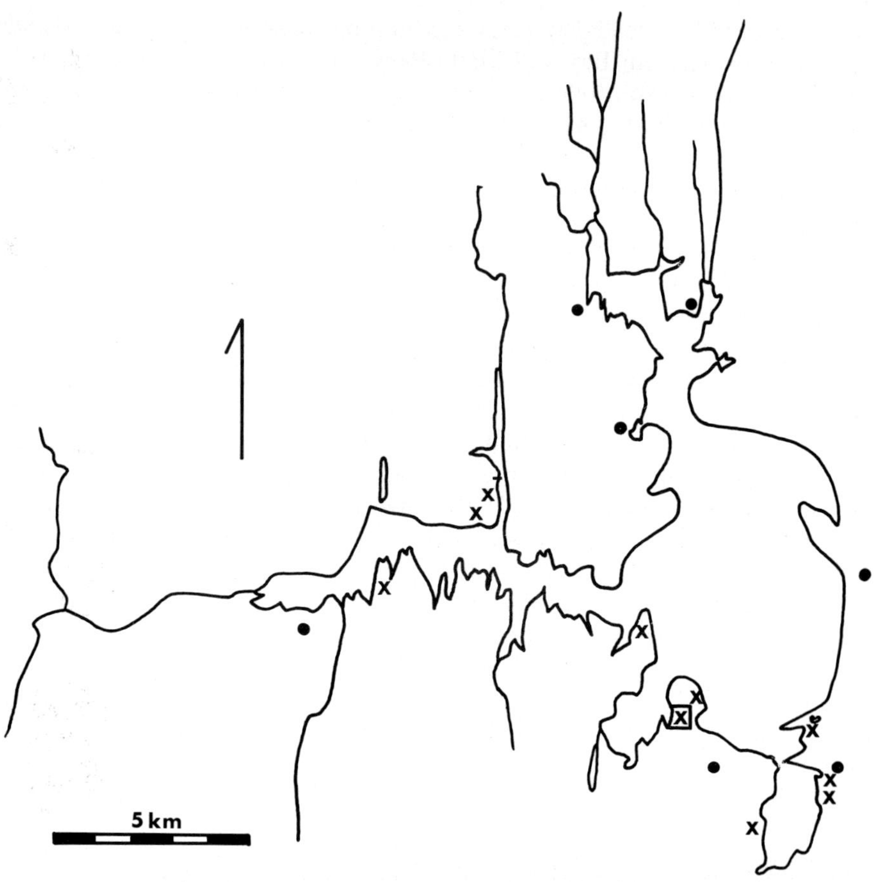

Fig. 1. The distribution of crocodile and fish eagle nests at Lake Sibaya. ☒ Arena of crocodile nests, 1970/71. **X** Individual crocodile nests, 1970/71. ● Site, or approximate site, of fish eagle nests, 1970–1976.

eastern shore of the northern arm as well as on the northern shore of the western arm adjacent to Nyanene forest. The low success rate of the 30 nests under observation, nearly all of which were remote from human habitation, is nevertheless disturbing.

The female crocodile guards the nest throughout the incubation period of 84 to 96 days Pooley, 1969a) although she may enter the lake on hot days, probably for thermoregulation. The main predators of eggs (and hatchlings) are leguaans which dig up unattended nests. The exposed eggs attract other predators such as large birds and the water mongoose *Atilax paludinosus.* Pooley (1969a) found that 22 out of 65 nests examined on the eastern shores of Lake St Lucia had been opened and destroyed by water leguaans. At Ndumu Game Reserve predation rarely exceeded 20% of the nests. Leguaans and water mongooses also prey on crocodile eggs elsewhere in Africa (Stevenson-Hamilton, 1947; Cott, 1961; Pienaar, 1966; Modha,

1967). Other causes of juvenile mortality are excessive rainfall or heat, trampling and parasites (Pooley, 1962).

The hatchlings emit a sharp yelp in response to movement on the ground above the nest when they are ready to hatch. This noise can be heard up to 4 m away. In many instances the young are unable to extricate themselves from the nest, unlike marine turtles, and need the assistance of the parent. Pooley (1974, 1977) demonstrated in a fascinating series of experiments and observations that the female responds to the yelps and digs up the nest with her snout. About 75% of the hatchlings use a caruncle, or egg tooth, to crack open the shell, but the parents may assist hatching by gently rolling the eggs in their mouths and cracking them open. The female carries the young from the nest to water in her buccal pouch. The hatchlings participate actively in this mouth-transport process by yelping and flicking their tails, apparently to attract the female's attention, and by climbing into her mouth. The young stay in close proximity to the female once they have reached the lake edge, and are guarded for up to five weeks. The latter observation was confirmed on a batch of young in a small pan adjacent to Lake Sibaya in 1974. This group consisted of 18 individuals initially but decreased to five after 30 days. Two adult crocodiles were in attendance throughout this period.

Soon after the young crocodiles leave the nursery, cool weather sets in. Pooley (1969b) found that crocodiles dig burrows up to 4 m long in which they hibernate. The burrows are also used for shelter from predators and inclement weather during summer.

Small crocodiles (<1.5 m) feed along the shore and are ecologically separated from larger individuals which mainly feed offshore. Closely observed juveniles in the field and in captivity fed on insects (Odonata, Hemiptera, Coleoptera, Orthoptera), frogs (*Hyperolius*, *Rana* but not *Bufo*), *Potamon sidneyi* and carrion. The food of intermediate size individuals (1.5 to 3 m) was only determined from the examination of the remains of prey on the shore, and consisted almost entirely of *Clarias gariepinus.* A sample of 54 catfish skulls was collected from known crocodile feeding areas and the total length of the catfish prey calculated from a skull length : total length ratio. The modal length of the prey was 520 mm (range 230 to 670 mm), which is slightly smaller than the modal size in the *C. gariepinus* population in the lake (♀560 mm, ♂580 mm). On three occasions crocodiles were seen capturing catfish during a spawning run in shallow water. *Clarias* also formed an important part of the diet of *Crocodylus niloticus* in Mweru Wa Ntipa and the Upper Zambezi and Kafue Rivers in Zambia, whereas *Tilapia, Synodontis, Protopterus* and *Barbus* were the dominant fish prey in other systems (Cott, 1961).

Other crocodile prey collected at Lake Sibaya included the remains of *Sarotherodon mossambicus, Varanus niloticus* and the grey-headed gull *Larus cirrocephalus.* Their prey undoubtedly includes other animals, many of which would be caught and eaten in the water. Cott (1961) has shown that the adult Nile crocodile is a versatile opportunist which thrives on prey as diverse as insects, molluscs, aquatic birds, reptiles and mammals, as well as being a cannibal. Large crocodiles (>3.0 m) feed increasingly on terrestrial vertebrates (reptiles and mammals) and at Lake Sibaya included in their diet two domestic goats, a cow and the carcasses of two young hippopotami.

Crocodiles are able to catch *Clarias gariepinus*, a relatively slow-moving fish, but what of their efficiency as predators on the more elusive cichlids? Crocodiles are well-known for their habit of removing cichlids from gillnets, and readily feed on cichlids in captivity so we can assume that they are a preferred prey. Pooley (1962) found that young crocodiles could not catch live cichlids in large pools, but readily fed on this prey in very shallow water when the prey was easy to catch. Undisturbed crocodiles observed from a 7 m tower on the terrace in Lake Sibaya made no attempt to catch shoaling cichlids, or even nest-guarding male *Sarotherodon mossambicus*. Whether or not they are capable of capturing free-swimming cichlids remains an open question, but one would expect that they would maximise their efficiency by preying on these elusive fish under optimal conditions for predation *e.g.* in shallow water, at night and when the fish shoals have disbanded, in the same manner as *C. gariepinus*.

Aves (Plates 11–22)

The diverse avifauna of Maputaland (472 species, Bruton & Cooper 1979) has, in common with the other vertebrates, strong tropical affinities. Most Maputaland species inhabitat the densely wooded area of Ndumu Game Reserve (402 species) but high numbers are also associated with coastal dune forest and aquatic habitats in the Kosi (285) and Sibaya (273) catchments, as well as the more arid thornveld of Mkuzi Game Reserve (273 species). The Lebombo montane area has at least 288 species.

Of the 273 species at Lake Sibaya, 62 are closely associated with the lake through their breeding, feeding or roosting habits. The distribution of these species in four main habitats at Lake Sibaya is given in Table 4. The abundance of birds correlates directly with the diversity of other taxa, as determined by Allanson *et al.* (1974). Most bird species are found in sheltered bays, which have the richest communities of plants and animals (as well as cover for secretive species), and least in the more depauperate exposed shore and open water habitats. The massive concentrations of shallow-water feeders and waders, such as flamingos and pelicans, which are such a feature of Lake St. Lucia, and to a lesser extent Kosi and Pongola, are not found at Lake Sibaya, although occasional flamingos have been sighted.

An indication of the relative abundance of birds associated with Lake Sibaya was obtained during two counts around the whole lake shore over three days in March 1970 (low lake level) and May 1976 (high lake level). The order of abundance found during these surveys (Table 5) was confirmed during other detailed counts along parts of the lake shore.

The most numerous birds were the reed and white-breasted cormorants. Both species are social roosters and breeders. Reed cormorants typically

Table 4. Checklist and habitat preferences of the birds of Lake Sibaya. A = Sheltered bays, B = Sheltered shores, C = Exposed shores and D = Open water.

	A	B	C	D
Podicipidae				
Podiceps cristatus (Linnaeus, 1758) Great crested grebe				×
Podiceps ruficollis (Pallas, 1764). Cape dabchick	×	×		×
Phalacrocoracidae				
Phalacrocorax carbo (Linnaeus, 1758). White-breasted cormorant	×	×	×	×
Phalacrocorax africanus (Gmelin, 1789). Reed cormorant	×	×	×	×
Anhingidae				
Anhinga rufa (Lacepede & Daudin, 1802). Darter	×	×	×	×
Ardeidae				
Ardea cinerea Linnaeus 1758. Grey heron	×	×		
Ardea melanocephala Vigors & Children, 1826. Black-headed heron		×		
Ardea goliath Cretzschmar, 1826. Goliath heron	×	×		
Ardea purpurea Linnaeus, 1766. Purple heron	×	×	×	
Casmerodius albus (Linnaeus, 1758). Great white egret		×		
Egretta garzetta (Linnaeus, 1766). Little egret	×	×		
Egretta intermedia (Wagler, 1829). Yellow-billed egret	×	×		
Bubulcus ibis (Linnaeus, 1758). Cattle egret	×	×	×	
Ardeola ralloides (Scopoli, 1769). Squacco heron	×			
Butorides striatus (Linnaeus, 1758). Green-backed heron	×	×		
Ardeola rufiventris (Sundevall, 1850). Rufous-bellied heron	×			
Ixobrychus minutus (Linnaeus, 1766). Little bittern	×	×		
Nycticorax nycticorax (Linnaeus, 1758). Night heron	×	×	×	
Scopidae				
Scopus umbretta Gmelin, 1789. Hamerkop		×	×	
Ciconiidae				
Ephippiorhynchus senegalensis (Shaw, 1800.) Saddlebill		×		
Threskiornithidae				
Threskiornis aethiopicus (Latham, 1790). Sacred ibis	×	×		
Hagedashia hagedash (Latham, 1790) Hadedah	×	×	×	×
Plataleidae				
Platalea alba Scopoli, 1786. Spoonbill			×	
Phoenicopteridae				
Phoenicopterus ruber Linnaeus, 1758. Greater flamingo			×	
Anatidae				
Plectropterus gambensis (Linnaeus, 1766) Spurwing goose		×	×	
Alopochen aegyptiacus (Linnaeus, 1766) Egyptian goose	×	×		

Table 4 (*Continued*)

Checklist and habitat preferences of the birds of Lake Sibaya

	A	B	C	D
Nettapus auritus (Boddaert, 1783) Pygmy goose	×			
Anas undulata Dubois, 1837. Yellow-billed duck	×	×	×	
Anas erythrorhyncha Gmelin, 1789. Red-bill teal	×	×		
Dendrocygna viduata (Linnaeus, 1766) White-faced duck	×	×		
Thalassornis leuconotus Eyton, 1838 White-backed duck	×	×		
Aquilidae				
Milvus aegyptius (Gmelin, 1788) Yellow-billed kite	×	×	×	
Haliaëtus vocifer (Daudin, 1800) Fish eagle	×	×	×	×
Pandion haliaëtus (Linnaeus, 1758) Osprey			×	×
Circus ranivorus (Daudin, 1800) African marsh harrier	×	×		
Rallidae				
Limnocorax flavirostris (Swainson, 1837) Black crake	×			
Porphyrio porphyrio (Linnaeus, 1766) Purple gallinule	×			
Gallinula chloropus (Linnaeus, 1758) Moorhen	×			
Fulica cristata Gmelin, 1789. Red-knobbed coot.	×			
Jacanidae				
Actophilornis africanus (Gmelin, 1789) African jacana	×			
Rostratulidae				
Rostratula benghalensis (Linnaeus, 1758) Painted snipe	×			
Charadriidae				
Charadrius marginatus Vieillot, 1818 White-fronted sandplover	×	×		
Charadrius tricollaris Vieillot, 1818 Three-banded sand plover	×			
Hoplopterus armatus (Burchell, 1822) Blacksmith plover	×	×		
Scolopacidae				
Tringa nebularia (Gunnerus, 1767) Greenshank	×	×		
Tringa hypoleucos (Linnaeus, 1758) Common sandpiper	×	×		
Calidris ferruginea (Pontoppidan, 1763) Curlew sandpiper		×		
Recurvirostridae				
Recurvirostra avosetta Linnaeus, 1758. Avocet	×	×		
Himantopus himantopus (Linnaeus, 1758). Stilt		×	×	
Burhinidae				
Burhinus vermiculatus (Cabanis, 1868) Water dikkop		×		

Table 4 (*Continued*)

Checklist and habitat preferences of the birds of Lake Sibaya

	A	B	C	D
Laridae				
Larus cirrocephalus Vieillot, 1818 Grey-headed gull			×	×
Sternidae				
Hydroprogne caspia (Pallas, 1770) Caspian tern			×	×
Chlidonias leucoptera (Temminck, 1815) White-winged lake tern			×	×
Sterna paradisaea Pontoppidan, 1763 Arctic tern			×	×
Bubonidae				
Scotopelia peli Bonaparte, 1850. Fishing owl	×			
Alcedinidae				
Ceryle rudis (Linnaeus, 1758) Pied kingfisher	×	×	×	
Megaceryle maxima (Pallas, 1769) Giant kingfisher	×	×	×	
Corythornis cristata (Pallas, 1764) Malachite kingfisher	×	×	×	
Halcyon albiventris (Scopoli, 1786) Brown-hooded kingfisher	×			
Ploceidae				
Ploceus cucullatus (Müller, 1776) Spotted-backed weaver	×			
Ploceus subaureus A. Smith, 1839 Yellow weaver	×			
Amblyospiza albifrons (Vigors, 1831) Thick-billed weaver	×			
Total:	45	39	23	12

formed large roosting aggregations e.g. on Mandosi peninsula where flocks numbering 20 to 179 birds (mean 99, $n=25$) were counted in 1970 and 1971. White-breasted cormorants usually formed smaller groups (5 to 76, mean 24, $n=25$).

Phalacrocorax carbo nests in trees overhanging the lake whereas *P. africanus* commonly uses *Phragmites* and *Cyperus* beds. In May 1976 76 *P. carbo* nests were in use in flooded *Casuarina equisitifolia* trees on the northern shore of the south basin. A colony of 25 *P. carbo* nests in a large *Ficus* tree at Banda Banda produced three to four (rarely two or five) eggs per nest, but juvenile mortalities were high. An unexpected predator was *Varanus niloticus*, which is an expert climber. One or two (rarely three) young were reared. The adult pair feed the young large quantities of fish which are regurgitated into the fledgling's mouth. There is considerable wastage which, combined with copious faeces, provides food for an assemblage of fishes below the nesting colony. These fishes consist mainly of *Clarias gariepinus* and *Aplocheilichthys katangae*, but *Pseudocrenilabrus philander* and *Tilapia rendalli* are also present. A sample of food was collected from two *P. carbo* fledglings, which readily regurgitate their stomach contents

Table 5. The fifteen most frequently seen birds during two whole-lake counts at Lake Sibaya (small, camouflaged or secretive species could not be counted accurately).

	Low lake level – March 1970	High lake level – May 1976	Total	% of total
1. Reed cormorant	289	51	340	23.2
2. White-breasted cormorant	93	241	334	22.8
3. Yellow-billed egret	31	258	289	19.8
4. Cape dabchick	100	26	126	8.6
5. Pied kingfisher	50	39	89	6.2
6. African jacana	9*	46	55	3.8
7. Purple heron	34	9*	43	2.9
8. Common tern	35	0	35	2.4
9. Hadedah ibis	3	29	32	2.2
10. Black-winged stilt	25	0	25	1.7
11. Malachite kingfisher	12	12	24	1.6
12. Fish eagle	8†	13	21	1.4
13. Black crake	4*	17	21	1.4
14. Pygmy goose	8*	7*	15	1.0
15. Darter	0*	14*	14	1.0
			1463	

* these counts are probably underestimates
† 13 to 17 fish eagles were consistently counted in subsequent years at high lake levels.

when threatened. The sample consisted of 19 *Glossogobius giurus* and nine *Croilia mossambica.*

Gobies also dominated the diet of eight *Phalacrocorax carbo* shot at Lake Sibaya in May & October 1976 (Table 6). The elongate cichlid *Pseudocrenilabrus philander*, constituted about 17% of the small sample. The diets of *Phalacrocorax africanus* and another piscivore, *Anhinga rufa*, were not determined directly, but field observations indicated that they fed more frequently on cichlids than *P. carbo.*

Three kingfishers are commonly associated with the lake and a fourth, *Halcyon albiventris*, is mainly terrestrial but, contrary to current opinion, also feeds on stranded fish. Pied kingfishers are important predators of juvenile cichlids, gobies, *Gilchristella aestuarius* and *Barbus* spp. They may band together and kill fairly large prey – the largest which I personally observed was an 85 g *Sarotherodon mossambicus.* The less common giant kingfisher feeds mainly on cichlids, but also on *Potamon sidneyi.* The diminutive malachite kingfisher takes very small prey (surface feeding fishes, such as *Aplocheilichthys katangae*, and emerging aquatic insects).

The most common inhabitant of sandy beaches is the white-fronted sandplover which regularly breeds at Lake Sibaya. Small diurnal waders include the black-winged stilt, avocet, greenshank and, rarely, spoonbill. Prominent among the large wading birds, which are mainly crepuscular, are the purple heron, great white heron and goliath heron. Squacco, rufous-bellied and night herons are secretive species whereas the hamerkop is often abroad during the day. When the lake level rises and extensive areas of shore are flooded, the composition of the littoral community charges. The

Table 6. Stomach and oesophagus contents of eight *Phalocrocorax carbo* shot at Lake Sibaya in May and October 1976.

		Prey species	No. of prey	Mean prey length (mm)	Prey length range (mm)
Large adult	May 76	*G. giurus*	60	84	57–109 ($n^*=55$)
	May 76	*P. philander*	9	51	37–60 (9)
Large adult	May 76	empty			
Large adult	May 76	*G. giurus*	55	83	65–101 (48)
Large adult	May 76	*G. giurus*	31	84	74–101 (31)
		P. philander	8	55	45–61 (8)
Subadult	Oct 76	*S. mossambicus*	2	—	140,195
Subadult	Oct 76	*S. mossambicus*	1	—	180
		Tilapia spp.	1	—	155
		P. philander	2	(heads only)	—
		G. giurus	3	—	—
Subadult	Oct 76	*P. philander*	2	(heads only)	
Subadult	Oct 76	empty			
Total		*G. giurus*	149(86% of total)	84	57–109 (134)
		P. philander	21	52	37–61 (17)
		S. mossambicus	3	172	140–195 (3)
		Tilapia spp.	1	—	155
			174		

* n = number of prey which could be measured.

Plate 12. Pied kingfisher, *Ceryle rudis*. Photo: P. J. Ginn.

small waders (clean sand feeders) are replaced by large scavengers, such as yellow-billed egrets and hadedahs.

Several terns and gulls are found at Lake Sibaya as well as on the marine coast. Grey-headed gulls and Caspian terns are common at times, and feed mainly on gobiid and juvenile cichlid fishes. White-winged lake tern, common tern and Arctic tern visit the lake sporadically, sometimes in large flocks (Plate 18). Other Sibaya bird species which feed on the marine coast

Plate 13. Giant kingfisher, *Megaceryle maxima*. Photo: P. J. Ginn.

Plate 14. Avocet, *Recurvirostra avosetta*. Photo: P. J. Ginn.

Plate 15. Grey heron, *Ardea cinerea*. Photo: R. A. Jubb.

Plate 16. Black-headed herons, *Ardea melanocephala* and cattle egrets, *Bubulcus ibis*. Photo: D. Wolff © Wildlife Society of Southern Africa.

Plate 17. Little egret, *Egretta garzetta* Photo: P. J. Ginn.

Plate 18. Terns gathered on the shores of Lake Sibaya. Photo: M. N. Bruton.

Plate 19. Darter, *Anhinga rufa*. Photo: P. J. Ginn.

Plate 20. Red-knobbed coot, *Fulica cristata*. Photo: P. J. Ginn.

of shore are flooded, the composition of the littoral community changes. The are fisheagle, pied kingfisher, reed cormorant, darter, white-fronted sand-plover and yellow-billed kite.

The most common anatids are the pygmy goose, Egyptian goose and spurwing goose. The latter two species feed on land, whereas the former feeds on water plants and invertebrates in sheltered bays. The distinctive avifauna of sheltered bays also includes the African jacana, a lily-trotting species which feeds on insects and snails, and several skulking birds (black crake, purple gallinule, moorhen and little bittern).

Three species of weavers (Table 4) construct their nests in colonies on emergent plants along sheltered shores, but feed terrestrially. Two grebes occur in open water habitats: the Cape dabchick is one of the most common birds on the lake, whereas only two pairs of great-crested grebe have been recorded.

The only raptor which is an important piscivore is the fish eagle, *Haliaëtus vocifer*. Yellow-billed kites (Plate 22) are common in summer, and feed mainly on frogs, reptiles and snails. The osprey and fishing owl are rare.

Seven pairs of fish eagles, and occasional solitary birds, were recorded at Lake Sibaya. Their nests were evenly spaced around the lake shore (Fig. 1). Their observed diet at Lake Sibaya was mainly live and stranded *Sarotherodon mossambicus* and *Clarias gariepinus*. Live prey are snatched from the water surface. Scales and skulls were collected from the remains of prey to determine prey size, using scale radius: total length (for the cichlid) and skull-length : total length (for the clariid) ratios.

A sample of 28 *S. mossambicus* had a mean total length of 15 cm (range 12 to 17 cm), whereas the mean size of 32 *C. gariepinus* prey was 50 cm (range 32 to 60 cm).

The cichlid prey correspond to the modal size of adults in the lake population, whereas the catfish were comprised mainly of small to intermediate size adults. In Lake Naivasha fish eagles preyed on tilapia in the range 200–400 g (Brown & Hopcraft, 1973). *C. gariepinus* is an important prey of the fish eagle in many other areas (Brown, 1960; Jubb, 1968; Junor, 1968; Tomkinson, 1975; Whitfield & Heeg, 1977). In Ndumu Game Reserve the mean mass of 11 prey (*S. mossambicus*, *C. gariepinus*, *Hydrocynus vittatus*, *Tilapia rendalli*) caught naturally was 1.41 kg, but artificially supplied *C. gariepinus* weighing 3.65 kg were also taken (Tomkinson, 1975).

Fish eagles hunt during the day, and capture their prey from the upper 1 m of the water column. Their main prey, *C. gariepinus* and adult *S. mossambicus*, rarely enter shallow or surface waters during the day. *C. gariepinus* are obligate air breathers, but their rises to the water surface are so rapid that the fish eagle has little chance to effect a capture. The catfish does, however, feed at the water surface, usually at night, but also during the day. On other occasions they have been seen swimming slowly or lying motionless near the water surface, neither breathing or feeding, apparently basking. They are particularly vulnerable to fish eagle predation at these times. As noted previously, *C. gariepinus* forages in shallow terrace waters at night at low lake level, but feeds in these areas throughout the 24 h cycle when terrace waters are deep.

Plate 21. Fish eagle, *Haliaëtus vocifer*. Photo: P. J. Ginn.

Plate 22. The yellow-billed kite, *Milvus aegypticus*. Photo: C. Bruton.

The conspicuous black nest-guarding males of *S. mossambicus* would appear to be particularly vulnerable to fish eagle predation, but their nests are usually excavated in water deeper than 1 m. The only shallower nests belonged to very small males, or to large males which dug nests under the cover of fallen *Acacia karroo* trees on the lake edge.

The density of fish eagles at Lake Sibaya (one per 8.4 km shoreline) is considerably lower than that reported for lakes in East Africa (1.3 per km, L. Albert, Green, 1964; 1.91 per km, L. Edward, Eltringham, 1975). Their low density at Lake Sibaya reflects the low biomass of fish and the low overall production in the lake. The daily consumption of fish by fish eagles has been estimated at about 1 kg/bird, including wastage (Green, 1964). A population of 15 fish eagles would remove approximately 5500 kg of fish from the lake each year, at this rate of consumption.

The white-breasted cormorant is reported to consume about 400 g of fish per day (Whitfield & Blaber, 1979). Their numbers at Lake Sibaya fluctuated widely but usually exceeded 80 birds. Their consumption of fish (approximately 12,000 kg per annum), combined with that of the reed cormorant and darter, therefore far exceeds that of the fish eagle, and any other bird, at the lake. Furthermore, Cott (1961) has shown on the basis of large samples collected from many different systems in Africa, that the two cormorant species and the darter, *Anhinga rufa*, each consume ten times more fish than the crocodile.

Birds are the dominant semi-aquatic vertebrates at Lake Sibaya, and feed at all levels of the food pyramid on detritus, plants, invertebrates, fishes and littoral vertebrates. Fishes, mainly gobiids, removed by the two cormorant species are probably the major outgoing energy pathway from the lake. In addition, the consumption of aquatic and littoral invertebrates and frogs by the many sheltered habitat species is likely to contribute significantly to energy export.

Mammalia

Although the Mozambique plain in Maputaland was inhabited by a great diversity and abundance of mammals in former times, hunting, game elimination during the 'Nagana' campaign and rapid deterioration of the habitat have decimated the population, and most of the large mammals are now confined to Ndumu and Mkuzi Game Reserves.

Only six mammal species are known to be consistently associated with Lake Sibaya (Table 7): two mongooses, an antelope, the hippopotamus and two rodents. The coastal dune forest is inhabited by at least 25 species, but none can be regarded as part of the lake fauna. The dwarf mongoose *Helogale parvula* and banded mongoose *Mungos mungo* are diurnal feeders in the forest although the former occasionally eat crabs and frogs on the lake shore. In 1958, Tinley made a clear sighting of a clawless otter *Aonyx capensis* at the mouth of the Magembe river in the northern arm. This aquatic

Table 7. The mammals associated with Lake Sibaya, and their preference for three main habitat types.

	Adjacent swamps	Sheltered bays	Exposed shores
Carnivora			
Viverridae			
Ichneumia albicauda G. Cuvier, 1829	×		×
White-tailed mongoose			
Atilax paludinosus G. Cuvier, 1829			×
Water mongoose			
Artiodactyla			
Hippopotamidae			
Hippopotamus amphibius Linnaeus, 1758	×	×	×
Hippopotamus (Plate 23)			
Bovidae			
Redunca arundinum (Boddaert, 1785)	(open grassland)		
Reedbuck			
Rodenta			
Otomyidae			
Otomys irroratus (Brants, 1827)	×	×	
Vlei otomys Muridae			
Dasymys incomtus (Sundevall, 1846)	×	×	
African marsh rat			

mammal, which feeds on crabs, fish, frogs and rodents (Roberts, 1951; Smithers, 1971) has not been sighted or reported at Sibaya again although they do occur further south at Lake St Lucia (Kröger & Forrest, 1978). They are characteristically inquisitive and friendly and usually reveal their presence, but these traits may also lead to their extermination.

The white-tailed mongoose is a solitary, nocturnal predator of small animals, especially toads, crabs, reptiles and insects (Smithers, 1971). There are only two recent records of this species from Lake Sibaya (Bruton, 1978). Water mongooses occur widely along undisturbed shores where they feed on *Potamon sidneyi* and other shore animals. Reedbuck occur in open grassland and swamp areas on the southern shores of the western arm. Their preference for grassland close to permanent water is probably related to the availability of new grass shoots in areas subject to hippopotamus grazing, occasional light burning and changes in lake level. The two rodents are secretive and rarely encountered.

An early report on the hippopotami of Lake Sibaya (Tinley, 1958a) provides a valuable contrast with the author's observations and records of Natal Parks Board in recent years. Nothing was known of the population of

Plate 23. A herd of hippopotami, *Hoppopotamus amphibius*. Photo: D. Wolff © Wildlife Society of Southern Africa.

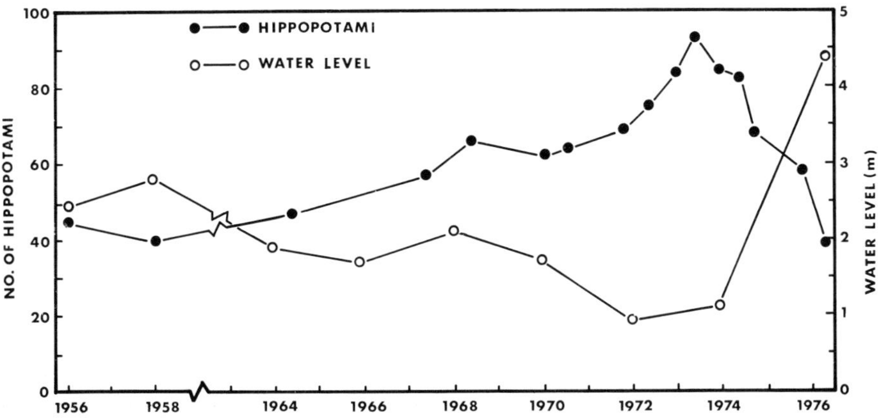

Fig. 2. The number of hippopotami in Lake Sibaya at different lake levels from 1956 to 1976. Hippopotamus counts were obtained from Natal Parks Board records and the author's own observations. Lake level records are from an Ott recorder at Lake Sibaya Research Station, and from the back-calculated data of Pitman & Hutchison (1975) (after Bruton, 1978).

hippopotami in the lake prior to 1940. Intensive hunting took place in Lake St Lucia to the south in the nineteenth and early twentieth centuries and there can be little doubt that the Sibaya population would have been reduced at the same time.

The extent of grassland areas in eastern Maputaland has increased at the expense of coastal forest and closed woodland (Acocks, 1975) due to incessant burning and tree-felling by local inhabitants. The carrying capacity of the Lake Sibaya catchment for hippopotami, in terms of grazing, may therefore have increased due to man's disturbance. Several African villages which were established close to the lake edge prior to 1940 have since been abandoned due to the increase in hippopotamus numbers.

Information gathered by Tinley from local Africans indicated that a regular increase in hippopotamus numbers has taken place since about 1940. This increase was due not only to natural breeding but also to migrations from the Kosi and Pongola systems to Lake Sibaya. At present no hippopotamus movement takes place between Sibaya and Pongola, but infrequent migrations to Kosi, Mosi swamp and Mgobezeleni still occur.

The number of hippopotami in Lake Sibaya during the period for which reliable records are available (1956 to 1976) ranged from 39 to 93 (Fig. 2). This number varied inversely with lake level so that most hippopotami were counted in the main lake at low lake levels. Although some sampling error may be involved – hippopotami are more difficult to count at high lake levels, especially from the air – the more important explanation is that part of the population moves into and inhabits adjacent swamps and lakelets when water levels are high. The reason for this movement is not clear, but the greater depth of inshore terraces in the lake (which reduces the area suitable for lying up) and the larger size of adjacent swamps (which provide a more extensive daytime refuge when their water level is high) may be contributing factors. Similarly, Field (1968) found that hippopotami in the Queen Elizabeth National Park, Uganda, occupied temporary inland wallows during the rains, and returned to permanent water during the dry season.

The maximum number (93) recorded in 1973 represents more than a doubling of the population since 1958 despite the removal of at least 35 dangerous or nuisance animals between 1963 and 1976 for management purposes (Natal Parks Board records, unpublished). The average annual increase of 2.9% between 1958 and 1973 (3.8% between 1964 and 1973) is low compared with figures obtained for the hippopotamus population in the Luangwa River, Zambia (average 5.5%, Ansell, 1965a). Attwell (1963) has noted that the hippopotamus is a resourceful animal which can make a spectacular comeback under protection provided there is sufficient food, space and cover.

In 1958 the largest herd consisted of eight individuals (Tinley, 1958a). The maximum herd size increased to 15 (occasionally 21) during the period 1970 to 1975 when relatively high numbers of hippopotami inhabited the main lake. The largest herds were found in the central western arm and south west basin. The average size of 83 herds counted between 1956 and 1976 was 6.3 (unpublished records of Natal Parks Board, and personal observations). Three to five solitary hippopotami also occurred in the lake.

Hippopotami mainly inhabited the northern and western arm and sheltered bays in Lake Sibaya (Fig. 3). They were rarely encountered in water deeper than 10m. Lying up areas, 0.5 to 2 m deep, were situated in sheltered sandy inshore areas with sparse stands of *Scirpus littoralis* and *Typha latifolia*. One to five hippopotami were present in Vasa pan immediately north of Lake Sibaya from 1970–76, and irregular sightings were also made in the numerous small pans and swamps to the north-east and south-west of the main basin, as well as in Matigane lake.

Hippopotami may breed throughout the year (Mentis, 1972; Skinner *et al.*, 1975) but seasonal peaks from November to July have been demonstrated in the Kruger National Park (Pienaar *et al.*, 1966). The bulls establish a dominance heirarchy in the herd, and perform vigorous aggressive displays (Verheyen, 1954; Physick, 1972). Courtship and mating have been observed in Zululand by Scotcher (1973). The female remains with the herd until an advanced stage of pregnancy has been reached, when she isolates herself in a sheltered bay or swamp. The gestation period lasts about 240 days (Laws & Clough, 1966). One young is usually born, in sheltered shallow water away from the herd. A dense swamp composed of *Phragmites mauritianus* and *Typha latifolia* on the western shore of the south basin (Fig. 3) was used by at least three nursing cows between 1970 and 1976. The cow remained with the calf during the day but fed in adjacent grasslands and fields at night while the calf was hidden in a tangle of sedges and reeds. Cleared lying-up areas in dense forest close to the lake may also be used by nursing females at Lake Sibaya (Tinley, 1958a). Lactation lasts for 10–12 months, and the calving interval is usually two years although conceptions may take place earlier under conditions of excellent nutrition

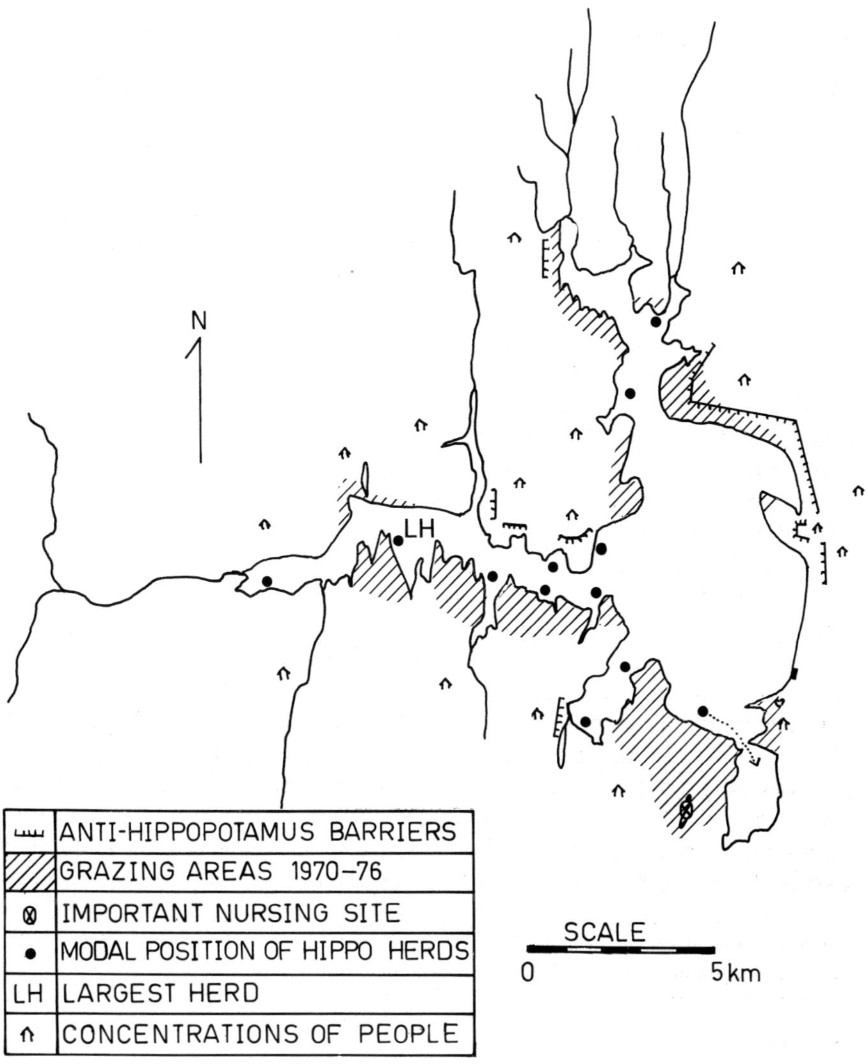

Fig. 3. The distribution of people, anti-hippopotamus barriers and hippopotamus grazing areas, herds and nursing sites at Lake Sibaya from 1970 to 1976.

(Laws & Clough, 1966). Young bulls reach puberty after 7–8 years after which they are evicted from the herd (Skinner *et al.*, 1975).

The hippopotamus is diurnally aquatic and nocturnally terrestrial. Luck & Wright (1964) have shown that there is a considerable loss of water directly through the skin, and as a result they are usually found near water. Water is also used as a buoyant medium and refuge. At night, hippopotami leave the lake to feed, following well-defined paths to the grazing areas. Their feeding movements rarely take them further than 3 km from Lake Sibaya.

The hippopotamus is herbivorous and feeds mainly on grass and sedges

(Verheyen, 1954; Bere, 1959; Ansell, 1965b; Jackson & Garlton, 1966; Pienaar *et al.*, 1966; Field, 1968, 1972; Lock, 1972; Mackie, 1976). Other items in their diet include aquatic plants (reeds, bulbs of *Nymphaea capensis, Pistia stratiotes*) and the fruit, flowers and bark of certain trees. Mr R. H. Taylor (pers. comm., 1978) of Natal Parks Board has found that the following grasses and sedges form an important part of the diet of *H. amphibius* at Lake St Lucia: *Ischaemum arcuatum, Acroceras macrum, Stenotaphrum secundatum, Rhynchospora brownii, Echinochloa pyramidalis, Sporobolus virginicus, Paspalum vaginatum, Nymphaea capensis* (water lily), *Digitaria macroglossa, Setaria sphacelata* and *Themeda triandra.* These are probably the most important food species at Lake Sibaya. At the Kosi lakes and on the Pongola floodplain, hippopotami grazed the grasses *Imperata cylindrica, Urelytrum squarrosum, Ischaemum arcuatum* and *Setaria* spp., as well as bulbs of *Nymphaea capensis* and young shoots of the reed *Phragmites mauritianus* (Tinley, 1958b, 1958c). Other plants which form an important part of the diet of *H. amphibius* elsewhere in Africa, and occur at Lake Sibaya, are given in Table 8.

H. amphibius usually feeds on grass less than 15 cm tall. The leaves or stalk are grasped by the sharp-edged bony lips and pulled off, sometimes causing the plant to be uprooted. They are capable of grazing very close to the ground. The canines and incisors are used for fighting and threat behaviour, and not for grazing. Intensive grazing results in the formation of a mosaic of short-cropped areas or lawns which are interspersed with islands of tall, relatively unpalatable grasses and shrubs. On the lawns, hippopotamus grazing favours the development of creeping stoloniferous plants, and those species which can colonize bare ground and withstand trampling, such as tussocky grasses, and discourages the growth of erect species (Field, 1968, Lock, 1972, in Uganda). Although the population

Table 8. Some plants in the diet of *Hippopotamus amphibius*, which occur at Lake Sibaya (according to Breen & Jones, 1971).

	Reference
Typhaceae	
Typha capensis Rohrb	Pienaar *et al.* (1966)
Potamogetonaceae	
Potamogeton spp.	Personal observations
Gramineae	
Themeda triandra Forsk	Verheyen (1954), Pienaar *et al.* (1966), Field (1972)
Panicum maximum Jacq.	Pienaar *et al.* (1966), Field (1968, 1972), Mackie (1976).
Panicum repens L.	Verheyen (1954), Curry–Lindahl, in Field (1972)
Phragmites mauritianus Kunth	Attwell, 1963
Cyperaceae	
Cyperus spp.	Pienaar *et al.* (1966), Modha, in Field (1972), Mackie, 1976
Araceae	
Pistia stratiotes L.	Verheyen (1954), Field (1972)

density of hippopotami at Lake Sibaya is not high, feeding is concentrated in small areas and grazing mosaics and other heavily-grazed areas are commonly encountered.

The weight of the stomach contents of culled Sibaya hippopotami has not been determined, but in Uganda average values of 34 to 38 kg dry weight were obtained which may in part be the accumulation of two night-feeding periods (Field, 1972). Thurston *et al.* (1968) demonstrated that fermentative digestion takes place in the presence of a diverse ciliate fauna, and Arman & Field (1973) that digestion of roughage is highly efficient. The need for rumination and fine mastication of the food has apparently been replaced by a long retention time. The latter state requires a very large stomach which could only evolve in a large animal which does not rely on speed to escape from predators, and which has constant access to water as a buoyant medium.

The most important grazing areas were on the gradually-sloping southern shores of the main basin and western arm (Fig. 3). There were few African habitations near the lake in these areas whereas in the north and east Africans lived close to the lake, usually on high ridges. Anti-hippopotamus trenches, log barriers, and wire fences have been constructed along some shores (Fig. 3) but they are poorly maintained. Noise-making contraptions – long pieces of wire on which are hung an assortment of metal objects – are also strung along the edges of fields (Tinley, 1958a; personal observations). Stray animals, usually solitary individuals, or nursing females, occasionally enter fields and villages at night to feed on ground nuts and maize, and may damage crops and endanger human lives. Persistent nuisance animals are shot by the conservation authorities. No hippopotami were shot at Lake Sibaya by local Africans between 1970 and 1976, but three well-maintained hippopotamus fall-traps and seven heavy snares were found during this period on the shores of the northern arm. Hippopotami were responsible for at least one human death at Lake Sibaya between 1970 and 1976.

The grazing method of the hippopotamus, the large food requirements of the population and its restriction to an area within about 3 km of permanent water (at Lake Sibaya) make it potentially destructive to grasslands. The term 'overgrazing' implies harm to both the pasture and individual plants as well as to the soil. Vesey-Fitzgerald (1965) has noted however that heavy grazing by a variety of herbivores tends to keep pasture in an early (more palatable) stage of growth, and must not be confused with overgrazing. In this way, hippopotami may improve the grazing for other species *e.g.* reedbuck. However the paucity or absence of other herbivores which complement the feeding habits of the hippopotamus and maintain a 'grazing mozaic' has probably lead to the deterioration of grasslands in the Sibaya area. Controlled burning may in this case remove some of the unpalatable adult grasses and scrub and encourage the regrowth of favoured species, but the natural balance cannot be restored until other herbivores are reintroduced.

As the hippopotamus grazes on land and usually defaecates in the water, there is a drain of nutrients to the lake with no major return pathways (Laws & Clough, 1966; Laws, 1968). Viner (1975) estimated the contribution of

Table 9. The nutrients released by *Hippopotamus amphibius* faeces into Lake George, according to Viner (1975). The approximate contribution of the faeces of 93 *H. amphibius* in Lake Sibaya has been calculated on the basis of Viner's mean values.

	Total nutrients available in dung		Dissolved and suspended nutrients released from dung.	
	mean kg/day/hippopotamus		mean kg/day/hippopotamus	
	Viner (1975)	x93	Viner (1975)	x93
Carbon	1.9	176.7	0.55	51.15
Nitrogen	0.06	5.58	0.3	27.9
Phosphorus	0.172	16.0	0.012	1.12
Iron	0.019	1.77	0.007	0.65
Silica			0.033	3.07
Sodium			0.051	4.74
Potassium			0.041	3.81
Calcium			0.02	1.86
Magnesium			0.016	1.49

minerals by hippopotamus excretion to Lake George (Table 9). Values for total minerals are more relevant to the amount transferred from land as the majority of this material is fibrous and falls to the bottom where it decomposes slowly. The dissolved and suspended fractions are more useful when considering the lake, as they represent the more readily metabolizable portion. As a rough approximation, the contribution of minerals from hippopotamus excreta to Lake Sibaya has been calculated using the average of Viner's (1975) maximum and minimum figures. Based on these calculations, substantial amounts of carbon and nitrogen, and lesser amounts of other minerals, would be transferred into the water by a population of 93 hippopotami (Table 9).

The transfer of nutrients from land to water is probably the most important ecological effect of the hippopotamus in Lake Sibaya, especially as the lake is bedded on low nutrient sands and has no major inflow of nutrients and a low primary productivity (as shown by Allanson & Hart, 1975).

By their movements, hippopotami open channels through dense floating and rooted hydrophytes, as well as through reed and sedge barriers on the water edge and in swamps. These channels provide a passage for small aquatic animals, and interconnect bays and swamps at high lake levels. Hippopotami also stir up bottom sediments and release nutrients into the water (van Ingen, 1950). Their continued presence in shallow terrace areas may prevent fishermen from interfering with breeding tilapia in summer, but they also damage tilapia nests. Hippopotamus carcasses provide a massive source of protein for scavengers, especially crocodiles and catfish. The current hippopotamus culling programme of the KwaZulu Government at Lake Sibaya makes provision for the removal of nuisance animals on condition that the meat is distributed equally to people living around the lake.

Discussion

The various marine transgressions and geological changes which characterised the early history of Maputaland, while still continuing, have allowed a relatively stable system of aquatic and terrestrial habitats to become established on the coastal plain during the past 40–50,000 years (King, pers. comm., 1975). Once this stabilisation had taken place, there were two main sources of vertebrate animals which could colonize the lowlands – the many subtropical/temperate species are non-breeding winter visitors. Besides the tropical species, the transition zone in Maputaland has several evident that far more tropical species have invaded Maputaland than subtropical or temperate forms. To this end the continuity of a tropical climate down the low-lying Mozambique Plain, which effectively connects northern Zululand and the tropics, would have been a vital factor. The relative failure of more local subtropical and temperate faunae to colonize the lowlands may be related to their lesser dispersion abilities compared with tropical forms, and to their lower tolerance of warm climates, but these factors need to be investigated. Among the most mobile of the vertebrates, the birds, many subtropical/temperate species are non-breeding winter visitors. Besides the tropical species, the transition zone in Maputaland has several endemics among the Amphibia (Poynton, 1964), reptiles (Bruton & Haacke, 1975), birds (Lawson, 1967) and mammals (Roberts, 1936), as well as fishes (see chapter on fishes), and molluscs (van Bruggen & Appleton, 1977). Poynton (1967) has suggested that the relatively unstable environmental conditions in a transition zone may result in the formation of isolated populations and hence favour cladogenesis and the development of species with relatively limited ranges compared with the extensive ranges of tropical species.

The non-piscine vertebrates of Lake Sibaya provide a vital link between the lake 'microcosm' and the terrestrial biotope. Many species occupy the transition zone between lake and land, and apparently have well-defined feeding niches. Others, such as *Varanus niloticus*, are eurytopic and euryphagic. Many Amphibia, and the crocodile, have different ecological roles as juveniles and adults.

The complexity of the vertebrates' involvement in the food web of Lake Sibaya is summarised in Fig. 4. The Amphibia, snakes, small crocodiles and some small birds are important secondary consumers which feed on small (mainly invertebrate) herbivores. The leguaan, large crocodiles and large birds are mainly tertiary consumers of these small predators, as well as herbivorous fish, and also act as scavengers. Hippopotami transfer nutrients in the opposite direction, from land to lake.

Acknowledgements

I am grateful to J. C. Poynton, D. G. Broadley, R. H. Taylor, A. C. Pooley, W. D. Haacke and R. C. Hart who made useful comments and criticisms,

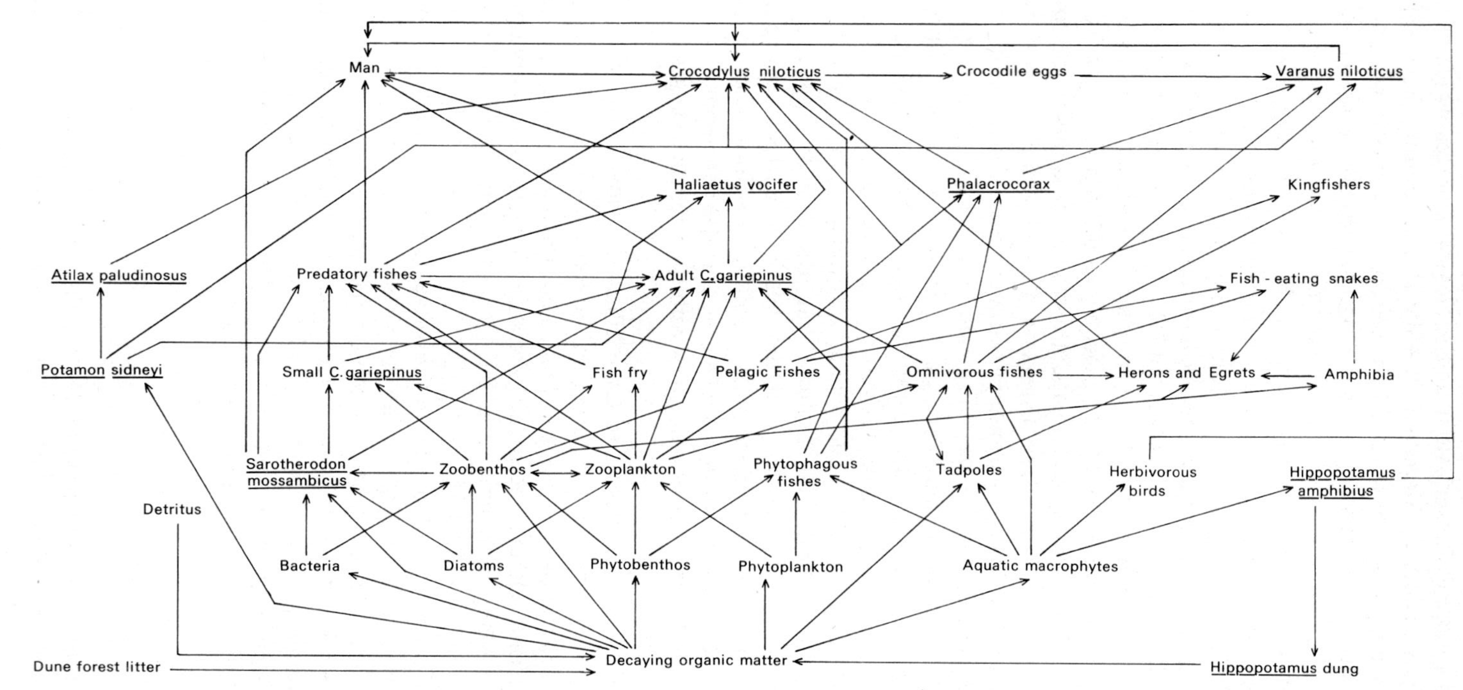

Fig. 4. Food web indicating the main trophic interrelationships of the vertebrates of Lake Sibaya.

and to B. R. Allanson for comments on the manuscript. Field assistance was provided by A. Bruton, S. H. Kröger and C. Bruton. Drafts and the manuscript were typed by C. Bruton and H. Tomlinson. This research would not have been possible without the full co-operation of the Department of Agriculture and Forestry, KwaZulu Government Service, and Natal Parks Board.

References

Acocks, J. P. H. 1975. Veld types of South Africa. Mem. Bot. Surv. S. Afr. No. 40: 1–128.

Allanson, B. R., M. N. Bruton & R. C. Hart. 1974. The plants and animals of Lake Sibaya, KwaZulu, South Africa: A checklist. Rev. Zool. afr. 88: 507–532.

Allanson, B. R. & R. C. Hart. 1975. The primary production of Lake Sibaya, KwaZulu, South Africa. *Verh. Internat. Verein. Limnol.* 19: 1426–1433.

Ansell, W. F. H. 1965a. Hippo census on the Luangwa River 1963–1964. The Puku, No. 3: 15–27.

Ansell, W. F. H. 1965b. Feeding habits of *Hippopotamus amphibius* Linn. The Puku, No. 3: 171.

Arman, P. & Field C. R. 1973. Digestion in the hippopotamus. E. Afr. Wildlife J. 11: 9–17.

Attwell, R. I. G. 1963. Surveying Luangwa hippo. Puku, No. 1: 29–49.

Bere, R. M. 1959. Queen Elizabeth Park, Uganda: the hippopotamus problem and experiment. Oryx 5: 116–124.

Breen, C. M. & I. D. Jones. 1971. A preliminary list of Angiosperms collected in the vicinity of Lake Sibayi. Trans. R. Soc. S. Afr. 39 (3): 235–245.

Brown, L. H. 1960. The African fish eagle *Haliaetus vocifer* especially in the Kavirondo Gulf. Ibis 102: 285–297.

Brown, L. H. & J. B. D. Hopcraft. 1973. Population structure and dynamics in the African fish eagle *Haliaetus vocifer* (Daudin) at Lake Naivasha, Kenya. E. Afr. Wildlife J. 11: 255–269.

Bruggen, A. C. van 1967. Preliminary note on the distribution of the land molluscs of Tongaland (NE Zululand), In: EM van Zinderen Bakker (ed.), Palaeocecology of Africa, 2: 87–90. Cape Town.

Bruggen, A. C. van 1969. Studies on the land molluscs of Zululand with notes on the distribution of land molluscs in southern Africa. Zool. Verh., Leiden, 103: 1–116.

Bruggen, A. C. van & C. C. Appleton. 1977. Studies on the ecology and systematics of the terrestrial molluscs of the Lake Sibaya area of Zululand, South Africa. Zool. Verh., Leiden 154: 1–44.

Bruton, M. N. 1978. Recent mammal records from eastern Tongaland in KwaZulu, with notes on hippopotamus in Lake Sibaya. Lammergeyer 24: 19–27.

Bruton, M. N. & K. H. Cooper. 1979. Studies on the ecology of Maputaland. Rhodes University Press, Grahamstown.

Bruton, M. N. & W. D. Haacke. 1975. New reptile records from the tropical transition zone of south-east Africa. Lammergeyer 22: 23–32.

Cott, H. B. 1961. Scientific results of an enquiry into the ecology and economic status of the Nile crocodile (*Crocodilus niloticus*) in Uganda and northern Rhodesia. Trans. zool. Soc. Lond. 29(4): 211–356.

Eltringham, S. K. 1975. Territory size and distribution in the African fish eagle. J. Zool. 175: 1–13.

Field, C. R. 1968. A comparative study of the food habits of some wild ungulates in the Queen Elizabeth National Park, Uganda. Symp. Zool. Soc. Lond. No. 21: 135–151.

Field, C. R. 1972. The food habits of wild ungulates in Uganda by analyses of stomach contents. E. Afr. Wildlife J. 10: 17–42.

FitzSimons, V. F. M. 1943. The lizards of South Africa. Transvaal Museum Memoir No. 1. Trustees of the Transvaal Museum, Pretoria.

FitzSimons, V. F. M. 1962. Snakes of southern Africa. Purnell. Cape Town.

Gans, C. & A. C. Pooley. 1976. Research on crocodiles? Ecology 57 (5): 839–840.

Green, J. 1964. The numbers and distribution of the African fish eagle *Haliaetus vocifer* on the eastern shores of Lake Albert. Ibis 106: 125–128.

Jackson, G. & J. S. Garlton. 1966. The flora and fauna of Lolui Island, Lake Victoria. J. Ecol. 53: 573–597.

Jubb, R. A. 1968. More about fish eagles. Piscator 72:25–26.

Junor, F. J. R. 1968. The African fish eagle. Brief observation on the weights of fish that can be carried by the fish eagle, *Haliaëtus vocifer.* Piscator 72: 22–24.

Köppen, W. 1931. Grundriss der Klimakunde (ed. 2): i–xii, 1–388. Berlin.

Kröger, S. H. & G. W. Forrest. 1978. First record of clawless otter from Lake St. Lucia. Lammergeyer 26: 64.

Laws, R. M. 1968. Interactions between elephant and hippopotamus populations and their environments. E. Afr. agric. For. J. 33: 140–147.

Laws, R. M. & G. Clough. 1966. Observations on reproduction in the hippopotamus *Hippopotamus amphibius* Linn. In 'Comparative Biology of Reproduction in Mammals' (I. W. Rowlands, ed.), pp. 117–140. Academic Press, London.

Lawson, W. J. 1967. The affinities of the avifauna of Tongaland, In: EM van Zinderen Bakker (ed.), Palaeoecology of Africa, 2: 90–92. Cape Town.

Lock, J. M. 1972. The effects of hippopotamus grazing on grasslands. J. Ecol. 60: 445–467.

Luck, C. P. & P. G. Wright. 1964. Aspects of the anatomy and physiology of the skin of the hippopotamus (*H. amphibius*). Q. Jl exp. Physiol. 49: 11–13.

Mackie, C. 1976. Feeding habits of the hippopotamus on the Lundi River, Rhodesia. Arnoldia Rhod. 7 (34): 1–16.

McLachlan, G. R. & R. Liversidge. Roberts' Birds of South Africa. Cape & Transvaal Printers, Cape Town.

Mentis, M. T. 1972. A review of some life history features of the large herbivores of Africa. Lammergeyer No. 16: 1–89.

Modha, M. L. 1967. The ecology of the Nile crocodile (*Crocodylus niloticus* Laurenti) on Central Island, Lake Rudolph. E. Afr. Wildlife. J. 5: 74–95.

Physick, R. 1972. Hippopotamus behaviour. Lammergeyer No. 15: 78.

Pienaar, U. de V. 1966. The reptiles of the Kruger National Park. National Parks Board of Trustees, Pretoria.

Pienaar, U. de V., P. van Wyk & N. Fairall. 1966. An experimental cropping scheme of hippopotami in the Letaba River of the Kruger National Park. Koedoe 9: 1–33.

Pitman, C. R. S. 1974. A guide to the snakes of Uganda. Wheldon & Wesley, Glasgow.

Pitman, W. V. & I. P. G. Hutchison. 1975. A preliminary hydrological study of Lake Sibaya. Hydro. Res. Unit Rep. No. 4/75: 35pp.

Pooley, A. C. 1962. The Nile Crocodile *Crocodilus niloticus.* Notes on the incubation period and growth rates of juveniles. Lammergeyer 2 (1): 1–55.

Pooley, A. C. 1969a. Preliminary studies on the breeding of the Nile crocodile *Crocodylus niloticus*, in Zululand. Lammergeyer No. 10: 22–44.

Pooley, A. C. 1969b. The burrowing behaviour of crocodiles. Lammergeyer No. 10: 60–63.

Pooley, A. C. 1969c. Some observations on the rearing of crocodiles. Lammergeyer No. 10: 45–57. Addendum: 58–59.

Pooley, A. C. 1970. A preliminary checklist of the reptiles found within the Mkuzi and Ndumu Game Reserves in northern Zululand. Lammergeyer 3 (2): 41–55.

Pooley, A. C. 1974. How does a baby crocodile get to water? Afr. wild life J. 28 (4): 8–11.

Pooley, A. C. 1977. Nest opening response of the Nile crocodile *Crocodylus niloticus.* J. Zool. 182: 17–26.

Poynton, J. C. 1964. The Amphibia of Southern Africa: a faunal study. Ann. Natal Mus. 17: 1–334.

Poynton, J. C. 1967. Zoogeographical problems in northern Natal, In: E. M. van Zinderen Bakker (ed), Palaeoecology of Africa, 2: 85–87. Cape Town.

Poynton, J. C. & D. G. Broadley, 1978. The herpetofauna. In 'Biogeography and ecology of southern Africa' (Ed. M. J. A. Werger) W. Junk, The Hague: 925–948.

Poynton, J. C. & V. A. Wager, 1979. The Amphibia of Maputaland. In: 'Studies on the ecology of Maputaland' (Ed. M. Burton & K. Cooper). Rhodes University Press, Grahamstown.

Roberts, A. 1936. Report on a survey of higher vertebrates of north eastern Zululand. Ann. Transv. Mus. 18 (3): 163–252.

Roberts, A. 1951. The Mammals of South Africa. Central News Agency. Cape Town.

Schulze, B. R. 1947. The climates of South Africa according to the classifications of Köppen and Thornthwaite. S. Afr. geogr. J. 29: 32–42

Scotcher, J. S. B. 1973. Observations on the breeding behaviour of hippopotamus. Lammergeyer No. 19: 35–36.

Skinner, J. D., J. A. Scorer & R. P. Millar. 1975. Observations on the reproductive physiological status of mature herd bulls, batchelor bulls, and young bulls in the hippopotamus *Hippopotamus amphibius amphibius* Linnaeus. Gen. comp. endocrinol. 26: 92–95.

Smithers, R. H. N. 1971. The mammals of Botswana. Trustees of the National Museums of Rhodesia, Museum Memior No. 4: 340 pp.

Stevenson-Hamilton, J. 1947. Wild life in South Africa. Cassell & Co. Ltd.

Stuckenberg, B. R. 1969. Effective temperature as an ecological factor in southern Africa. Zool. Afr. 4 (2): 145–197.

Thurston, J. P., C. Noirot-Timothee & P. Arman. 1968. Fermentative digestion in the stomach of *Hippopotamus amphibius* (Artiodactyla: Suiformes) and associated ciliate protozoa. Nature, Lond. 218: 882–883.

Tinley, K. L. 1958a. A preliminary report on the ecology of Lake Sibayi, Tongaland. Unpublished report to Natal Parks Board, Pietermaritzburg.

Tinley, K. L. 1958b. A preliminary report on the ecology of the Pongola and Mkuze Flood Plains, Tongaland. Unpublished report to Natal Parks Board, Pietermaritzburg.

Tinley, K. L. 1958c. A preliminary report on the ecology of the Kosi Lake System, Tongaland. Unpublished report to Natal Parks Board, Pietermaritzburg.

Tomkinson, A. J. 1975. Notes on the mass-carrying ability of the African fish eagle. Lammergeyer 22: 19–22.

Van Ingen, T. C. 1950. Report by the fish culturist. Rep. Game Dep. Uganda 1948: 46–47.

Verheyen, R. 1954. Monographie èthologique de l'hippopotame (*Hippopotamus amphibius* Linne). Explor. Parc natn. Albert, Brussels, pp. 5–90.

Vesey-Fitzgerald, D. F. 1965. The utilization of natural pastures by wild animals in the Rukwa Valley, Tanganyika. E. Afr. Wildlife J. 3: 38–48.

Viner, A. B. 1975. The supply of minerals to tropical rivers and lakes (Uganda) pp. 227–261. In 'Coupling of Land and Water Systems' (Ed. A. D. Hasler). Ecological Studies 10: 1–309. Springer-Verlag, Berlin.

Wager, V. A. 1965. The frogs of South Africa. Purnell. Cape Town.

Whitfield, A. K. & S. J. M. Blaber, 1979. Feeding ecology of piscivorous birds at Lake St. Lucia, Part 3: swimming birds. Ostrich 50: 10–20.

Whitfield, A. B. & J. Heeg. 1977. On the life cycles of the cestode *Ptychobothrium belones* and nemotodes of the genus *Contracaecum* from Lake St. Lucia, Zululand. S. Afr. J. Sci. 73: 121–122.

10 The utilization and conservation of Lake Sibaya

M. N. Bruton

Introduction

The remote situation of Lake Sibaya in an area proclaimed for rural development since 1936 has limited the extent to which the lake has been utilized on an intensive basis. As a result the lake and its catchment are at present largely unspoilt. The rapidly expanding rural population of Zululand, and the increasing need for food and water supplies and recreational areas of the emerging KwaZulu nation, could, however, soon change this situation. The pattern of utilization of Lake Sibaya is therefore likely to undergo a rapid transition from subsistence use by a small population of rural people to sophisticated manipulation of the ecosystem for the benefit of large resident and visiting groups.

Past Utilization

The amaThonga who inhabit the vicinity of Lake Sibaya (Plates 1 and 2) differ linguistically and culturally from the other African people of south-east Africa. The tribe is divided into several groups of which the vaka-Tembe, Matshabane, Mabaso and Zikhali clans inhabit the Lake Sibaya area (Thorrington-Smith *et al.*, 1978). The amaThonga established contact with the outside world through the anchorage at Delagoa Bay far earlier than most Bantu groups. According to Denoon (1972) they had been trading with the Arabs and later the Portuguese since the sixteenth century. The coastal belt of southern Mozambique and northern Zululand where they settled was, however, low-lying and unhealthy, and not at all suited to stock-farming so the amaThonga explored a variety of other ways of earning a living. They hunted and snared wild game, made extensive use of indigenous fruits and vegetables, and fished extensively in the coastal lagoons, lakes and rivers, which is unusual among the southern Bantu. They also practised shifting agriculture but with limited success due to the infertility and salinity of the marine sands of the Mozambique plain.

Present Utilization

The Fishes

Due to their penchant for fishing, and the need to live near water, the amaThonga are mainly concentrated near lakes, rivers and the sea. Their fishing methods differ according to the nature of the adjacent water body

Plate 1. An amaThonga family at Lake Sibaya. Photo: C. Bruton.

(Junod, 1962; Tinley, 1964). In the rich alluvial floodplain pans of the Pongola river, 'fonya' thrust baskets (illustrated in Chapter 12). 'umono' valve baskets, gillnets and handlines are used. In 'fonya drives' large groups of people form a line across a pan and drive the fishes into the shallows where they are caught by thrusting the basket into shallow water. The main species caught are *Sarotherodon mossambicus, Tilapia rendalli* and *Clarias gariepinus.* On the Pongola floodplain the whole life of the people is intimately related to the fluctuating water levels in the flood-plain pans and dependent cycles of crop cultivation and fish harvesting (Tinley, 1964; Coke & Pott, 1970).

Guide fences and palisade traps are used extensively in the lagoon at Kosi to sample migrating estuarine fishes. The main species caught are *Gerres rappi, Pomadasys commersoni* and *Acanthopagrus berda* (Blaber, 1978).

Due to their high productivity and/or connection to the sea, the Pongola and Kosi systems have diverse fish populations (50 and 125 species respectively) compared with that of Lake Sibaya (18 species). Furthermore, the greater depth and clearer water of Lake Sibaya necessitate different methods of fish capture. Fishing effort at Lake Sibaya is divided into four categories:

a) all-year fishing, in deep and shallow water from the shore, using handlines, rod and line (Plate 3) and 'umono' valve baskets. The main fishes

Plate 2. An amaThonga 'nduna' or headman at Lake Sibaya. Photo: C. Bruton.

caught in deeper water are *Clarias gariepinus*, *Sarotherodon mossambicus* and *Tilapia rendalli*. Cichlids and clariids also dominated the catches in shallow inshore areas (Table 1). The catfish are normally fried or dried, whereas the smaller fishes are incorporated into a stew.

b) Summer fishing for juveniles and small species which migrate into the warm shallow water during the day. 'Umono' valve baskets are used either with or without reed barricades and trenches to guide the fishes into the trap

Table 1. Accumulated catches of 33 'umono' valve baskets set by local Africans in marginal vegetation on the southern shore of Crocodile Point, Lake Sibaya, at various times between 1973 and 1975.

	No. of fish	% of total
Cichlidae		
Pseudocrenilabrus philander	850	39.5
Tilapia sparrmanii	580	27.0
Sarotherodon mossambicus	362	16.8
Tilapia rendalli	142	6.6
Clariidae		
Clarias gariepinus	49	2.3
Other (mainly small cyprinodontids)	167	7.8
	2150	

Plate 3. Home-made rods are used to catch *Clarias gariepinus*, *Sarotherodon mossambicus* and *Tilapia rendalli* throughout the year at Lake Sibaya. Photo: A. Bruton.

(Plates 4 & 5). Crude handnets made of matted reeds and palm leaves are also used to herd fishes into shallow water and onto the shore. At low lake levels the main species caught is *Sarotherodon mossambicus*, but at high lake level other cichlids as well as *Clarias gariepinus*, *C. theodorae*, *Glossogobius giurus*, *Barbus paludinosus* and *Aplocheilichthys katangae* may be caught in large numbers.

c) Summer fishing for breeding adult fishes in shallow water. In September and October adult male *Sarotherodon mossambicus* move into terrace waters in preparation for nest-construction. At this time they readily take bait and large catches are made by fishermen using handlines. On 23 November 1973 three young Africans were observed to catch 105 adult *S. mossambicus* in one day, but catches of 10–15 fish per fisherman/day are more common (Plate 6). Few of the local people have the inclination to fish intensively. Once the male *S. mossambicus* have established nests they take bait less readily and the fishermen resort to spears which are thrown from above the surface but with little success. No use is made of underwater masks or spears at Lake Sibaya, although this equipment is used sparingly at Kosi estuary by Africans. Mouth-brooding female *S. mossambicus* are clubbed and speared at Lake Sibaya when they swim close inshore to release the advanced fry. In general, the tilapia fishery in the lake is, however, poorly developed.

Spawning *C. gariepinus* are speared or clubbed during their nocturnal inshore spawning migrations in summer. The yield on five nights along the

Plate 4. Unbaited 'umono' valve baskets are set in channels in flooded marginal areas to catch juveniles and small species of fish. Photo: M. Bruton.

Plate 5. A young boy inspects an 'umono' fish trap made out of *Scirpus* stems. Cyprinids, cyprinodontids and small cichlids are caught in these traps. Photo: C. Bruton.

Plate 6: Adult male *Sarotherodon mossambicus* caught at their nests by youths using rod and line at the beginning of the breeding season (September 1970). Photo: M. Bruton.

east shore of the main basin ranged from 29 (by one fisherman) to 163 (by a group). The majority of the catfish caught are dried (Plate 7) for later consumption. Approximately 550 kg of *C. gariepinus* is caught in Lake Sibaya each year during spawning runs.

Smaller numbers of adult breeding *Tilapia rendalli*, *T. sparrmanii* and *Pseudocrenilabrus philander* are also caught in summer. *Ctenopoma multispinis*, which migrates overland on a spawning run after heavy rain, are occasionally collected in large numbers, but they are primarily used as bait to catch marine fishes.

d) African fishermen in the vicinity of Lake Sibaya make large catches of fishes in the adjacent Indian Ocean, using rod and line. The main species caught are *Caranx sexfasciatus*, *C. williamsi*, *Argyrosomus hololepidotus*, *Lutjanus rivulatus*, *Diplodus sargus*, *Rhabdosargus sarba*, *Tylosurus leiurus* and *Trachinotus russellii*. The combined catches of four adult men in May and June 1970 consisted of 405 fish of which 368 (91%) were caught in the sea and the rest in Lake Sibaya. There is a trend for fishermen to fish in the lake mainly in summer and in the sea mainly in winter. In addition, African women collect several marine intertidal invertebrates for food, including the mussel *Perna perna*, red bait *Pyura stolonifera*, nudibranch *Hexabranchus marginatus*, and various species of limpets, sea urchins and prosobranch molluscs (Jackson, 1976; personal observations, Plate 8). No freshwater

Plate 7. The tails of *Clarias gariepinus* caught on a spawning run are sun-dried for later consumption. Photo: C. Bruton.

Plate 8. An amaThonga woman collecting *Pyura stolonifera* from the marine intertidal zone at Mbibi adjacent to Lake Sibaya. Photo: C. Bruton.

invertebrates were recorded in their diet at Lake Sibaya, but some terrestrial insects (coleopteran larvae and locusts) are occasionally eaten.

No boats are used by the African people on Lake Sibaya other than crude rafts to cross narrow inlets. Fear of crocodiles and hippopotami, and the great depth and heavy wave action of the lake, are given as the main reasons. This is in contrast to the situation at Kosi and on the Pongola floodplain where homemade punts are regularly used. A steam-powered paddleboat was used on Lake Sibaya in 1945 by Mseleni Mission Hospital staff (Theron, 1975).

Other Aquatic Vertebrates

Although hippopotamus fall-traps (Plate 9) and baited snare-traps for crocodiles (Plate 10) have been found at Lake Sibaya (see Chapter 9), there is no regular cropping of these animals by local inhabitants. Water leguaan *Varanus niloticus* and small birds are also snared occasionally but snaring is less common than in the Pongola floodplain and Kosi areas. The use of fire-arms is also uncommon at Sibaya. Hippopotami culled by the conservation authorities (Plates 11 & 12) provide a welcome addition of protein to the diet of people around the lake. Thirty-five hippopotami were culled at Lake Sibaya between 1963 and 1976.

Plate 9. An hippopotamus fall-trap excavated in a hippopotamus path adjacent to Lake Sibaya. The covering foliage had been burnt off by a grass fire. Sharpened vertical stakes are embedded in the floor of the trap. Photo: A. Bruton.

Plate 10. A crocodile trap consists of a strong enclosure made of sticks, a cable snare attached to a heavy log and positioned in the entrance on the left, and bait (dead puppies, kittens, tortoises and other animals). Photo: M. Bruton.

Water Plants

Bulbs of the water lilies *Nymphaea capensis* and *Nymphoides indica* are commonly collected from the lake and eaten. In addition the reed *Phragmites mauritianus* and the sedge *Typha latifolia* are used as building materials and for making fishtraps and guide fences.

A limit of 5 km was adopted to estimate the current human contact with Lake Sibaya and associated lakelets. Within this distance there were 631 huts and 1451 people but only 169 huts (389 people) are considered to be dependent on these waterbodies for everyday water supplies; the rest live closer to streams or small ponds but may use the main lake during the dry season or in dry years (Appleton & Bruton, 1979). In addition, a considerable number of people use water piped from the western arm of the lake to Mseleni Mission Hospital, but the volume used (an average of 1.27 megalitres per week in 1975, with a projected annual usage of 75 megalitres per annum in 1976) is negligible in proportion to the maintained volume of the lake (about 0.9 km^3; Hill, 1969).

From an economic point of view, the main asset of Lake Sibaya is that it forms a large unpolluted water supply in a country which is generally arid. The lake water is fresh, well-oxygenated and clean, with elevated chloride levels (Allanson & van Wyk, 1969). Samples taken from African watering sites revealed that the water is suitable for drinking without treatment as indicated by the low chemical oxygen demand (12.6 to 28.0 mg/l, $n = 35$), and low levels of nitrate (N), phosphorus, phosphate (P), and acceptable levels of total dissolved solids and sulphates (Lubbe *et al*, 1973).

Plate 11. A culled hippopotamus is removed from Lake Sibaya near the research station. Photo: M. Bruton.

Plate 12. Hippopotami culled by the conservation authorities were examined, measured, sampled for various internal organs and then given to the amaThonga for use as food. Photo: C. Bruton.

In 1958 the South African Department of Forestry embarked on an extensive afforestation programme in eastern Maputaland (Marwick, 1973). Today, pine and eucalyptus plantations are established to the north and south of Lake Sibaya (Plate 13) and the harvest is being utilized in a variety of ways (Plate 14).

The total catchment area of the lake is estimated at 530 km^2, of which approximately 65 km^2 is taken up by the lake itself and 70 km^2 is covered by exotic plantations, mainly in the north-east (Fig. 1). Dense concentrations of trees undoubtedly have high water requirements (but also shade the soil) and they may affect the hydrological balance in shallow aquifer systems, as claimed by Tinley (1971a) for Lake St Lucia. This possibility also exists at Lake Sibaya, but present indications are that longterm rainfall patterns are a more important determinant of lake level fluctuations in this endorheic system (Pitman & Hutchison 1975; Allanson, this volume).

The forest stations at Mbazwane and Manzengwenya obtain their water supplies from the Mbazwane river and a well near Vasa pan respectively, both of which are outside the catchment of Lake Sibaya.

Attempts to establish pulp mills at Lake Sibaya have been forestalled due to the pollution risk of pulp effluent. Another suggestion to utilize water from Lake Sibaya as an additional freshwater source to 'flush out' Lake St Lucia during saline phases (Tinley, 1971a) has not received any support. Any draw-down of Sibaya water would severely affect the littoral zone, the

Plate 13. Pine plantations at Mbazwane adjacent to Lake Sibaya. Photo: M. Bruton.

Plate 14. Pine logs being sorted at Mbazwane prior to transfer to pulp mills in southern Zululand. Photo: M. Bruton.

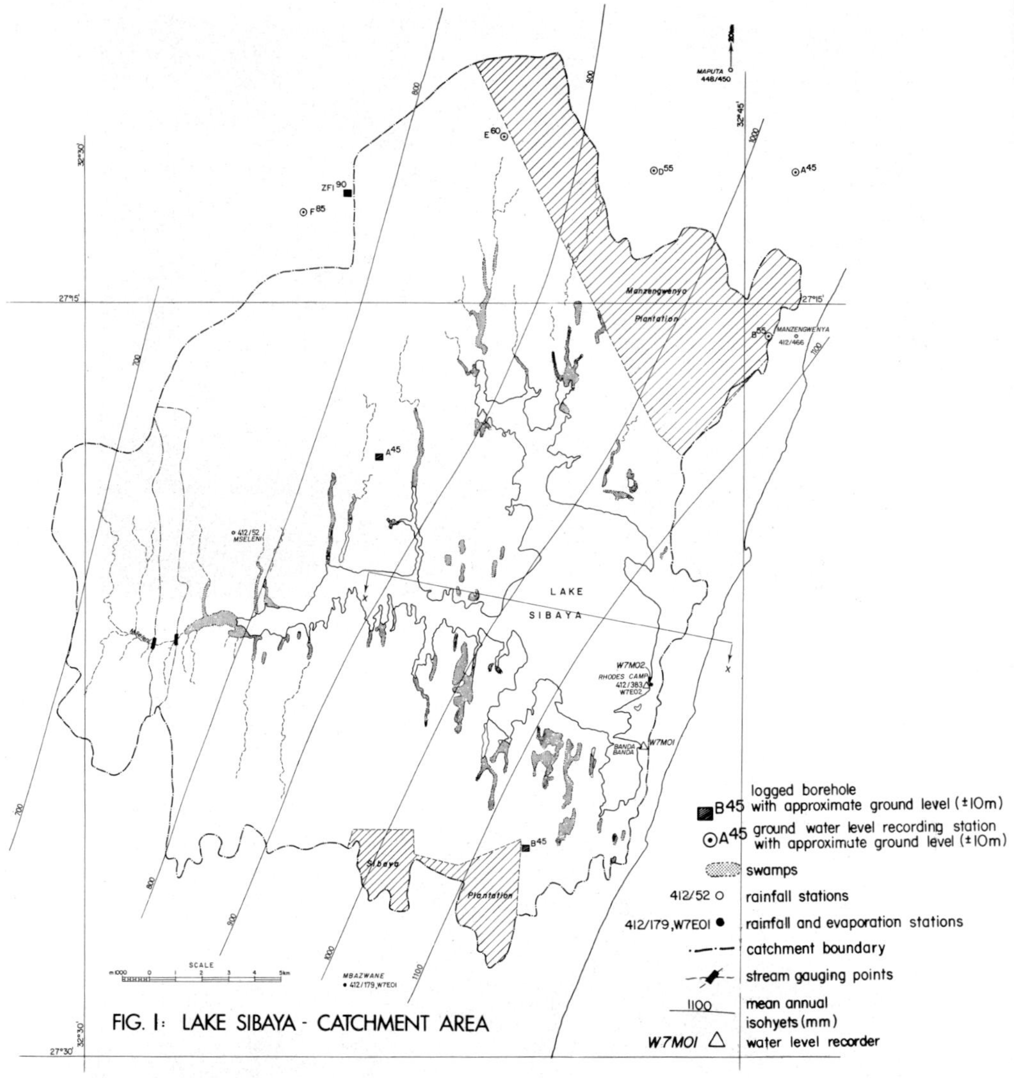

Fig. 1. The catchment of Lake Sibaya showing the extent of exotic tree plantations in the north-east and south (from Pitman & Hutchison, 1975).

richest part of the lake, and draw-down to sea level would reduce the surface area by 84% to only 15 km^2 (Allanson, 1975). In this connection it is important to realize that in any evaluation of the effect of future abstraction, the lake and its catchment must be treated as an integrated aquifer system, sensitive to long term changes in regional rainfall, and connected to the ocean by seepage under the dunes.

The only irrigation practised by the amaThonga at Lake Sibaya is a crude form of inundation of potato fields. The sandy soils of Maputaland are underlain by impervious Cretaceous strata of marine origin which are potentially saline under irrigation. These soils are therefore unsuitable for intensive irrigation and are best used for subsistence agriculture with irrigation aid only during the summer growing season.

Soils within the Catchment

More fertile soils occur in association with swamp forest, coastal forest and coastal dune forest than in open, more arid areas of Maputaland. Fields are made inside the forests and maize, potatoes, sweet potatoes, groundnuts, millet, kaffircorn, casava, beans, bananas, pumpkins, calabashes and paw-paws are grown (Lubbe *et al.*, 1973; Tinley, 1976). The amaThonga also make extensive use of wild fruits and vegetables, but this resource is dwindling due to the reduction of forests, as a result of burning, tree–felling for timber and firewood, and the establishment of fields (Plate 15 & 16). Twenty-eight different kinds of green leaves, as well as mushrooms and 44 kinds of wild fruit have been recorded in their diet (Lubbe *et al.*, 1973). During times of food scarcity many people, especially those in inland areas e.g. Mseleni, rely almost entirely on these natural products as a source of food.

Intoxicating beverages are made from *Sclerocarya caffra*, millet, kaffircorn and sugarcane. The ilala palm, *Hyphaene natalensis* is tapped extensively in eastern Maputaland for 'ubusulu' wine (Plate 17), which is very nutritious

Plate 15. Maize fields established in closed *Syzygium cordatum* woodland adjacent to Lake Sibaya. Photo: M. Bruton.

Plate 16. Fields established in place of coastal forest on the northern shores of Lake Sibaya. Photo: M. Bruton.

and may be an important protein supplement, particularly in inland areas, as well as provide essential vitamin B, riboflavine and nicotinic acid (Campbell, 1969; Torres, 1973). Ilala palms are distributed at an average density of about 92 plants per hectare over an area of 960 km^2 in northern Zululand, and their exploitation has become an important economic activity (Moll, 1972). In addition, the date palm *Phoenix reclinata* is tapped for cider.

The amaThonga keep few cattle and goats compared with the Zulus, largely due to the unsuitability of grasslands on the Mozambique plain for intensive grazing, and do not regard their livestock as an important source of food (Junod, 1962; Lubbe *et al.*, 1973). Although meat is favoured by all, cattle are mainly slaughtered for ceremonial purposes, and as a result the intake of meat is restricted to special occasions. Chickens are eaten more often, but eggs and milk are hardly ever consumed. Furthermore, women and children are traditionally forbidden from consuming eggs, milk and meat, but this taboo is not observed very strictly. These dietary habits stress the importance of natural products as a source of protein for the amaThonga and emphasize the need to manage fish and hippopotamus populations as a longterm food resource.

Diseases Associated with the Lake Sibaya Area

The warm wet lowlands of the Mozambique plain were once rife with diseases which seriously hampered settlement and exploration in the area. Today, only malaria and bilharzia present a serious threat, but other localized diseases are still found.

A strange joint disease affects a large proportion of the indigenous African population in the Mseleni area adjacent to Lake Sibaya. Although the aetiology of Mseleni Joint Disease has been studied in detail, a full explanation has not as yet been given (Burger *et al.*, 1973; Fellingham *et al.*, 1973; Lockitch, Fellingham & Elphinstone, 1973; Lockitch, Fellingham, Witmann *et al.*, 1973, Lubbe *et al.*, 1973). The disease affects most joints, but primarily the hips, and in serious cases the patient's ability to walk is curtailed. The prevalance rate is about 2.4% in children under 10, 11% in all males and 39% in all females. Evidence is presented which indicates that hereditary factors may be involved but the investigators also noted that those groups most affected by the disease (women and children) were largely deprived of protein foods due to poor availability and traditional taboos. The Mseleni people eat less fish than populations to the east between Lake

Plate 17. The sap of the ilala palm *Hyphaene natalensis* is tapped and fermented to form a potent brew called 'ubusulu'. Caps made out of *H. natalensis* leaves are used to shelter the cut stem from the sun. Photo: C. Bruton.

Sibaya and the sea, and less 'ubusulu' wine than people to the north. These observations stress the need for protein supplements in the form of fish and other animal products in the diet of rural people adjacent to Lake Sibaya.

Nagana, a type of sleeping sickness in cattle which decimated domestic herds last century, was virtually eliminated this century following Sir David Bruce's discoveries in Maputaland that the trypanosome is conveyed by the bite of infected tsetse flies (MacArthur, 1955). The fly is infected in the first place by feeding on game animals, which harbour the trypanosome in their blood, and serve as a reservoir of infection. This finding was followed by a drastic game eradication campaign in northern Zululand when thousands of antelope were destroyed, and the Game Reserves at Mkuzi and Hluhluwe were threatened. Thereafter trapping and spraying campaigns were undertaken to destroy the flies themselves. According to Mr E. B. Kluge who was in charge of the latter operation, most of the spraying was directed at *Glossina pallidus*, which occurred primarily in Game Reserves to the south of Maputaland, but nuclei of *G. austeni*, which are less troublesome, occurred until 20 years ago at Lake Sibaya as well as at Sodwana and Tete pan on the Pongola floodplain. Few cases of trypanosomiasis in cattle have been diagnosed in recent years in Maputaland.

Malaria is still widespread in the coastal lowlands despite regular spraying programmes in African kraals. Visitors to the area are strongly advised to take prophylactic drugs. The vectors, anopheline mosquitoes, are abundant in the warm months and breed primarily in ponds and swamps. The fishes *Tilapia sparrmanii*, *Aplocheilichthys katangae* and *A. myaposae*, which include mosquito pupae in their diet, probably reduce vector populations in larger water bodies.

The epidemiology of human and bovine schistosomiasis in the Lake Sibaya area is discussed by Appleton & Bruton (1979). Three species of *Schistosoma* occur in Maputaland: *S. haematobium*, the cause of human urinary bilharzia, and *S. mattheei*, the cause of bovine bilharzia, are widespread (Fig. 2) as is their snail intermediate host *Bulinus* (*Physopsis*) *globosus*. The third schistosome, *S. mansoni*, the cause of human intestinal bilharzia, is rare. Its intermediate host, *Biomphalaria pfeifferi*, has a discontinuous distribution, occurring only in deeper open waterbodies, including Lake Sibaya, but not in shallow ponds and swamps. The absence of *S. mansoni* from Maputaland in general (Azevedo *et al.*, 1961) and Lake Sibaya area in particular, despite the presence there of *B. pfeifferi* which are susceptible to the parasite, presents an intriguing situation. Appleton (1977) has demonstrated that the fecundity of *B. pfeifferi* is impaired at high water temperatures, and in shallow water bodies subject to marked elevation in water temperature during the hot season colonisation is unsuccessful. Such work helps us to understand the significance of Wright's (1966) stress upon ecological factors as the fundamental determinants of trematode infections in molluscs. In the Lake Sibaya area there is little human contact with the deep, temperature-stable bays inhabited by *B. pfeifferi*, whereas cattle and humans regularly utilize the shallow ponds where *B.* (*Ph.*) *globusus* and *S. haematobium* are common. According to local villagers,

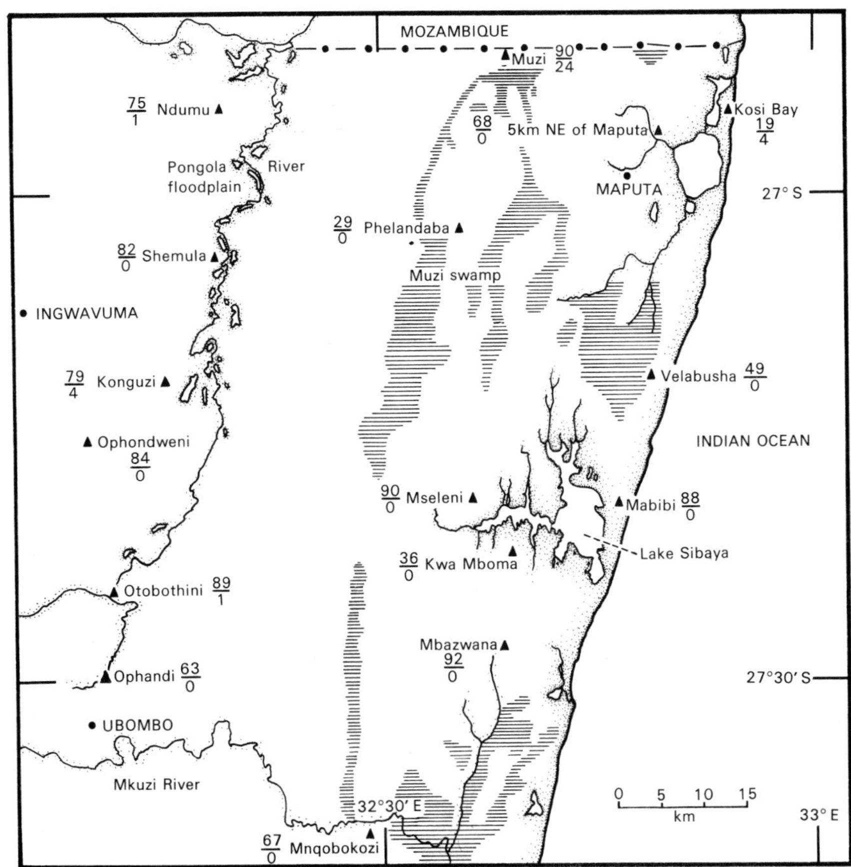

Fig. 2. Map of Maputaland showing the percentage prevalence of *Schistosoma haematobium* (numerators) and *S. mansoni* (denominators) at various localities (after Appleton & Bruton, 1979).

their limited contact with deeper lakelets and bays is due largely to the danger of attack by crocodiles and, to a lesser extent, hippopotami. Some human contact points on the lake are protected from crocodiles by barricades of branches in the water and anti-hippopotamus trenches are a common feature of shores adjacent to African villages. Should the present crocodile and hippopotamus populations be reduced, as is likely to happen if the human population is significantly increased and stricter conservation practises are not introduced, it seems probable that human contact with deep ponds and bays where *B. pfeifferi* does occur, would intensify and that if *S. mansoni* was introduced, it would survive.

The main infection points for *S. haematobium* and *S. mattheei* are shallow ponds and streams, but transmission also takes place in the trenches dug by the amaThonga in association with sweet potato and potato fields, which are flooded in summer (Appleton & Bruton, 1979).

During the past 4 to 5 years Lake Sibaya has entered an inundative phase and the rising water has flooded extensive areas of low-lying grassland. This has resulted in the main sandy-beach human contact points being converted into possible transmission foci of *S. haematobium* as *B.* (*Ph.*) *globosus* is common in the flooded areas. This 'pioneering' by *B.* (*Ph.*) *globosus*. is comparable with that by other bulinid intermediate hosts in some African man-made lakes and which was followed by a sometimes drastic spread of *S. haematobium* transmission (Jordan, 1975; Waddy 1975). Analysis and interpretation of rainfall data for southern Africa by Dyer (1976) and the hydrological analysis provided by Pitman Hutchison (1975) suggest however that the inundative phase in Lake Sibaya may not be permanent. In addition, the flooded grasses will eventually die and be replaced by a sandy beach if the shoreline remains stable.

Proposed Utilization

The Fishes

Of the eighteen species of fish known from Lake Sibaya, twelve are small and have no commercial value (Table 2) although they do support a subsistence fishery when caught in large numbers. Four of the larger fishes (*Marcusenius macrolepidotus, Clarias theodorae, Tilapia rendalli* and *Labeo molybdinus*) are uncommon and constitute a small proportion of the catch.

Table 2. The fishery potential of the fishes of Lake Sibaya.

	Maximum total length in Lake Sibaya (mm)	Maximum weight in Lake Sibaya (g)	Fishery potential
Gilchristella aestuarius	70	<10	4
Marcusenius macrolepidotus	250	250	4
Barbus paludinosus	90	<10	2
Barbus viviparus	80	<10	2
Labeo molybdinus	300	330	4
Clarias theodorae	170	100	4
Clarias gariepinus	1080	9000	1
Aplocheilichthys katangae	50	<10	2
Aplocheilichthys myaposae	50	<10	2
Sarotherodon mossambicus	300	360	1
Tilapia sparrmanii	140	115	2
Tilapia rendalli	180	223	3
Pseudocrenilabrus philander	125	25	2
Ctenopoma multispinis	130	35	3
Glossogobius giurus	140	20	2
Croilia mossambica	100	<10	4
Silhouettea sibayi	50	<10	4
Hepsetia breviceps	80	<10	4

1: High fishery potential
2: Small species, only important in subsistence fishery
3: Uncommon species, only important in subsistence fishery
4: rare in catches, no fishery potential.

Sarotherodon mossambicus and *Clarias gariepinus* are the most common large species, and constitute the only exploitable populations.

S. mossambicus of all sizes are readily eaten by the local Africans and could form the basis of a small fishery (Bruton, 1975). Fishing methods should be used which do not destroy breeding fishes, or interfere unduly with recruitment of young into the adult population. Fishing should therefore be carried out after the end of the breeding season when adult fish are feeding again and in relatively good condition. Furthermore, methods should be chosen which permit fishermen to work in reasonable safety from crocodiles and hippopotami, and which are feasible economically.

S. mossambicus are most abundant in shallow water from September to May. Juveniles feed throughout the year, but adults breed from September to January and feed from January to May. Netting from January to May would give the highest yields of non-breeding fishes in good condition. Netting between May and September would be fruitless as few fishes enter shallow water in the cool season.

In the initial stages of a Sibaya fishery a beach seine net (30 m wide, 5 m long, 3 m deep, 10 mm bar nylon mesh), which can be operated by a few men from the shore without the need for a boat, would be most suitable for catching *S. mossambicus* and other cichlids. Seine netting over nesting areas should be avoided as the net causes extensive damage to nests. Homemade fishtraps are also most effective and no purpose would be achieved by introducing commercially-made traps.

Suitable seining beaches close to human habitations occur on the north-east shore of the main basin, western and northern shore of the south-west basin and northern shore of the western arm.

A pilot *S. mossambicus* seine fishery conducted over three weeks at three netting sites in April and May 1974 following the recommendations listed above (given in detail by Bruton, 1975) gave an average catch of 351 g per pull, and a total catch per day after 6 pulls of only 2.1 kg. The average standard length of *S. mossambicus* caught was 104 mm (weight 29 g). Thus, even though netting was performed under optimal conditions, the catches were very low. Assuming that each net is used on 15 days per month from January to May, the total catch would be 157 kg, per netting site, giving a total of 471 kg for three sites. Local fishermen are unlikely to maintain a higher rate of netting, so the expected yield per netting season would be about 500 kg. A similar seine netting programme carried out from February to May in 1970, 1971 and 1972 gave yields of 480, 540 and 510 kg. When the lake entered an inundative stage from 1974 yields from inshore seine netting increased to over 700 kg in the period January to May due to the increased diversity and abundance of cichlids in inshore areas. Furthermore, the density of juvenile and subadult cichlids was high in early summer and netting could be carried out throughout the warm season without undue interference with breeding by adults. These yields are nevertheless very low compared with the Pongola system where catches exceeding 900 kg in weight can be made during a single 'fonya' drive (Coke & Pott, 1970).

S. mossambicus is one of the most successful pond culture fishes in the world. Cultivation in freshwater and brackish impoundments takes place on all continents, but a particularly lucrative industry has developed in south-east Asia (Vaas & Hofstede, 1952). A fishfarming enterprise aimed at supplementing the diet of the local amaThonga would be an economical proposition if the high initial costs of construction were met by outside finance and suitably qualified farm managers could be found. Recommendations on the culture of *S. mossambicus* in pools are made by Vaas & Hofstede (1952), Pannikar & Tampi (1954), Natal Parks Board (1964), and by Bruton (1975) with particular reference to Lake Sibaya. The culture or introduction of exotic fishes is not recommended.

C. gariepinus are readily eaten by local Africans and, because of their large size, they are the most sought after freshwater fish at Lake Sibaya. Shore-based capture methods (e.g. rod and line, spears) are only moderately successful except during spawning runs when massive catches are made. Any commercial exploitation of spawning catfish must be avoided but traditional methods at the present intensity should be maintained due to the importance of catfish in the diet of the amaThonga. *C. gariepinus* may be caught using gillnets but yields are not high (approximately 12 kg per 100 m net per night with a stretch mesh of 50 mm), even in the productive inshore zone. Gillnets are also wasteful as they produce large numbers of decaying fish if they are not serviced regularly, as often happens on a windswept lake. Furthermore, their strict size selectivity tends to cause damage to a particular part of the population. Pike (1969) agrees that Lake Sibaya could not support a *Clarias* gillnet fishery.

Longlines have been used successfully to catch *C. gariepinus* in Lake Sibaya (Bruton, 1978). Each longline consists of a simple wooden reel with a buoy, wire handle and clip, and 30 kg dacron line, a steel trace, sinker and hook (Fig. 3). The longlines are clipped onto loops on a surface line at 10 m

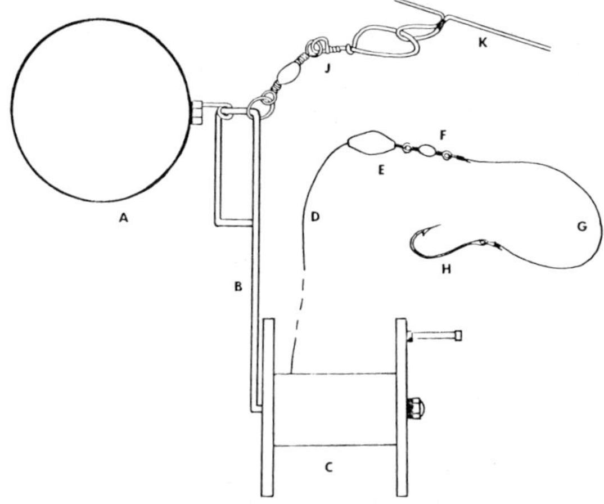

Fig. 3. A longline unit. A. Polystyrene float. B. Wire handle. C. Hardboard reel. D. 36 kg breaking strain dacron line (varying lengths on reel). E. 14 g running sinker. F. Swivel. G. 40 kg steel trace. H. Brass hook. J. Swivel with snap clip. K. Nylon surface line with loop. (from Bruton, 1978).

Table 3. The catch per unit effort (one 120 unit set laid for 6 hours) of *C. gariepinus* in offshore habitats in Lake Sibaya using longlines in 1974 and 1975 (after Bruton, 1978).

Habitat	Catch per unit effort	No. of longline settings (120 unit sets)
Terrace	14.4	33
Sheltered bay	8.3	16
Steep slope	5.1	20
Gradual slope	4.4	14
Profundal	1.3	23

intervals and the hooks, baited with tilapia flesh, are lowered to the bottom. *C. gariepinus* hooks itself by playing against the buoy, but they can surface for breathing and even continue feeding while on the line (diving observation). Small rafts could be used for setting and checking the longlines. This apparatus is inexpensive, and easy to use and maintain. Catches are variable, being lower in deep water (Table 3), but approximately one catfish should be caught on each set of 10 lines in six hours. Longlines catch *C. gariepinus* in the total length range 200 to 1080 mm but most catfish caught measure between 400 and 700 mm. Three fishermen operating in different parts of the lake using 40 longlines two nights a week from September to March would obtain a yield of about 2000 kg.

The pond culture of *C. gariepinus* at Lake Sibaya is also a possibility, as the breeding biology and early development of this and related species in natural as well as artificial waterbodies is now well-known (see reviews by Bruton, 1979, and in press).

Fish have always been an important source of food for the amaThonga, but an increased demand for this commodity has become apparent in recent years. Lake Sibaya cannot support an intensive fishery as found in the estuarine lagoons at Kosi and in the flood-plain pans of the Pongola river, but the lake could provide a regular supply of fish for local people if the relatively dense littoral populations are carefully exploited.

Hippopotamus and Crocodile Populations

Calculations based on the area and availability of suitable grazing indicate that the Lake Sibaya catchment could support a hippopotamus population of more than 150 animals (R. H. Taylor, pers. comm., 1979). A culling rate of 6% pa (the approximate annual rate of increase of *H. amphibius* populations in the wild, Ansell, 1965) would provide an annual yield of 9 hippopotami to supplement the diet of local people. A regular supply of hippopotamus flesh to local inhabitants would ensure that this mammal retains its rightful place in the aquatic ecosystem, with considerable benefit to the lake and its fauna.

There is also a strong case for retaining breeding populations of the Nile crocodile in Lake Sibaya, despite the potential threat to human lives. The crocodile plays an important role as a predator of tertiary consumers, especially *Clarias gariepinus*, and as a scavenger. Furthermore, this species has been drastically reduced throughout its range in Africa, and only stringent conservation methods will restore their numbers (Cott & Pooley, 1972).

Besides their natural roles, both the hippopotamus and the crocodile would be important tourist attractions at Lake Sibaya if the lake is incorporated into a Game Reserve.

Conservation and Recreation

The coastal lakes and estuarine lagoons of the southern Mozambique Plain, in association with the Indian Ocean, the marine beach, coastal dune forest and swamp forest, form a spectacular scenic area of considerable conservation value and recreational potential. Some of these coastal waterbodies have been protected for many years e.g. Lake St Lucia which is part of a Nature Reserve established in 1897, and Lake Nhlange which lies adjacent to Kosi Bay Nature Reserve, established in 1950.

The early expeditions of 1947, 1948 and 1949 organized by the Natal Society for the Preservation of Wildlife and Natural Resorts created considerable interest in the Kosi lakes, Lake Sibaya and the adjacent marine coast, and the first attempts to establish a coastal nature reserve in the area can be traced back to the reports of that expedition. Several recent reports have emphasized that Maputaland is one of the last unspoilt wilderness areas along the South African coast, and noted that economic commonsense and prudence demand the protection of the delicate ecosystems which support the diverse life on which a prosperous tourist industry could be based (Heydorn & Hughes, 1969; Tinley, 1971a; Heydorn, 1972; Heydorn & Wallace, 1973; Hughes, 1973; Grindley, 1975; Grindley *et al.*, 1976; Bruton, 1976; Moll, 1978; Thorrington-Smith *et al.*, 1978; Bruton & Cooper, 1979). While some of these reports were concerned only with the marine environment, others (e.g. Tinley, 1971a; Bruton, 1979; Moll, 1978; Thorrington-Smith *et al.*, 1978; Bruton & Cooper, 1979) proposed a holistic approach in which valuable marine, coastal (including coastal lake) and inland areas were integrated into a single large nature reserve, the Maputaland National Park.

Discussion

The utilization and conservation of Lake Sibaya needs to be viewed from the perspective of the lake as a unique independent microcosm as well as a part of an extensive watershed on a coastal plain. The lake itself provides protein in the form of fish and hippopotami to supplement the predominantly carbohydrate diet of local inhabitants, but the fish yield is very low compared with other aquatic systems in Maputaland. In contrast, Lake

Sibaya is a far more extensive source of freshwater for irrigation, domestic use and recreational activities than other waterbodies in the area. The establishment of population foci within the catchment of Lake Sibaya should however be considered carefully due to the risk of pollution of this deep endorheic system. Similarly uncontrolled use of fertilizers may result in eutrophication problems.

The inclusion of Lake Sibaya in a large Maputaland National Park would ensure that this sensitive and valuable ecosystem is conserved for the longterm benefit of the local people, as well as provide a further dimension to the recreational activities of visitors to the Park.

Although the more important organisms and trophic pathways of Lake Sibaya have been studied, there are still important gaps in our knowledge which must be filled if the lake is to be utilized and conserved in the most efficient way. A continued scientific presence in the area should therefore be an essential part of the management plan for the National Park.

A potentially troublesome aspect of the development of more efficient land use as well as a tourist industry in Maputaland is the continued presence there of various diseases. Malaria is a constant threat though modern prophylactic drugs are now available, but bilharzia is widespread and dehabilitating. Molluscicides are ineffective against a snail vector as widespread as *Bulinus* (*Physopsis*) *globosus*, so efforts should be made to break the life cycle of the parasite by providing piped water, storage tanks and water taps for humans, and drinking troughs for cattle. Reduced contact with deeper water bodies would decrease the chance of the introduction of *Schistosoma mansoni*, whose presence would seriously hamper recreational development at Lake Sibaya. The creation of a 'green belt' about 2 km wide around Lake Sibaya, with access only at certain pre-selected points, would reduce the risk of human intestinal bilharzia as well as protect the lake from domestic pollution (Allanson, 1975). Mseleni Joint Disease is a serious problem in a localized area but until the aetiology is known all that can be done is to improve the diet of the groups concerned and investigate the disease further.

The apparent incompatibility of water sports, hippopotami and crocodiles in Lake Sibaya can be overcome by confining water contact activities to the south basin, and excluding the dangerous animals from there. Boating, fishing and guided tours could be conducted on the other parts of the lake. Without these concessions to recreational demands it will not be possible to establish a sufficiently large, economically viable National Park in Maputaland.

The development of Maputaland needs to be planned in such a way that abundant natural resources are utilized on a sustained yield basis, the biotic diversity of the area is retained, and further work opportunities are created. The main natural resources are the fishes of the Pongola pans, Kosi lakes, Lake St Lucia and Indian Ocean, the freshwater of Lake Sibaya, the variety of indigenous fruits and timber, and the fertile soils associated with alluvial floodplains and wooded areas. An adequate management plan of such a diverse area would require careful multidisciplinary planning. Details of land use allocation in Maputaland could be determined using McHarg's (1969) intrinsic suitability mapping technique, as A'Bear *et al.*, (1977) have done in a modified way for the eastern shores of Lake St Lucia. This method involves the evaluation of the suitability of key natural resources for various usages and provides an invaluable data base for monitoring changes in natural resources, determining the success of land use management programmes and selecting suitable sites for development.

Acknowledgements

I am grateful to R. H. Taylor, C. C. Appleton, K. H. Cooper and C. W. Hardwich for useful discussions, and to the KwaZulu Government Service for permission to work in Maputaland. This research was conducted under the auspices of the Institute for Freshwater Studies of Rhodes University. The manuscript was prepared in the J. L. B. Smith Institute of Ichthyology. Ms E. M. Tarr and Mr P. Russell are thanked for assistance with the illustrations, and Mrs H. Tomlinson for typing the manuscript. My wife Carolynn provided extensive field assistance.

References

A'Bear, D. R., A. M. Little, A. J. Phelan & R. H. Taylor. 1977. St Lucia eastern shores: an evaluation of natural resources. Paper read at the Wildlife Management Association Symposium, University of Pretoria, July 1977. 18 pp.

Allanson, B. R. 1975. The utilization of Lake Sibaya. Report to the Department of Bantu Administration and Development, Pretoria. 7 pp.

Allanson, B. R. & J. D. van Wyk. 1969. An introduction to the physics and chemistry of some lakes in northern Zululand. Trans. R. Soc. S. Afr. 38 (3): 217–240.

Ansell, W. F. H. 1965. Hippo census on the Luangwa River 1963–1964. The Puku, no. 3: 15–27.

Appleton, C. C. 1977. The influence of temperature on the life cycle and distribution of *Biomphalaria pfeifferi* (Krauss, 1848) in south-eastern Africa. Int. J. Parasitol. 7: 335–345.

Appleton, C. C. & M. N. Bruton. 1979. The epidemiology of schistosomiasis in the vicinity of Lake Sibaya, with a note on other areas of Tongaland (Natal, South Africa). Ann. trop. Med. Parasit. (in press)

Azevedo, J. F. de, L. do C. M. de Medeiros, M. M. da C. Faro, M. de L. Xavier, A. F. Gandara & T. de Morais. 1961. Os moloscos de agua doce do ultramar Português II – Moloscos de Mozambique. Junta de Investigacoes do Ultramar No. 88. Lisboa, 394 pp.

Blaber, S. J. M. 1978. Fishes of the Kosi system. Lammergeyer 24: 28–41.

Bruton, M. N. 1975. The utilization of the fishes of Lake Sibaya. Report to the KwaZulu Government, Institute for Freshwater Studies, Rhodes University. 13 pp.

Bruton, M. N. 1976. Proposal for a marine reserve in Tongaland. Report to the Marine Reserves Commission of the Department of Sea Fisheries, 7 pp.

Bruton, M. N. 1978. The habitats and habitat preferences of *Clarias gariepinus* (Pisces: Clariidae) in a clear coastal lake (Lake Sibaya, South Africa). J. limnol. Soc. sth. Afr. 4(2): 81–88.

Bruton, M. N. 1979. The breeding biology and early development of *Clarias gariepinus* (Pisces: Clariidae) in Lake Sibaya, South Africa, with a review of breeding in species of the subgenus *Clarias* (*Clarias*). Trans. zool. Soc. Lond. 35: 1–45.

Bruton, M. N. In press. Guidelines for the culture of *Clarias gariepinus* in southern Africa. South Afr. J. Wildl. Res.

Bruton, M. N. & K. H. Cooper, (eds). 1979. Studies on the ecology of Maputaland. Rhodes University Press, Grahamstown.

Burger, F. J., C. D. Elphinstone, S. A. Fellingham, P. C. Grey & Z. A. Hogewind. 1973. Mseleni joint disease: biochemical survey. S. Afr. Med. J. 47: 2331–2338.

Campbell, G. G. 1969. A review of scientific investigations in the Tongaland area of northern Natal. Trans. R. Soc. S. Afr. 38 (4): 305–316.

Coke, M. M. & R. Mc. C. Pott. 1970. The Pongola floodplain pans. Natal Parks Board, Pietermaritzburg. 34 pp.

Cott, H. B. & A. C. Pooley. 1972. The status of crocodiles in Africa. IUCN Publications New Series, Supplementary paper no. 33. 98 pp.

Denoon, D. 1972. Southern Africa since 1800. Longman, London.

Dyer, T. G. J. 1976. Expected future rainfall over selected parts of South Africa. S. Afr. J. Sci. 72: 237–239.

Fellingham, S. A., C. D. Elphinstone & W. Wittmann. 1973. Mseleni joint disease: Background and prevalence. S. Afr. Med. J. 47: 2173–2180.

Grindley, J. R. 1975. The need for the establishment of marine reserves in South Africa. Proceedings of the Council for the Habitat conference on coastal conservation, Durban. 9 pp.

Grindley, J. R., K. H. Cooper & A. V. Hall. 1976. Proposals for marine nature reserves for South Africa. Wildlife Society of Southern Africa, Durban. 16 pp.

Heydorn, A. E. F. 1972. Tongaland's coral reefs – an endangered heritage. Afr. wild life. 26 (1): 20–23.

Heydorn, A. E. F. & G. R. Hughes. 1969. Urgent need for marine reserves. Afr. wild life. 23 (4): 270–278.

Heydorn, A. E. F. & J. H. Wallace. 1973. The role of the fauna of Kosi lake system and the associated coastal regions in terms of the possible proclamation of a nature reserve. Report to the Department of Bantu Administration and Development, 13 pp.

Hill, B. J. 1969. The bathymetry and possible origin of Lakes Sibayi, Nhlange and Sifungwe in Zululand (Natal). Trans. R. Soc. S. Afr. 38 (3): 205–216.

Hughes, G. R. 1973. The possible role of sea turtles in the utilization of the Maputaland coast. Report to the Department of Bantu Administration and Development, 11 pp.

Jackson, Lynette F. 1976. Aspects of the intertidal ecology of the east coast of South Africa. Invest. Rep. oceanogr. Res. Inst. No. 46: 1–72.

Jordan, P. 1975. Schistosomiasis – epidemiology, clinical manifestations and control. In: Man-made lakes and human health, eds: N. F. Stanley & M. P. Alpers: 35–50. Academic Press, London.

Junod, H. A. 1962. The life of a South African tribe. Vols. I & II. University Books, New York.

Lockitch, G., S. A. Fellingham & C. D. Elphinstone. 1973. Mseleni joint disease: A radiological study of two affected families. S. Afr. Med. J. 47: 2366–2376.

Lockitch, G., S. A. Fellingham, W. Wittmann, P. D. de Villiers, I. S. de Wet & G. T. du Toit. 1973. Mseleni Joint Disease: the pilot clinical survey. S. Afr. Med. J. 47: 2283–2293.

Lubbe, A. M., C. D. Elphinstone & S. A. Fellingham. 1973. Mseleni joint disease: food and water supplies. S. Afr. Med. J. 47: 2225–2233.

MacArthur, W. 1955. An account of some of Sir David Bruce's researches, based on his own manuscript notes. Trans. R. Soc. trop. Med. Hyg. 49 (5): 404–412.

McHarg, L. 1969. Design with Nature. New York, Natural History Press.

Marwick, C. W. 1973. Kwamahlati – the story of forestry in Zululand. Dept. of Forestry, Pretoria. 72 pp.

Moll, E. J. 1972. The distribution, abundance and utilization of the lala palm, *Hyphaene natalensis*, in Tongaland, Natal. Bothalia 10 (4): 627–636.

Moll, E. J. 1978. The vegetation of Maputaland. Trees in South Africa 29 (4): 31–59.

Natal Parks Board. 1964. Fish farming with tilapia in Natal. Natal Parks Board, Pietermaritzburg. 16 pp.

Pannikar, N. K. & P. R. S. Tampi. 1954. On the mouth-breeding cichlid, *Tilapia mossambica* Peters. Indian J. Fish. 1: 217–230.

Pike, T. 1969. Composition, relative abundance and size range of fish populations in Lake Sibaya. News Lett. limnol. Soc. sth. Afr. 13 (suppl.): 38–43.

Pitman, W. V. & I. P. G. Hutchison. 1975. A preliminary hydrological survey of Lake Sibaya. Hydrol. Res. Unit Rep. No. 4/75: 1–35.

Theron, A. 1975. Out of the ditch. The Mseleni story. Theron & Theron, Pretoria.

Thorrington-Smith, C. J. Rosenberg, & G. McCrystal. 1978. The preliminary development plan for KwaZulu. Shuter & Shooter, Pietermaritzburg.

Tinley, K. L. 1964. Fishing methods of the Thonga tribe in north-eastern Zululand and southern Mozambique. Lammergeyer 3 (1): 9–39.

Tinley, K. L. 1971. Lake St Lucia and its peripheral sand catchment. Wildlife Society of Southern Africa, Durban. 62 pp.

Tinley, K. L. 1976. The ecology of Tongaland. Wildlife Society of Southern Africa, Durban. 140 pp.

Torres, J. L. R. 1973. Summary of the political, social and economic organization of the Tonga people living on the Pongola floodplains. Internal report, University of Natal, Durban. 6 pp.

Vaas, K. F. & A. E. Hofstede. 1952. Studies on *Tilapia mossambica* Peters in Indonesia. Contr. Ind. Fish. Res. St. No. 1, 1–88.

Waddy, B. B. 1975. Research into the problems of man-made lakes, with special reference to Africa. Trans. R. Soc. trop. Med. Hyg. 69: 39–50.

Wright, C. A. 1966. The pathogenesis of helminths in the Mollusca. Helminth. Abstr. 35: 207–224. 35: 207–224.

1 The lake as an ecosystem

B. R. Allanson

This chapter represents an attempt to assemble together those "bits" of information from the previous chapters and other relevant literature which are likely to provide some insight into the structure and functioning of the lake as an ecosystem. Certainly it is not difficult to give, as Fig. 1 shows, the principal components of this ecosystem and the paths of principal interaction. What is difficult to determine is the magnitude of these interactions and how their rates vary under the influence of a diverse array of climatological, edaphic and biological factors. Notwithstanding, it is possible to draw on the main framework of interaction to indicate which principal events control the operation of the system and which may be sensitive to man-made perturbations.

Climate and the Lake Shore

There is no doubt that this largely endorheic lake undergoes a cyclical pattern of volume increase and decrease, which by virtue of the shape of its basin (B. J. Hill, this volume) is reflected in a marked change in lake level. The periodicity of this cycle is of the order of 20–30 years, and, as B. R. Allanson (this volume) has reported, is matched closely by the pattern of precipitation in the summer rainfall region of southern Africa. This change, coupled with wind-driven currents, is responsible for the shape and dimension of the littoral shelf, particularly on the eastern and western shores. Our study of the lake has taken place during a period of increasing level, so we are unaware of the effect these factors have upon the dimension and stability of the littoral zone. Some indication of the extent of the shore terrace is given in Plate 1. Wave erosion of the unstable elevated shoreline transports sand onto the terrace, increasing its height and creating the wave-cut platforms commonly seen around the lake. This slow cyclical pattern of inundation and regression, meaning in effect that the dimension of the littoral terrace cyclically changes, brings with it a number of major edaphic and biological changes.

At low lake levels the sand beach, terrace and macrophyte beds are well defined, the latter with a downward limit set by light attenuation at a depth of 7–8 m. The terrace above the macrophyte beds is essentially a hostile environment, with no cover from avian predators for benthic feeding fish during the daytime, but its illuminated sands are a rich source of benthic algae and interstitial and surface detritus used extensively by a variety of cichlid fry and juveniles. Bowen (1976a & b, 1978, 1979) has argued with considerable effect that the nutritional quality of the benthic food

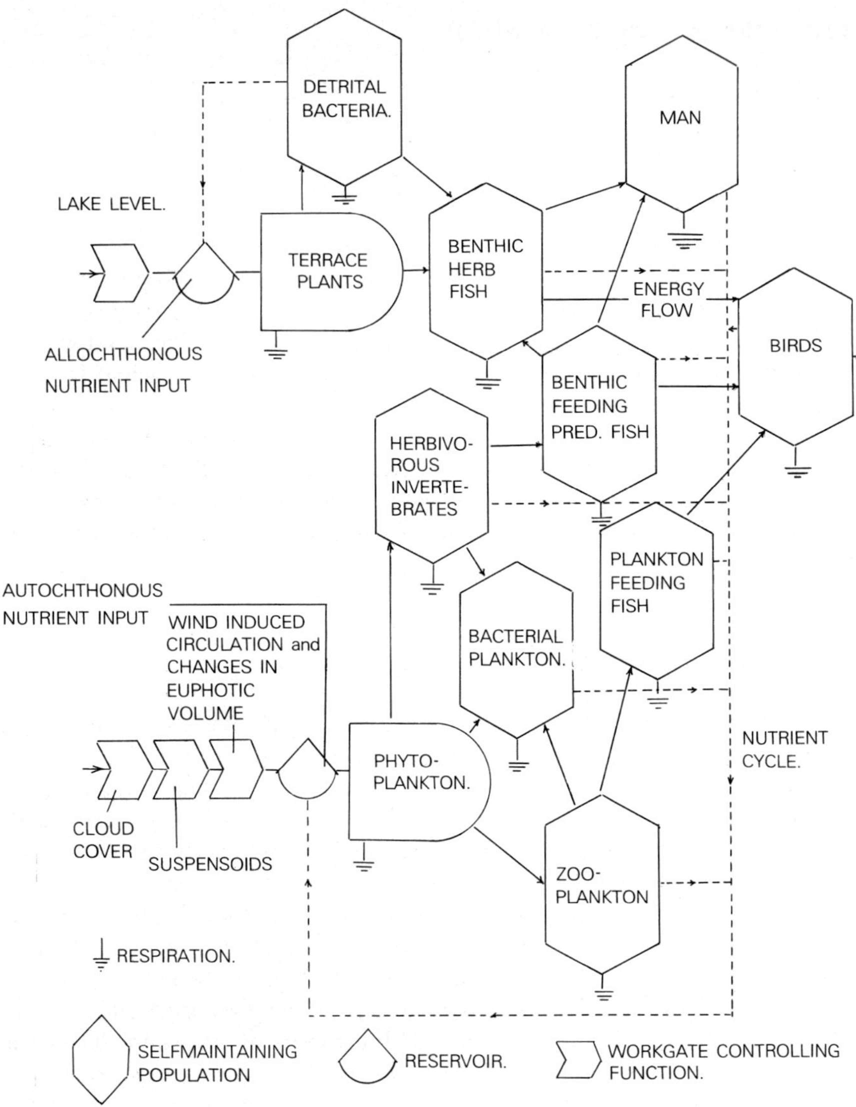

Fig. 1. The principal biological components of the Lake Sibaya ecosystem expressed in energy circuit language. See Odum, H. T. (1972) An energy circuit language for ecological and social systems. Chapter 4: 140–210, Systems Analysis and Simulation in Ecology, Vol. II, Ed. B. C. Patten. Academic Press, New York.

resources in the lake depends not only upon diatom distribution but also, and more importantly, upon the bacterial content of the surface and interstitial detritus. The detritic material originates largely, if not entirely, from grass and forest debris which accumulates along the shoreline. With the rise in lake level this allochthonous component of the benthic food resource is greatly increased. This fact coupled with the marked increase in cover, provided by flooded marginal wet lands and the increase in the sedge community, is expected to influence the number of young fish successfully reared and possibly the status of the adult cichlid community, particularly *Sarotherodon mossambicus.*

We see here a direct effect of a climatological cycle upon the trophic resources of the lake littoral. It is significant that the expected increase in soluble reactive phosphate following upon flooding of the marginal swamps and *Acacia* fringe did not occur. Phosphate ions liberated into the water from dead and dying plants will be taken up rapidly by the developing algal and bacterial biomass. Howard-Williams & Allanson (1978) have demonstrated the rapidity with which phosphate is lost from senescent hydrophyte material and taken up by an epiphyte association in another coastal lake, Swartvlei, and an equivalence with the inundated littoral plants of Lake Sibaya can reasonably be expected. Not all the allochthonous material following inundation remains in the littoral subsystem. A good deal is passed quite quickly, by wave action, into deeper water where decay continues.

C. Howard-Williams (this volume) has demonstrated that an additional and important influence of lake level rise is the rapid destruction of the macrophyte beds of the low level terrace. Lake level changes tend to be steplike in nature, with rapid increases occurring over a period of weeks or months during the cycle, followed by quite long periods of level stability. The integration of these increments eventually reaches a point when subsurface irradiation falls below the compensation point for the hydrophyte community, and it decays. This effect is illustrated in Fig. 2.

Plate 1. The flooded terrace showing stumps of *Acacia karroo*. The position of the terrace break is shown by arrow 1. Arrow 2 indicates the approximate position of the shoreline at mean level.

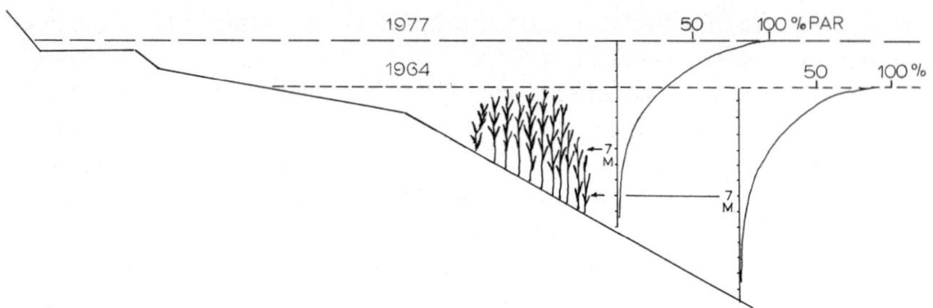

Fig. 2. The elevation of the compensation point for hydrophytes in Lake Sibaya following flooding.

Howard-Williams (this volume) has also shown that water depth, while explaining a goodly proportion of hydrophyte biomass variance in the lake, is comparatively unimportant to the main hydrophyte species of the terrace slope, namely *Potamogeton schweinfurthii.* He has argued that an increase in turbidity plays an important role in the changes brought about in the submerged macrophyte beds. I have hinted at the importance of turbidity changes (Allanson, this volume) in the lake, but our information is not sufficiently quantitative to support or reject the view expressed by Howard-Williams. Nevertheless I remain convinced that the change in the pattern of subsurface irradiance with increasing water depth, coupled with small changes in turbidity, certainly not high enough to be measured by significant changes in Secchi disc transparency, are sufficient to destroy the macrophyte beds of the main basin.

The products of their decay will form part of an allochthonous nutrient and energy input into the lake ecosystem, but, as in the inundated littoral, no significant elevation in soluble reactive phosphorus is measured. Indeed if my data reported in Chapter 4, Table 8 are to be relied upon there is, if anything, a decrease in soluble reactive phosphorus at high lake levels, which may represent a step in the establishment of a new SRP equilibrium at the peak of the high level phase.

Light and the Fixation of Carbon

The sequelae of long term climatological events are diverse and find expression in the physical, chemical and biological subsystems pertaining to the littoral and pelagic zones of the lake. The effect of volume increase upon the fixation of carbon needs specific consideration.

As far as our data show there is no pronounced seasonal peak in primary production, but the months of November and December appear to be periods when fixation in terms of $mgC \cdot m^3 \cdot hr^{-1}$ is at its highest. However it is difficult to distinguish seasonal changes from daily meteorological events, as it may well have been that the measurements were made during periods of variable cloud cover. This would certainly influence the pattern as Fig. 5 in Chapter 5 might suggest.

The thermal instability of the lake prevents the sequestration of nutrients in a hypolimnion which, as a result of either a pattern of monomictic turnover or internal seiche activity as in very deep tropical lakes like Lake Malawi, become available to the phytoplankton flocks in the euphotic zones. This suggests the existence of a slow continuous regeneration of nutrients from senescing cells at all levels of the water column, such that the standing crop is at all times in equilibrium with its chemical environment. In the absence of any material input of nutrients from surface runoff or from the subterranean aquifers, this would appear to be the only mechanism maintaining the algal standing crop, certainly at or near low lake levels. This means, in essence, that stimuli to rapid growth of algal populations are in general lacking, although Hart & Hart (1977) have reported marked increases in *Synedra acus* between August and October and in *Melosira granulata* between February and May. Bluegreens, particularly representatives of the genera *Anabaenopsis* and *Anabaena*, showed peaks in January and February.

The other photosynthetic components must now be considered. C. Howard-Williams (this volume) records that the macrophytes of the lake are responsible for only a moderate fraction of the dry matter production, in fact some 27,876 tons of dry matter per year, which is 11% of that produced by the phytoplankton community. With the reduction in macrophyte beds this amount will become even smaller during the late stages of inundation, and be replaced very largely by the production of the epipsammic and periphyton community of the extensive littoral terrace. The energy fixation potential of the pelagic photosynthetic algal community is impressive and in very large lakes with weakly developed shorelines and steep sides it must represent the major, if not the only, significant source of energy fixation. However, in lakes of the morphology of Lake Sibaya, the shallow illuminated terrace contributes significantly to the energy fixing systems of the lake, as the data in Table 1 show.

As might be expected, there is a marked increase in the contribution of epipsammic production at high lake levels. Clearly this cannot be ignored, nor has it, as the studies of herbivore utilization of the terrace production by M. N. Bruton (this volume), S. H. Bowen (Bowen *op. cit*), and M. Caulton (1978) have shown.

Another less obvious though possibly important effect of the elevation of lake level is upon the proportion of the euphotic volume at the extremes of lake levels as we know them. These proportions are given in Table 2.

At low lake levels the higher proportion of the euphotic volume to total lake volume ensures that the algal community is maintained very largely within photosynthetically active depths. At high levels the volume of dark water increases. This condition is comparable to those lakes in which the thermocline occurs well below euphotic depth. But whereas in such lakes the thermocline becomes a barrier to further downward circulation, in Lake Sibaya the bottom sediments are reached. It is therefore not surprising, in a

Table 1. A comparison of pelagic and episammic production in Lake Sibaya during the high and low lake level phases.

	Low level	High level
Area of littoral shelf <3 m depth	10 km^2	16 km^2
Area of main basin excluding terrace area	27.2 km^2	24.9 km^2
Net fixation of carbon in tonnes.day^{-1}		
On shelf	16	26
In pelagic zone	46	50
Percentage of total energy fixed day^{-1} by epipsammic community	26%	34%

warm transparent and well oxygenated lake with full circulation, that the primary production of the volumetrically large euphotic zone is efficiently intercepted by a complex and rich benthic invertebrate and bacterial community. These hydroclimatological conditions may also control to some extent the structure of the phytoplankton. Hutchinson (1967) differentiates a *meroplankton* to which the taxa *Melosira* and *Synedra* belong. The dominance of such components in the phytoplankton of the lake is obviously linked to the accessibility of large areas of bottom sediments within the euphotic volume.

Such conditions existed in the lake during low levels when the majority of our work was done. The long term effect of the decrease in relative euphotic volume during high lake levels are as yet unstudied, although physicochemical sampling in January 1977 certainly did not indicate that the benthos of the lake, particularly in the deeper levels, was being subjected to increasing stressful conditions, in particular reduced dissolved oxygen tensions. A striking example of the influence of a rapid increase in level, coupled with a marked decrease in the depth and volume of the euphotic zone, upon hydroclimate and benthos is given by Boltt & Allanson (1975) for a somewhat similar coastal ecosystem, Lake Nhlange some 60 km to the

Table 2. A comparison of the proportion of euphotic volume as percentage of total lake volume at low and high lake levels.

	Low levels	High levels
	1964	1977
Cool season*	71	51
Hot season*	82	68

* Depth of 1% PAR = 8 & 11 m respectively; and the proportion is determined from the volumes delineated by these depths.

north. The inflow of humic water from the southern swamps reduced Secchi disc transparency to <1 m. Dissolved oxygen concentrations dropped sharply below 13 m to between 0–1 mg l^{-1} with anaerobic conditions developing at the sediment–water interface. Notwithstanding full circulation within the water column, it took 3 years before sufficiently high oxygen levels could be maintained at this interface to allow a recolonisation of benthic fauna. It may well be that in the time scale of the events leading to the maximum elevation of lake level in Lake Sibaya, 5–10 years, the opportunity existed for the ecosystem to adapt to, principally, the input of allochthonous organic material and the relative decrease in the volume of the euphotic zone. However, the decrease in euphotic volume is never sufficiently marked to significantly influence the growth of the phytoplankton.

Nutrients and their Cycling

I have already hinted at the reasons behind the low concentration of soluble reactive phosphate (SRP) in the water of the lake. Unfortunately our knowledge of the concentrations of the various phosphorus and nitrogen moeities and their distribution within the lake is simply not adequate. Surface inflow is insignificant compared with that derived from the subterranean aquifers of the catchment. It seems likely, although not positively established, that flow from the aquifers, along with a component of cyclical salt from the sea, is responsible for the elevated chloride concentration in the lake which is so essential to the successful maintenance of a variety of estuarine taxa.

If, as has been frequently suggested, the low nutrient status of the lake waters derives from its catchment, then our attention must be given to describing the links (Fig. 1) in the internal recycling of phosphorus, and the extend to which nitrogen is fixed by the bacterial and bluegreen components of the lake flora. The importance of the allochthonous litter input must likewise be assessed.

Utilization of Primary Production

As far as we know the lake does not have a phytoplanktonic feeding fish. Those herbivores among its piscifauna are largely if not exclusively benthic algal or hydrophyte feeders. From R. C. Hart's work (Hart 1973, Hart & Allanson 1975, and Hart & Hart 1977 and Hart this volume) on the zooplankton of the lake it becomes clear that the principal utilizer of pelagic phytoplankton production is the calanoid copepod *Pseudodiaptomus hessei*. But even here constraints are imposed: this copepod filters nannoplankton and bacterioplankton at greater rates than net phytoplankton, and because of its diel feeding rhythmicity feeding takes place predominantly at night.

The importance of nannoplankton (<20 μm) in the primary production processes of the lake has been established by Allanson & Hart (1975), but a relevant question at this stage is; how efficiently is this resource utilized in the lake? Given that the integral of productivity during the hot season can be as high as 1847 $mgC.m^{-2}.day^{-1}$, and that the principal grazer *P. hessei*, with a mean biomass of 28.5 $mgC.m^{-2}$, is firmly locked onto or into the sediments by virtue of a remarkable endogenous rhythm during daylight hours, it follows that during the productive period, namely daylight hours, the phytoplankton community is being little utilized. While this is of little consequence to the utilization by zooplankton, one is left with the impression that a large fraction of the daily production is not used by grazers and is available for degradation by bacteria. The importance of bacteria in the economy of lakes is well established, but the points at which they contribute to the flow of energy will vary from system to system. In general, however, the decay of phytobenthos, phytoplankton and hydrophytes will maintain a bacterial biomass, and the whole degrading complex will contribute to the pool of DOM and detritus. This is a view essentially similar to that of Sorokin (1965).

While our understanding of these events in the pelagic zone of the lake is poor, events in the littoral are known with a greater degree of certainty. Bowen (1976a & b, & 1979) has obtained evidence which points strongly to the importance of bacteria in the diet of the cichlid *S. mossambicus*, which possesses the metabolic equipment to both lyse and digest bacterial cells. These together with diatoms make up the primary food intake of this cichlid on the terrace. S. H. Bowen sees in his model a dependence by the bacterial floc community upon DOM secreted by the benthic diatom community; and that, for the provision of the right quality of food, both trophic elements are necessary. This supports Sorokin's *(op. cit)* point that 'bacterial biosynthesis in some respects is nearer to primary than to secondary production, for the basic process of the initial formation of bacteria in the water medium proceeds mostly by the utilization of organic matter which is practically lost for the direct utilization by animals'.

In short the role of bacteria as a food resource and not merely as a means to an end, namely the formation of detritus, particularly in those lakes tending towards oligotrophy, needs re-evaluation. The transfer of energy to the major herbivore on the terrace is summarized in Fig. 3.

These interrelationships have been taken a step further in our understanding of energy flow, particularly within the terrace subsystem, largely by the work of Caulton (1978). His studies on the bioenergetics of herbivorous cichlids, particularly *Tilapia rendalli*, and the application of these results to *Sarotherodon mossambicus* using the data from Bowen (1976b) provide an opportunity of extending the analysis of utilization to the terrace-feeding population. In Fig. 4 the components of the energy input and utilization in a 20 gm *S. mossambicus* are shown.

Unfortunately our estimates of real density of *S. mossambicus* are very inaccurate, but assuming a density of one individual per m^2, the utilization of the terrace primary productivity, in terms of assimilated energy, will be of the order of 5% per day. If the density is set at 0.1 individuals per m^2, which seems more likely, utilization will be 0.5% per day.

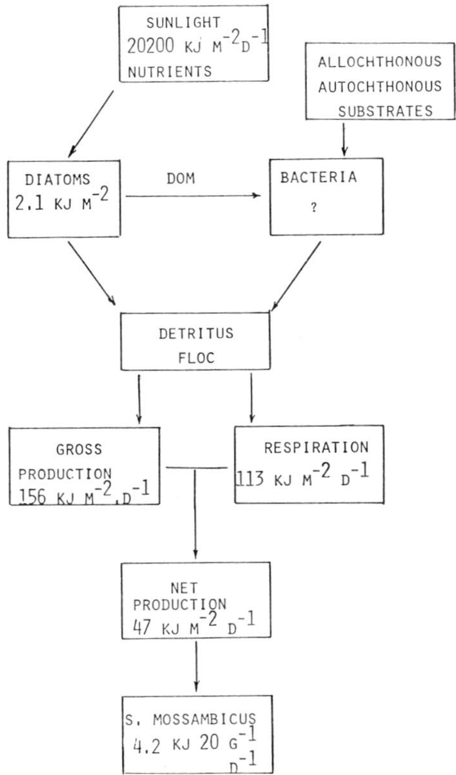

Fig. 3. The transfer of energy to the major herbivore on the terrace, *Sarotherodon mossambicus.*

M. N. Bruton (this volume) has recorded a yield per fisherman of between 10–15 fish per day from the terrace. This was primarily during the summer months (November–February) and represents one of the terminal points of terrace energy fixation.

A Preliminary Synthesis

In Lake Sibaya our attention has been towards the solution of largely descriptive problems, but it is clear from the review which the preparation of this volume has provided that some general principles about the lake as an ecosystem are beginning to emerge. To assist us in this task are the important ideas of, for example, Thienemann, Rawson and Neumann who attempted to isolate a principle set of factors which could be said to govern the productivity of lakes. Among those of particular relevance were morphometry, climate and edaphic features. Within the last 10–15 years these ideas have been developed by the work of Ryder (1965), and in 1974 Johnson demonstrated a series of log:log relationships between the morphoedaphic index (TDS/mean depth) and an index of production for

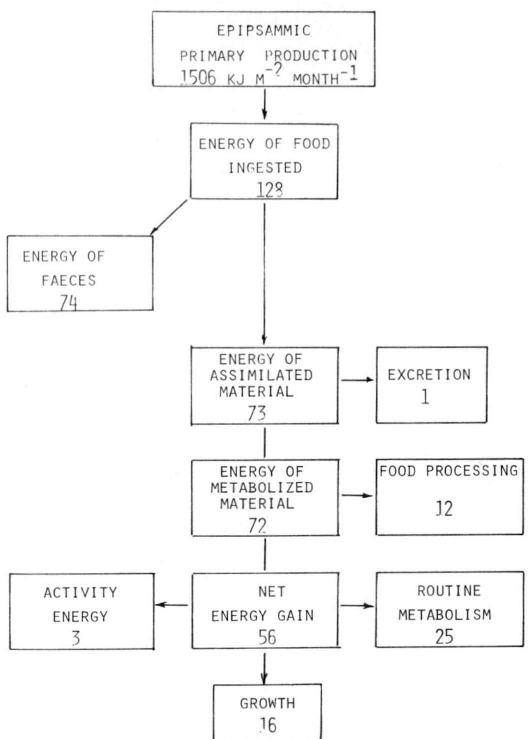

Fig. 4. The partition of ingested food energy by *S. mossambicus* when feeding on the terrace. Data from Caulton (1978) and Caulton unpublished. Units are KJ 20 g^{-1}·$month^{-1}$ with the exception of those for primary production.

benthos (standing stock: $kg.m^{-2}$ × turnover rate in days) for a wide variety of lakes. Applying this type of analysis to data from Lake Sibaya shows that its morphoedaphic index, 36, and production index, 32, expressed in a log : log plot, fall within the variance of Johnson's data for lakes on the Atlantic side of North America. This suggests that production in Lake Sibaya of the major benthic faunal taxa *Grandidierella lignorum* and *Apseudes digitalis* is related to lake depth, area and dissolved solids. It may well appear to us, with the wisdom of hindsight, to be almost self-evident. What we need now is a much deeper understanding of how this production is controlled by the environmental factors.

In attempts to achieve this, one assemblage of ideas keeps recurring to me, based upon three principal facts about the lake:

1. That the lake undergoes a clearly defined long term cycle of rising and falling, which, having regard to the morphometry of the lake, influences both the size of the littoral zone and the ratio of the euphotic zone to the non-euphotic or dark zone.

2. That the lake circulates freely throughout its depth under the stress of

wind, such that homothermal conditions are maintained during both hot and cool seasons.

3. That as a consequence of full circulation the phytoplankton community is distributed throughout the depth of the lake (Hart & Hart 1977), but with diatoms, and in particular *Melosira* spp tending to maintain higher densities in deep water than in the surface waters during the hot season.

These facts and their ecological sequelae are shown in Fig. 5, in which I have attempted to show by reference to area diagrams the changes in the volumes referred to in Table 2, and the principal events linking benthic and pelagic components.

A descriptive analysis of this open water part of the ecosystem might reasonably begin by noting that the lake lacks a significant phytoplankton-herbivorous fish or, for that matter, an important piscine zooplankton predator, albeit our knowledge of the trophic biology of *Gilchristella aestuarius* is poor. Consequently the utilization of the primary productivity of the large euphotic zone is left to a rather simply constructed invertebrate zooplankton dominated by *Pseudodiaptomus hessei.* This copepod herbivore does not by any means exhaust its food resource, so that much of the carbon fixed in the euphotic zone passes either by passive sinking or more rapidly by wind-induced turbulence to the bottom sediments, where it becomes the resource upon which a substantial micro- and macrobenthos depends. The success of this benthic primary consumer component may be judged by the predator structure which is built upon it. Some indication of the complexity of this food web is given in Fig. 6.

The question as to whether this represents an efficient utilization of the primary energy fixed, particularly when we consider the low potential of the lake fishery, cannot be answered until we can express quantitatively the utilization of the carbon flux in this ecosystem: clearly one of the principal unresolved problems still to be tackled.

We see in this instance a modification or development of Beauchamp's suggestion (Beauchamp 1958) that in the absence of plankton feeders it is reasonable to expect a loss of nutrients from the lake system; in fact an irreversible loss into the 'sink' of the sediments. I have argued that the hydroclimatological conditions in Lake Sibaya will combat this tendency, and by virtue of its relatively rich benthos the lake will continue to circulate and partition the nutrient resources. We have suggested in an earlier paper (Allanson & Hart 1975) and in Chapter 5 of this volume that in this comparatively deep lake the low algal standing crop (2–4 μg chlorophyll.l^{-1}) and production may be the result of a failure to regenerate nutrients within an hypolimnion. Burgis (1978) has argued that this is 'manifestly not true of Lake George where very high levels of gross photosynthesis are maintained throughout the year ... despite the fact that levels of inorganic N and P in the water are at all times very low'. While our argument is not without flaw, it does point to the possibility that in Lake Sibaya we are experiencing an inefficient routing of nutrient resources. How

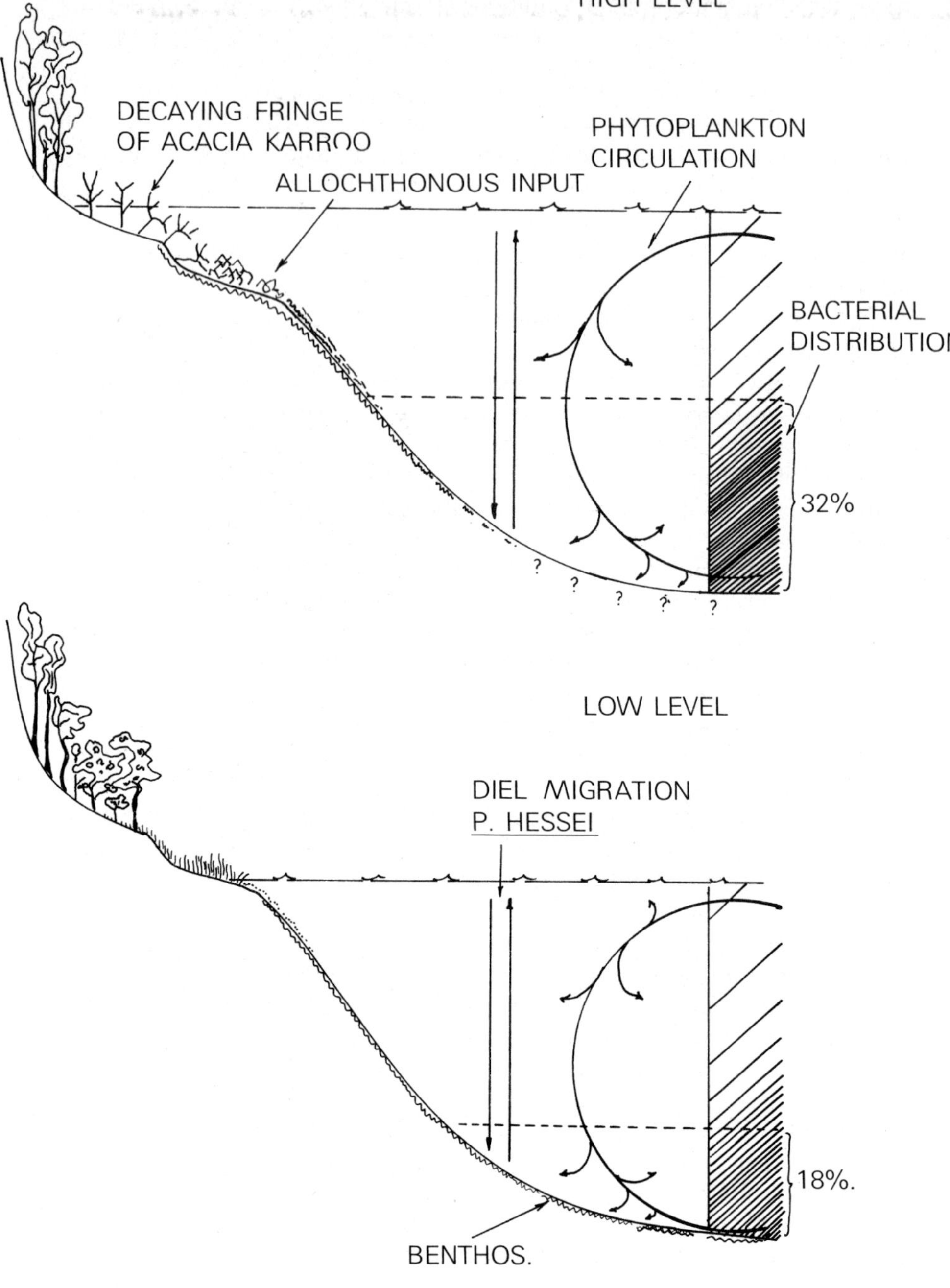

Fig. 5. Some of the principal events linking the pelagic and benthic components in Lake Sibaya at high and low lake level. This figure should be looked at in conjunction with Fig. 1. The limit of 'dark' water is indicated by a dashed horizontal line. The percentage given is based on volumes not vertical distance.

much is being incorporated into bacterial biomass both within and above the sediments for transfer into the microbenthos we do not know, but some preliminary data on bacterial growth in peptone broth at 30° (Hart 1973) does suggest that this may not be an insubstantial pathway.

In this regard it is important to stress the interdependence between bacterial and algal growth. S. H. Bowen's elegant studies on the quality of the food resources on the terrace of the lake, to which I have already referred, have established the importance of this interdependence in the growth of young *Sarotherodon mossambicus.* A combination of a decrease in food quality as the substrata become deeper, and its inability to forage widely over the deep bottom due to a depth limitation imposed by swim-bladder gas exchange physiology (Caulton & Hill 1973), limits this species to the lake littoral. The omnivorous cichlids, *Pseudocrenilabrus philander* and *Tilapia sparrmanii* are not constrained by a depth limit and utilize the whole of the benthic niche, both littoral and profundal.

Accepting that we know quantitatively very little of the flow of energy throughout the system, what data we do have suggests utilization as shown in Table 3.

The data were obtained from a variety of investigations during the hot, summer season. I have, therefore, expressed the incoming radiation on a daily basis: the figures are then more meaningful for Lake Sibaya than attempting a yearly total. The comparison which Table 3 gives between Lake Sibaya and Lake George is a useful one as it draws attention to the efficiency with which the net production is being utilized in both systems. Burgis (1978) on the basis of the accumulated data on Lake George argues

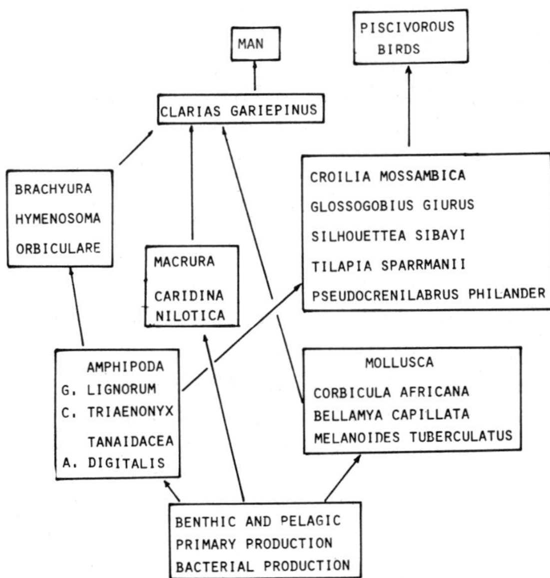

Fig. 6. The structure and interactions within the secondary consumer food web of the benthos in Lake Sibaya.

Table 3. Utilization of energy fixed in Lake Sibaya and Lake George, Uganda.*

	$KJ\,m^{-2}\,d^{-1}$		
	L. Sibaya	L. George	%
Solar radiation	20,200	19,700	100
Gross primary production†	152 (0.8)	226 (1.1)	100
Net primary production	43 (28)	64 (28)	100
Invertebrate herbivores			
Planktonic: *P. hessei* / Benthic: Amphipoda	1.1–1.3 (2.6–3.0)		
Piscine herbivores		1.7 (2.7)	

* Burgis and Dunn (1978)
The values given in brackets are percentages of the energy term above.
† The value for L. Sibaya is for phytoplankton, epipsammic algae and macrophytes.

that the small proportion (2.6%) of herbivore production principally represented by *Sarotherodon niloticus*, is an inefficient use of energy fixation. In Lake Sibaya the production of the invertebrate benthos is of the same magnitude. Unfortunately no production data are available for *S. mossambicus* in Lake Sibaya but as this lake has never sustained a viable cichlid fishery it is not unreasonable to suppose that production values very much lower than those reported for *S. niloticus* are to be expected. This confirms an earlier view that we are experiencing an inefficient routing of energy through the invertebrate benthos in Lake Sibaya. The tertiary routes which are followed are given in Fig. 6, and are principally via the predatory crab *Hymenosoma orbiculare* to *Clarias gariepinus* and Man. An equally important route is via the predatory gobiid and cichlid fish to birds.

Compared with Lake George with a herbivore : plant ratio of 3.2 the biomass pyramid for Lake Sibaya is somewhat more conventional. The herbivores are some 27% of the plant biomass – a value not greatly dissimilar from that of Loch Leven, 47%. Although these values are describing an instantaneous situation (Burgis and Dunn *op cit*), they do demonstrate that the autochthonous input of biomass is the primary energy resource; allochthonous inputs are additive to an already balanced situation.

In this chapter I have put together a number of ecological facts in the hope that some principles might emerge: success has been moderate largely because of the fragmented nature of much of our data. Clearly the need to develop research programmes to modify and strengthen the existing framework of knowledge about the lake is a first priority.

Acknowledgement

This chapter has benefited from discussions with Dr J. Talling, FRS; the errors are mine. Dr R. C. Hart and Dr M. N. Bruton commented upon the text of this chapter.

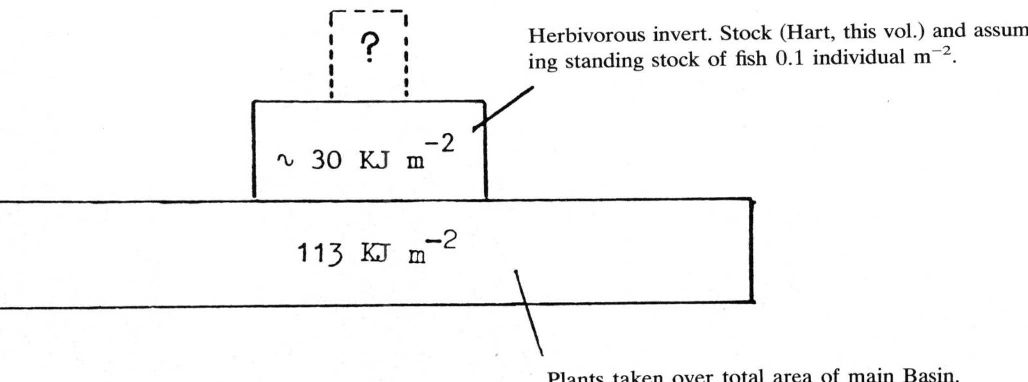

Fig. 7. A biomass pyramid for Lake Sibaya.

References

Allanson, B. R. & R. C. Hart. 1975. The primary production of Lake Sibaya, Kwazulu, South Africa. Verh. Internat. Verein. Limnol. 14: 1426–1433.

Allanson, B. R. & J. D. van Wyk. 1969. An introduction to the physics and chemistry of some lakes in Northern Zululand. Trans. R. Soc. S. Afr. 38: 217–240.

Beauchamp, R. S. A. 1958. Utilizing the natural resources of Lake Victoria for the benefit of fisheries and agriculture. Nature, Lond. 181: 1634–1636.

Blaber, S. J. M. & A. K. Whitfield. 1977. The biology of the burrowing goby *Croilia mossambica* Smith (Teleostei, Gobiidae). Environ. Biol. Fishes 1 (2): 197–204.

Bowen, S. H. 1976. Mechanism for digestion of detrital bacteria by the cichlid fish *Sarotherodon mossambicus* (Peters). Nature, Lond. 260 (5547): 137–138.

Bowen, S. H. 1976. Feeding ecology of the cichlid fish *Sarotherodon mossambicus* in Lake Sibaya, Kwazulu. Ph.D. thesis Rhodes University, Grahamstown, S. Africa.

Bowen, S. H. 1978. Benthic diatom distribution and grazing by *Sarotherodon mossambicus* in Lake Sibaya, South Africa. Freshwater Biol. 8: 449–453.

Bowen, S. H. A nutritional constraint in detritivory by fishes: the stunted population of *Sarotherodon mossambicus* in Lake Sibaya, Ecol. Monogr. 49: 17–31 49, 1979

Bruton, M. N. & B. R. Allanson. 1974. The growth of *Tilapia mossambica* Peters (Pisces: Cichlidae) in Lake Sibaya, South Africa. J. Fish Biol. 6: 701–705.

Burgis, M. J. 1978. Case studies of lake ecosystems at different latitudes in the tropics. The Lake George ecosystem. Verh. Internat. Verein. Limnol. 20: 1139–1152.

Burgis, M. J. & I. G. Dunn. 1978. Production in three contrasting ecosystems. In: Ecology of fresh water fish production. Ed. S. D. Jenkins. Chapter 6. Blackwells, Oxford.

Caulton, M. S. 1978. The effect of temperature and mass on routine metabolisms in Sarotherodon (Tilapia) mossambicus (Peters). J. Fish Biol. 12: 195–201.

Caulton, M. S. & B. J. Hill. 1973. The ability of *Tilapia mossambica* Peters to enter deep water. J. Fish Biol. 5: 783–788.

Hart, R. C. 1973. A contribution to the biology of *Pseudodiaptomus hessei* (Mrázek) (Copepoda: Calanoida) in Lake Sibaya, South Africa. Ph.D. dissertation, Rhodes University, Grahamstown.

Hart, R. C. & B. R. Allanson. 1975. Preliminary estimates of production by a calanoid copepod in subtropical Lake Sibaya. Verh. Internat. Verein. Limnol. 19: 1434–1441.

Hart, R. C. & R. Hart. 1977. The seasonal cycles of phytoplankton in subtropical Lake Sibaya: a preliminary investigation. Arch. Hydrobiol. 80 (1): 85–107.

Howard-Williams, C. & B. R. Allanson. 1978. Swartvlei Project Report Part III: The limnology of Swartvlei with special reference to production and nutrient dynamics in the littoral zone. Institute for Freshwater Studies Special Report No. 78/3: 1–280. Rhodes University, Grahamstown.

Hutchinson, G. E. 1967. A treatise on limnology. II. Introduction to lake biology and the limnoplankton. John Wiley & Sons, Inc., New York.

Johnson, M. G. 1974. Production and productivity. *In:* The benthos of lakes. R. O. Brinkhurst (1974). Macmillan Press Ltd., London.

Ryder, R. A. 1965. A method for estimating the potential fish production of north temperate lakes. Trans. Am. Fish. Soc. 94: 214–218.

Sorokin, Ju. I. 1965. On the trophic role of chemosystems and bacterial biosystems in water bodies. In: C. R. Goldman (Ed.) Primary productivity in aquatic environments. Mem. Ist. ital. Idrobiol. 18 suppl.: 189–205. University of California Press, Berkeley.

12 A history of the research station at Lake Sibaya: its people and development

A. G. Bruton

The development of research activities by the Institute for Freshwater Studies at Lake Sibaya progressed through three well-defined periods. During the years 1965 to 1968 work was carried out on an expeditionary basis from Rhodes University, Grahamstown, using tented camps on the lake shore and small boats with a minimum of sophisticated equipment. With the founding of a permanent research station by Professor B. R. Allanson in January 1968 a centre was established from which postgraduate students and research limnologists could carry out projects on a subtropical pristine lake ecosystem.

Early History

The first expedition to Lake Sibaya by Rhodes University staff and students took place during July 1965. A tented camp was established on the western shore of the lake and teams set out in a small fibreglass dinghy. The high winds and unpredictable weather characteristics of Lake Sibaya soon dictated the timing and extent of the various programmes. The first bathymetric survey, using a sounding weight and calibrated winch from the dinghy, was carried out almost exclusively at night to take advantage of calmer lake conditions.

The next expedition in January 1966 brought with it improvements in the equipment, mainly in the provision of a larger (5 metre) aluminium launch fitted with a 10 hp outboard engine. The provision of an echo-sounder enabled more complete bathymetric data to be gathered.

Subsequent expeditions under the leadership of Professor B. R. Allanson, the late Dr R. E. Boltt and Dr B. J. Hill, camped on the eastern shore of the lake (Plate 1) with access from a road which ran along the base of the coastal dune. These expeditions, which continued on a twice yearly basis, returned to this site in the lee of the dune and camped amongst the *Acacia karroo* trees which formed a fringe along the lake shore. From the tented camp at Sibaya trips were made to the Kosi lake system 60 km to the north.

It was soon found that the unique characteristics of the coastal lakes of this area and the many opportunities for further study signified the need for long term involvement. The two day trip from Rhodes University in Grahamstown was long and arduous and a limited number of projects could be tackled in the time available at the lake side. With these factors in mind Professor Allanson approached the office of the Chief Bantu Affairs Commissioner in Natal and put forward proposals for the establishment of a

Plate 1. Processing water samples at the tented camp, Professor Allanson and Mrs Gill Boltt, July 1967.

permanent research base on the shores of Lake Sibaya. The request was well received and during the closing months of 1967 a site was cleared and planned (Fig. 1) on the eastern shore of the lake in the vicinity of the old camping area. The first building to be erected was a prefabricated six roomed laboratory/residence (Plate 2), the loose panelled construction of which was ideally suited to the various roles it was to play in the development of the station. An engine room, campsite ablution block and a small workshop/storeroom followed. A 5 hp air-cooled diesel lighting plant of 3.0 kw, 220 volt output was installed in the engine room. The lake provided an ample supply of fresh water to the complex via a feed pipe extending into the lake, an electrical centrifugal pump and overhead tanks.

The buildings were supplied with basic facilities only. Such items as hot water gas appliances and laboratory equipment were fitted by Rhodes University technical staff. A campsite area was cleared to provide accommodation for short term visitors and trees planted to provide shade from the tropical sun.

The workshop was an essential facility at the field station. Due to the isolation of the station a comprehensive stock of raw materials, fittings and paint had to be maintained to ensure the efficient running of the complex. As the list of capital equipment grew so did the facilities in the workshop to cope with the repair and service of these items. The purchase of a powerful battery-charging plant, welding equipment and a second-hand lathe were steps in this development. The capital equipment list included four boats, four outboard engines, two electric water pumps, two 4 stroke centrifugal water pumps, the lighting plant and the station vehicle, all of which were maintained from the workshop. Due to a lack of suitable repair apparatus only minor repairs were undertaken on scientific instruments.

The first staff members to reside at the now established Lake Sibaya Research Station were Zoology Honours graduates from Rhodes University, John Minshull and his wife Jacqueline who arrived in January 1968. A local Thonga, Mr Gilbert Mhlambo, was hired at this time to assist with cleaning and maintenance and served on the station until his retirement in April 1975. The entire grounds of the station were grassed and permission was obtained from the local conservation authority to plant the garden with some rare and interesting indigenous plants. Temporary structures were made from locally available materials such as stout wooden poles, reeds and thatching grass to serve as outdoor storage areas. These structures were cheap and easy to rebuild when they became unsound.

Robert Hart and his wife Rei, both from Natal University in Pietermaritzburg, joined the staff of the research station during June 1969 and a new cottage was built on the northern perimeter of the grounds for their occupation.

John Minshull left the station at the end of 1969 and was succeeded as officer-in-charge by Robert Hart. Michael Bruton joined the staff in 1970 and another Thonga from the local village, Mr Nelson Mdletshe, was hired and worked at the station for six years during which time he became an invaluable field assistant and master fisherman. The inherent skill of the two

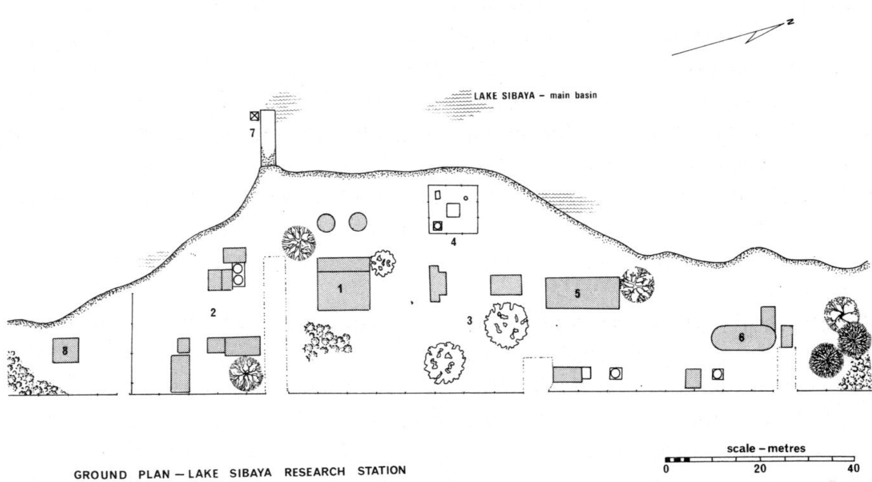

Fig. 1. Key to the Ground Plan – Lake Sibaya Research Station as at September 1975. 1. Laboratory and small guest flat. The two circular huts in front of this building serve as outdoor storage areas. 2. Service Complex: engine room, workshop, storerooms and water tanks. 3. Camping Area with toilet and shower block, indoor and outdoor cooking facilities. 4. Weather Station with evaporation tanks, rain guage, temperature and humidity recorders. 5. Residence: technical officer. 6. Residence: biologist. 7. Jetty and Lake Level Recorder. 8. Caretaker's Cottage.

Plate 2. The research station in an early stage of development (January 1969). Buildings visible are the laboratory, outdoor shelters and the campsite ablution block.

Thonga workmen in the local methods of construction of temporary buildings, fences and of fishing gear enabled these operations to be performed most economically and efficiently. Robert Hart left Sibaya in July 1972 to continue his studies at the Institute for Freshwater Studies base in Grahamstown and Michael Bruton, who was joined by his wife Carolynn in April 1972, began his four year spell as officer-in-charge. Christopher Appleton and his wife Margaret, of the Bilharzia Research Unit, Nelspruit, Transvaal, spent 11 months at the station during 1973 carrying out a field study on bilharzia host snails.

In each instance the staff made a personal and significant contribution to the facilities of the research station often beyond their normal line of duty. In this way the buildings, grounds and equipment were continually improved.

Until early 1974 the biologists were, with limited assistance from the Thonga labourers, responsible for the running and routine maintenance of the complex as well as the design and construction of their own sampling gear. It was found that an increasing amount of their time was spent on technical tasks and thus valuable research time was being wasted. The growth of the station and the increasing use made of its facilities necessitated the appointment of a staff member who could handle technical and administrative tasks and so enable the biologists to concentrate more closely on their individual research projects. Accordingly, in April 1974, Anthony Bruton and his wife Nola joined the staff and took over these duties. A two-bedroomed house was erected in the grounds by the KwaZulu Government to accommodate the new couple. The provision of this additional dwelling allowed for a modification of the original prefabricated house to include a self-contained flat for visiting scientists.

The increasing interest in Sibaya and its environs which was generated by the activities at Lake Sibaya Research Station and the expanded accommodation facilities thereafter ensured an almost continuous succession of visiting scientists at the station. In 1968, the first year of operation, 42 persons visited the station. In 1976, 202 persons passed through spending a total of 639 visitor days at the station. Of these visitors the majority were scientists from 31 different institutions, the balance being conservationists, planners and environmentalists. The conversion of one of the rooms of the laboratory building provided a spacious office-cum-meeting room where many meetings critical to the future of the Sibaya area were held between scientists, planners and senior government officials.

Inevitably the wives of the staff members became involved in the affairs of the station. The tasks which they performed covered such items as typing, bookkeeping, the routine gathering of meteorological data, laboratory assistance and visitor bookings. The initiation and upkeep of a herbarium collection and the management of the fast-growing library and reprint collection were also the responsibility of the ladies. Once the newly appointed technical officer had overcome an initial backlog of maintenance and repair work he was instructed in basic field sampling techniques. After this instruction several projects were carried out by the technical officer for research staff based in Grahamstown.

The major difficulties associated with the running of the remote field station were in respect of transport and communications. Although the Sibaya area is dotted with small Thonga villages, the nearest large settlement, a forest station with unreliable telephone facilities, was 20 km away. A close liaison with the Institute for Freshwater Studies headquarters was therefore impossible. The postal delivery made through a local store was slow and frequently delayed. The station was situated 80 km from the nearest macadamised road and during the rainy season the two access roads were frequently flooded or closed due to muddy conditions. During the dry season the dirt tracks were in poor condition increasing the maintenance costs of vehicles. The nearest centre with a comprehensive range of hardware and supplies was Empangeni, a three and a half hour drive under good conditions.

When the station was first established it was necessary to use four wheel drive vehicles to reach the site in its remote spot. An aged Land Rover seconded from the Rhodes University vehicle pool gave solid service until its engine expired in late 1973. This vehicle, then 15 years old, was fitted with a second-hand engine donated by the Mseleni Mission Hospital and so-equipped ran for another one and a half years, at which stage it was replaced. The roads had, in the interim, been improved and it was possible to use a two wheel drive vehicle in all but the worst conditions. The operation of the station was made considerably more effective by the generous donation of our total fuel and oil requirements by international oil companies. Although the facilities were run in strict accordance to the fuel conservation restrictions there is no doubt that, without this donation, financial constraints would soon have strangled research projects on the lake and its environs.

Research Equipment

The research equipment at the lake station comprised only the basic essentials necessary to the various programmes and in most cases more sophisticated analyses were carried out at the University. In the case of samples which deteriorated during transport to Grahamstown the necessary instrumentation for their processing was brought to the station on a temporary basis. The use of SCUBA (Self Contained Underwater Breathing Apparatus) (Plate 3) was someting of a pioneering venture in southern African limnology when it was first used in the lake during early expeditions. The following points will demonstrate the essential role which SCUBA played in research on Lake Sibaya:

a) general observation and broad surveys of visible plants, animals and substrate types.

b) detailed observations of fish behaviour, distribution of macrofauna and microfauna.

Plate 3. SCUBA divers A. T. Forbes and T. H. Wooldridge preparing to enter the water.

Plate 4. The Director of the Institute for Freshwater Studies, Professor B. R. Allanson and Miss Sandra Rudd conducting an investigation of the episammic algal community using the 'Hookah' diving apparatus.

c) quantitative collections of the benthic infauna and epifauna, fish faeces and diatom-rich material.

d) checking the efficiency of sampling apparatus such as grabs, substrate bins, longlines, seine and trawl nets.

e) repairs to the boats, laying of buoys and moorings, recovery of lost gear.

Later the purchase of a 'Hookah' diving apparatus (Plate 4) extended underwater investigations. This portable apparatus consisted of a small engine and belt-driven compressor with pressure gauges, safety valve and a sophisticated filtration unit. Two fifteen metre long hoses were fitted with conventional second stage aqualung type regulators. This unit, carried in a boat, was used to great effect primarily for investigations into the plants and animals of the shallow terraces of Lake Sibaya. Divers working in the warm clear waters of Lake Sibaya had to be constantly aware of the ever present danger from hippopotami and crocodiles.

The boats used in the early part of the programme were two solidly constructed 5 m aluminium launches. These two vessels, 'Tilapia' (plate 5) with a small forward cabin, powered by a 20 hp and later a 33 hp outboard engine, and 'Sibayi' (Plate 6) an open deck runabout with a 10 hp and later 20 hp outboard engine, gave sterling service under the most arduous conditions. Both vessels were fitted with removable davits to which hand-operated winches could be clamped and 'Tilapia' was equipped with a

Plate 5. 'Tilapia' equipped for a sampling trip on the lake. Equipment on the boat includes a current speed and direction recorder, an echo-sounder and a bathythermograph.

recording echo-sounder. 'Sibayi' with its large open deck area was an efficient craft for netting programmes. Buoys, which could be lighted at night by means of flashing lanterns, were laid in the lake at fixed sampling stations and along north-south and east-west transects. The position of the research station (Plate 7) on an exposed section of the main basin shoreline necessitated the construction of a wooden jetty and offshore mooring facilities.

Plate 6. The 5 m aluminium craft 'Sibayi' on a sampling trip to the upper reaches of the lake.

Plate 7. A view westwards of the research station, its jetty and the boat mooring, from the dune forest.

High winds were always a feature of the open lake and a severe restriction was placed on the type of weather in which one could safely operate in a 5 m boat. Moreover it was felt that a larger, more stable vessel would enable more sophisticated projects to be carried out on the lake. Consequently the vessel 'Croilia' (Fig. 2 and Plate 8) was launched on Lake Sibaya in June 1973 and 'Tilapia' was transferred to another Institute for Freshwater Studies project. The catamaran hull of 7.7 m length and 2.7 m beam with a displacement of approximately $1\frac{1}{2}$ tons, made 'Croilia' the ideal craft for work on Sibaya. The original design of this craft was copied by Dr Boltt, from that of a larger vessel operating on Loch Leven. A standard catamaran hull was cast in fibreglass and the deck arrangement subsequently designed in consultation with limnologists on the staff. A small cabin with table space

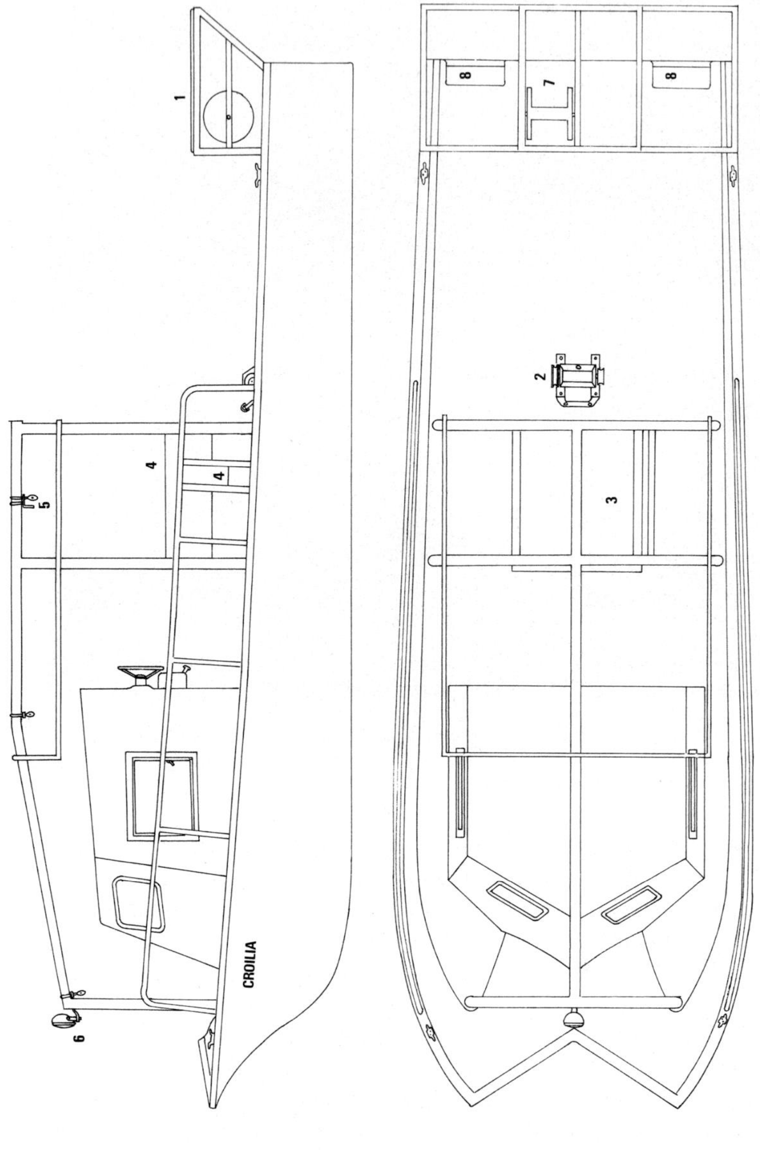

Fig. 2. Research vessel 'Croilia'. 1. Net setting desk. 2. Double capstan electric winch 12/24v. 3. Well (with removable cover). 4. Transom boards for the attachment of portable winches. 5. Anchor rope pulley, and heavy duty block. 6. Spotlight and forward anchor rope pulleys. 7. Anchor line reel. 8. Outboard engine attachment points.

Plate 8. 'Croilia', a 7.7 metre fibreglass catamaran. This craft was ideally suited to the conditions on Lake Sibaya.

and a seat with locker, was located well forward, and the helmsman's station attached to the cabin. Amidships a strong metal gantry above the central deck well allowed heavy equipment to be lowered into the lake by means of overhead pulleys. A double capstan winch allowed the easy recovery of heavy objects and was positioned so as to serve in raising the anchor. The 'Seascribe' recording echo-sounder was transferred to this craft which also had three positions to which portable calibrated limnological winches could be attached. A large 12 volt lead-acid battery located in the cabin supplied power to the winch, echo-sounder, cabin lamp and spotlight. A broad platform extending aft over the outboard engines served as a net setting area for the laying of gillnets and trawl nets. The engines fitted to the boat, a 33 hp and 25 hp were able to raise the boat to a planing attitude under limited load. In January 1975 a 40 hp engine was purchased to replace the smaller version, improving the performance of the craft. A heavy duty offshore mooring, consisting of four large transversely placed Danforth-type sand anchors, was placed by SCUBA divers to provide security during heavy weather. A small fibreglass dinghy was purchased for use as a tender. On a typical sampling trip to the further reaches of the lake the large craft would be anchored offshore and short trips to sampling stations would be made using the dinghy.

The capital equipment list grew as projects were expanded and funds became available. Standard limnological sampling gear was always at hand. A feature, though, of the equipment available at the research station was the fascinating array of improvised gear in the design of which people such as the late Dr Robin Boltt were masters. The isolated position of the station and often peculiar requirements of a sampling programme brought innovative thinking to the fore. This equipment, whose applications are described in other chapters, included the following:

1) electronic apparatus – multiple probe temperature recorders, an ultra low light level meter and later a pilot programme of sonic fish tagging.

2) a design of 'substratum bin' for the sampling of plankton during their diel movements (Hart, 1976).

3) fishing equipment such as longlines and a lightweight trawl net (Bruton, 1977).

4) impoundments for the observation of fish behaviour under semi-natural conditions. Such impoundments took the form of plastic lined pools in the grounds and later a concrete pool with observation tower and an infra-red lighting system (Bruton, 1977).

5) numerous items of apparatus were constructed for the randomized collection of the substrate and benthic epifauna and infauna (Boltt, 1969, Bowen, 1976).

A large variety of nets such as gillnets, seine nets and a selection of devices used by local Thonga fishermen were on hand. Frequent use was made of traditional Thonga fishing gear such as 'umono' traps and 'fonya' baskets (Plate 9) to supplement normal gear during sampling trips. The humidity of the semi-tropical Maputaland plain and the station's proximity to the sea dictated a rigid anti-corrosion programme. Field instruments could never be stored with dry cells in place as these would soon leak and cause damage to the circuits. Laboratory instruments such as the spectrophotometer and microscopes had to be kept under dehydrated conditions.

The Department of Water Affairs made a considerable contribution to the monitoring of physical parameters at Lake Sibaya. Their field officers maintained a close liaison with the staff at the research station. Two water level recorders were positioned on the eastern shore of the lake and a weather station was erected at the research station. Climatological data gathered on a daily basis included the lake level, rainfall, evaporation by Symons Tank and 'A' Pan, temperature, humidity, hours of sunshine and, for a limited period, wind strength. During the period of heavy rainfall and consequent rapid rise in lake level of 1976–77 the weather station was moved no less than three times in order to ensure continuous monitoring during this critical period. Later when the station was finally abandoned, a water level recorder was installed in the South Basin at a point where it could easily be serviced by forestry department employees. A wave height measuring apparatus was constructed at the station and, in conjunction with a timing apparatus and wind speed recorder, used to relate wind strength to wave height in the open lake. A flow rate measurement of two major

Plate 9. 'Fonya' thrust baskets, a traditional amaThonga fishing device, used for catching spawning *Clarias gariepinus* at Lake Sibaya.

streams discharging into Lake Sibaya was conducted by the research station staff over an extended period.

Flooding

Lake Sibaya Research Station was positioned at the base of a 167 m coastal dune with a slope of approximately 30°. The access road, serving the station, a village and forest station beyond, ran along the base of this dune.

A sharp rise in lake level (Plate 10) took place as a result of heavy rain during February and March 1975 and during that year the first steps were taken to combat the erosion of the grounds by wave action. Materials for this purpose were provided by the local forest station. When a further sudden rise in lake level occurred during the early months of 1976 efforts to minimize the damaging effects of wave action were redoubled and events took a more serious turn.

The construction of a dyke (Plate 11) across the entire lake frontage of the research station began. The wall was erected in the following manner: a double row of hardwood poles 3.6 m long were sunk into the sand in parallel lines using a high pressure jet of water from a portable water pump. A dry mix of sand and cement was made up and placed into hessian grain bags. The bags were placed in an interlocking arrangement between the rows of poles and the mixture then wetted. In places the wall reached a height of two

Plate 10. The research station at the peak of its development in July 1975. Beyond the laboratory building (partly obscured by trees in the foreground) are the two staff houses. Compare the water level to that shown on Plate 2.

metres and approximately 2500 grain bags were used in its construction. An area behind the wall was backfilled to reclaim ground. The impressive strength of this barrier was able to break the force of the waves thus keeping at bay erosion of the sandy substrate, and allowed the station to continue functioning in 1977.

The rains during the winter months of 1976 maintained a higher average than usual, preventing evaporation from reducing the lake level to a marked degree. At this stage it became obvious that if the early months of 1977

Plate 11. The flooded buildings, showing the sandbag wall and backfilled area (February 1977).

showed the same rainfall pattern as previous years, then the station would be flooded. Accordingly the equipment evacuation programme, which so far had been restricted to the more expensive and fragile items, was accelerated and all goods not actually engaged in research or essential maintenance schedules were moved to storage.

As the lake level rose during the closing months of 1976 it became obvious that the roads on either side of the station would become impassable due to the rise in lake level even before the station buildings became flooded. Again we are indebted to KwaZulu who through their Department of Forestry kept the roads in reasonable order until it became quite impossible to continue operating from the site.

The fabric of the station has deteriorated until in 1979 the shells of its buildings serve as nesting and roosting sites for the birds of the lake – *Naturam expellas furca, tamen usque recurret*!

References

Boltt, R. E. 1969. The benthos of some southern African lakes. Part II. The epifauna and infauna of the benthos of Lake Sibayi. Trans. R. Soc. S. Afr. 38: 249–269.

Bowen, S. H. 1976. Feeding ecology of the cichlid fish *Sarotherodon mossambicus* in Lake Sibaya, KwaZulu. Ph.D thesis, Rhodes University, Grahamstown.

Bruton, M. N. 1977. The biology of *Clarias gariepinus* (Burchell, 1822) in Lake Sibaya, KwaZulu, with emphasis on its role as a predator. Ph.D. thesis, Rhodes University, Grahamstown.

Hart, R. C. 1976. The substrate bin – a new sampling device for studying diel vertical migratory movements on to and off lake sediments. Freshwater Biol. 6: 155–159.

13 Bibliography

P. M. Eva

This bibliography spans the years 1952 to date, but very few of the publications refer to work done prior to the establishment of the Lake Sibaya Research Station. The work includes three M.Sc. theses and five Ph.D. dissertations, and covers a wide spectrum of organisms and physico-chemical components. Important fields of study not included in publications already printed, such as lake currents, epipsammic production and aquatic macrophytes have been included in chapters to this volume.

There are a number of publications cited which have a restricted distribution, such as reports of Regional Planning Commissions, and to Government Departments. Although not widely available, these contributions are considered to be an important part of the Lake Sibaya literature as they form the link between the academic and the applied whole, and in effect translate the academic findings to a form useful to those involved in making the decisions on the long term planning and utilization of the area.

Allanson, B. R. 1966. Lake Sibayi, Tongaland, Natal. A preliminary report from the Department of Zoology and Entomology, Rhodes University, Grahamstown.

Allanson, B. R. 1968. A new research station in South Africa. News Lett. Limnol. Soc. Sth. Afr. No. 10: 8–10.

Allanson, B. R. 1970. An examination of the value of ecological research in the control of Bilharziasis. Central African Journal of Medicine Suppl. to 16 No. 7: 14–19.

Allanson, B. R. 1973. Report on the utilization of Lake Sibaya. Secretary for Bantu Administration and Development, Pretoria, April 1973.

Allanson, B. R. 1974/1975. A further note on the primary production of Lake Sibaya. Institute for Freshwater Studies Reports and Reprints 1974/1975.

Allanson, B. R. 1976. Environmental factors in the limnology of coastal lakes in South Africa. Proceedings of the 1st Interdisciplinary Conference on Marine and Freshwater Research in Southern Africa. Port Elizabeth 5–10 July 1976.

Allanson, B. R. 1976. Ecology of water areas. Proceedings of the International symposium Planning for Environmental conservation, 4–6 September 1973. pp. 73–75. Published by the Organising Committee, International Symposium, Planning for Environmental Conservation, Department of Planning and the Environment, Pretoria.

Allanson, B. R. (in press) Coastal lakes. Ed. J. H. Day.

Allanson, B. R., M. N. Bruton & R. C. Hart. 1974. The plants and animals of Lake Sibaya, KwaZulu, South Africa: A checklist. Rev. Zool. afr. 88: 507–532.

Allanson, B. R. & R. C. Hart. 1975. The primary production of Lake Sibaya, KwaZulu, South Africa. Verh. Internat. Verein. Limnol. 19: 1426–1433.

Allanson, B. R., B. J. Hill, R. E. Boltt & V. Schultz. 1966. An estuarine fauna in a freshwater lake in South Africa. Nature, Lond. 209 (5022): 532–533.

Allanson, B. R. & S. Rudd. 1976/1977. Epipsammic production in the littoral of Lake Sibaya. Institute for Freshwater Studies Annual Reports and Reprints 1976/1977.

Allanson, B. R. & J. D. van Wyk. 1969. An introduction to the physics and chemistry of some lakes in Northern Zululand. Trans. R. Soc. S. Afr. 38: 217–240.

Anon. 1971. The 1947 Tongaland expedition. Natal Wildlife 12 (1): 21–25.

Anon. 1977. Institute for Freshwater Studies, Rhodes University. Scientiae 18 (4): 35–37.

Appleton, C. C. 1972/1973. The future bilharzia programme at Lake Sibaya. Institute for Freshwater Studies Annual Reports and Reprints 1972/1973.

Appleton, C. C. 1976. The influence of abiotic factors on the distribution of *Biomphalaria pfeifferi* (Krauss, 1848) (Planorbidae: Mollusca) and its life cycle in south-eastern Africa. M.Sc. thesis, Rhodes University, Grahamstown, January 1976.

Appleton, C. C. 1977. The influence of above-optimal constant temperatures on South African *Biomphalaria pfeifferi* (Krauss) (Mollusca: Planorbidae). Trans. R. Soc. trop. Med. Hyg. 71 (2): 140–143.

Appleton, C. C. 1977. The influence of temperature on the life-cycle and distribution of *Biomphalaria pfeifferi* (Krauss, 1948) in South-Eastern Africa. Int. J. Parasitol. 7: 335–345.

Appleton, C. C. 1977. The freshwater mollusca of Tongaland with a note on molluscan distribution in Lake Sibaya. Ann. Natal Mus. 23: 129–144.

Appleton, C. C. & M. N. Bruton. 1979. The epidemiology of schistosomiasis in the vicinity of Lake Sibaya with a note on the rest of Tongaland. Ann. trop. Med. Parasit. (in press).

Archibald, R. E. M. 1966. Some new and rare diatoms from South Africa. 2. Diatoms from Lake Sibayi and Lake Nhlange in Tongaland (Natal). Nova Hedwigia 12 (3+4): 477–495.

Blaber, Stephen J. M. & Alan K. Whitfield. 1977. The biology of the burrowing goby *Croilia mossambica* Smith (Teleostei, Gobiidae). Environ. Biol. Fishes 1 (2): 197–204.

Boltt, R. E. 1969. A contribution to the benthic biology of some Southern African lakes. Ph.D. dissertation, Rhodes University, Grahamstown.

Boltt, R. E. 1969. The benthos of some Southern African lakes. Part II. The epifauna and infauna of the benthos of Lake Sibayi. Trans. R. Soc. S. Afr. 38: 249–269.

Boltt, R. E. 1972/1973. Coastal lakes benthos. Institute for Freshwater Studies Annual Reports and Reprints 1972/1973.

Boltt, R. E., B. J. Hill & A. T. Forbes. 1969. The benthos of some southern African lakes. Part 1. Distribution of aquatic macrophytes and fish in Lake Sibayi. Trans. R. Soc. S. Afr. 38: 241–248.

Bowen, Stephen H. 1976. Mechanism for digestion of detrital bacteria by the cichlid fish *Sarotherodon mossambicus* (Peters). Nature, Lond. 260 (5547): 137–138.

Bowen, S. 1976. Feeding ecology of the cichlid fish *Sarotherodon mossambicus* in Lake Sibaya, KwaZulu. Ph.D. thesis, Rhodes University, Grahamstown.

Bowen, Stephen H. 1978. Benthic diatom distribution and grazing by *Sarotherodon mossambicus* in Lake Sibaya, South Africa. Freshwater Biol. 8: 449–453.

Bowen, Stephen H. 1979. A nutritional constraint in detritivory by fishes: the stunted population of *Sarotherodon mossambicus* in Lake Sibaya. Ecol. Monogr. 49: 17–31.

Breen, C. M. 1971. An account of the plant ecology of the dune forest at Lake Sibayi. Trans. R. Soc. S. Afr. 39 (3): 223–234.

Breen, C. M. & I. D. Jones. 1971. A preliminary list of angiosperms collected in the vicinity of Lake Sibayi. Trans. R. Soc. S. Afr. 39 (3): 235–245.

Bruton, M. N. 1971. Territorial behaviour in natural crocodile populations. Animal Behaviour Symposium, Zoological Society of Southern Africa, Durban, July 1971.

Bruton, M. N. 1973. A contribution to the biology of *Tilapia mossambica* Peters in Lake Sibaya, South Africa. M.Sc. thesis, Rhodes University, Grahamstown.

Bruton, M. N. 1974. Utilization of the fish fauna of Lake Sibaya. Bantu Investment Corporation of S.A. Ltd., August 1974. 9pp.

Bruton, M. N. 1974. Two new fish records from Lake Sibaya. News Lett. Limnol. Soc. sth. Afr. No. 22: 47–48.

Bruton, M. N. 1974. Lake Sibaya. Afr. wild life 28 (3): 13–15.

Bruton, M. N. 1975. Lake Sibaya – a land-locked estuary. Scientiae 16 (6): 18–28.

Bruton, M. N. 1975. First record of *Sarotherodon placidus* (Pisces: Cichlidae) from South Africa. Lammergeyer 22: 33–36.

Bruton, M. N. 1975. Key conservation areas in Tongaland. Report to the Town and Regional Planning Commission, Pietermaritzburg. Part I. 7pp. Part II. 2pp.

Bruton, M. N. 1975. Nyanene Forest at Lake Sibaya, with recommendations on its conservation. Report to the KwaZulu Department of Agriculture and Forestry, Ulundi. 4 pp.

Bruton, M. N. 1976. An outline of the proposed Tongaland Marine Reserve. Report to the National Marine Reserves Committee of the South African Department of Sea Fisheries, Cape Town. 8pp.

Bruton, M. N. 1976. On the size reached by *Clarias gariepinus.* J. Limnol. Soc. sth. Afr. 2 (2): 57–58.

Bruton, M. N. 1976. The utilization of the fishes of Lake Sibaya. Report to the KwaZulu Government. 11 pp.

Bruton, M. N. 1976. A proposal for a Marine Nature Reserve and research station in Tongaland. Report to the KwaZulu Conservation Trust and World Wildlife Fund. 14 pp.

Bruton, M. N. 1976. An outline of the proposed Tongaland Nature Reserve. Report to the proposed KwaZulu Conservation Trust. 5 pp.

Bruton, M. N. 1976. The coastal dune forest of north-eastern Zululand, with comments on conservation. Report to the KwaZulu Department of Agriculture and Forestry, Ulundi. 16 pp.

Bruton, M. N. 1976. Recommendations on the conservation and utilisation of eastern Tongaland. Report to the Town and Regional Planning Commission, Pietermaritzburg. 4 pp.

Bruton, M. N. 1977. The biology of *Clarias gariepinus* (Burchell, 1822) in Lake Sibaya, KwaZulu, with emphasis on its role as a predator. Ph.D. thesis, Rhodes University, Grahamstown.

Bruton, M. N. 1978. Recent mammal records from Eastern Tongaland in KwaZulu, with notes on hippopotamus in Lake Sibaya. Lammergeyer 24: 19–27.

Bruton, M. N. 1978. The habitats and habitat preferences of *Clarias gariepinus* (Pisces: Clariidae) in a clear coastal lake (Lake Sibaya, South Africa). J. Limnol. Soc. sth. Afr. (in press).

Bruton, M. N. 1979. The survival of habitat desiccation by air breathing clariid catfishes. Environ. Biol. Fishes (in press).

Bruton, M. N. 1979. Old two legs - the life of the sharptooth catfish. Afr. wild life (in press).

Bruton, M. N. 1979. The breeding biology and early development of *Clarias gariepinus* (Pisces: Clariidae) in Lake Sibaya, South Africa, with a review of breeding in species of the subgenus *Clarias* (*Clarias*). Trans. zool. Soc. Lond. 35: 1–45.

Bruton, M. N. 1979. The food and feeding behaviour of *Clarias gariepinus* (Pisces: Clariidae) in Lake Sibaya, South Africa, with emphasis on its role as a predator of cichlids. Trans. zool. Soc. Lond. 35: 47–114.

Bruton, M. N. 1979. The role of diel inshore movements by *Clarias gariepinus* for the capture of fish prey. Trans. zool. Soc. Lond. 35: 115–138.

Bruton, M. N. 1979. The effect of lake level changes on the nesting behaviour of *Sarotherodon mossambicus* (Pisces: Cichlidae) in Lake Sibaya. S. Afr. J. Zool. (in press).

Bruton, M. N. & B. R. Allanson. 1974. The growth of *Tilapia mossambica* Peters (Pisces: Cichlidae) in Lake Sibaya, South Africa. J. Fish Biol. 6: 701–705.

Bruton, M. N. & B. R. Allanson (in press). The growth of *Clarias gariepinus* (Pisces: Clariidae) in Lake Sibaya, South Africa. S. Afr. J. Zool.

Bruton, M. N. & R. E. Boltt. 1975. Aspects of the biology of *Tilapia mossambica* Peters (Pisces: Cichlidae) in a natural freshwater lake (Lake Sibaya, South Africa). J. Fish Biol. 7: 423–445.

Bruton, M. N. & W. D. Haacke. 1975. New reptile records from the tropical transition zone of south-east Africa. Lammergeyer 22: 23–32.

Campbell, G. D. & B. R. Allanson. 1952. The fishes of the 1947, 1948 and 1949 scientific investigations of the Kosi area, organized by the Natal Society for the Preservation of Wild Life and Natural Resorts. April Magazine Natal Society for the Preservation of Wild Life and Natural Resorts 1 (4): 1–8.

Campbell, G. G. 1969. A review of scientific investigations in the Tongaland area of northern Natal. Trans. R. Soc. S. Afr. 38 (4): 305–316.

Caulton, M. S. 1973. The ability of four species of southern African cichlid fishes to enter deep water. M.Sc. thesis, Rhodes University, Grahamstown.

Dussart, B. 1972. Les Copépodes du lac Sibayi (Natal). Bull. Mus. Hist. nat. Paris, 3e sér., No. 68, Zoologie 54: 869–873.

Farquharson, F. L. 1970. A new freshwater gobi (Pisces: Gobiidae) from Lake Sibayi, Zululand, South Africa. Ann. Cape Prov. Mus. (Nat. Hist.) 8 (10): 85–87.

Forbes, A. T. & B. J. Hill 1969. The physiological ability of a marine crab *Hymenosoma orbiculare* Desm. to live in a subtropical freshwater lake. Trans. R. Soc. S. Afr. 38: 271–283.

Hart, R. C. 1973. A contribution to the biology of *Pseudodiaptomus hessei* (Mrázek) (Copepoda: Calanoida) in Lake Sibaya, South Africa. Ph.D. dissertation, Rhodes University, Grahamstown.

Hart, R. C. 1976. The substrate bin – a new sampling device for studying diel vertical migratory movements on to and off lake sediments. Freshwater Biol. 6: 155–159

Hart, R. C. 1977. Feeding rhythmicity in a migratory copepod (*Pseudodiaptomus hessei* Mrázek). Freshwater Biol. 7: 1–9.

Hart, R. C. 1978. Horizontal distribution of the copepod *Pseudodiaptomus hessei* in subtropical Lake Sibaya. Freshwater Biol. 8: 415–421.

Hart, R. C. & B. R. Allanson. 1975. Preliminary estimates of production by a calanoid copepod in subtropical Lake Sibaya. Verh. Internat. Verein. Limnol. 19: 1434–1441.

Hart, R. C. & B. R. Allanson. 1976. The distribution and diel vertical migration of *Pseudodiaptomus hessei* (Mrázek) (Calanoida: Copepoda) in a subtropical lake in southern Africa. Freshwater Biol. 6: 183–198.

Hart, R. C. & R. Hart. 1977. The seasonal cycles of phytoplankton in subtropical Lake Sibaya: A preliminary investigation. Arch. Hydrobiol. 80 (1): 85–107.

Hill, B. J. 1965. Department of Zoology and Entomology, Rhodes University, expedition to northern Zululand. News Lett. Limnol. Soc. sth. Afr. 2 (2): 26.

Hill, B. J. 1969. The bathymetry and possible origin of Lakes Sibayi, Nhlange and Sifungwe in Zululand (Natal). Trans. R. Soc. S. Afr. 38: 205–216.

Hill, B. J. 1975. The origin of Southern African coastal lakes. Trans. R. Soc. S. Afr. 41: 225–240.

Minshull, J. L. 1968. A summary of the results obtained at the Lake Sibayi Research Station, for the period April to October 1968. Institute for Freshwater Studies Collection of Reprints and Reports 1968.

Minshull, J. L. 1969. Introduction to the food web of Lake Sibaya. Northern Zululand. News Lett. Limnol. Soc. sth. Afr. No. 13 (Suppl.): 20–25.

Pike, T. 1969. Composition, relative abundance and size range of fish populations in Lake Sibaya. News Lett. Limnol. Soc. sth. Afr. No. 13 (Suppl.): 38–43.

Pitman, W. V. & I. P. G. Hutchison. 1975. A preliminary hydrological study of Lake Sibaya. Hydrological Research Unit Report No. 4/75. 35 pp. University of the Witwatersrand, Johannesburg.

Rabie, A. L. 1966. Sibayimeer: Verslag oor besoek – Mei 1966. Hidrologiese Afdeling, Departement Waterwese, Pretoria.

Ribbink, A. G. J. 1976. A contribution to the understanding of the ethology of the cichlids of Southern Africa. Ph.D. dissertation, Rhodes University, Grahamstown.

Rudd, S. 1977. Diving ecology of Lake Sibaya, KwaZulu. E. Cape Naturalist No. 62: 20–22.

Scott, K. M. F. 1968. On some Trichoptera from Northern Zululand, South Africa. Proc. R. ent. Soc. Lond. (B) 37 (1–2): 1–8.

Tinley, K. L. 1958. A preliminary report on the ecology of Lake Sibayi, with special reference to the Hippopotamus. January to April 1958. Natal Parks, Game & Fish Preservation Board.

Tinley, K. L. 1964. Fishing methods of the Tonga tribe in north-eastern Zululand and southern Mozambique. Lammergeyer 3 (1): 9–39.

Tinley, K. L. 1976. The ecology of Tongaland. Report 1. Lake Sibayi. Natal Branch of the Wildlife Society. pp. 9–31.

Vahrmeyer, J. 1966. Notes on the vegetation of northern Zululand. Afr. wild life 20 (2): 151–161. June 1966.

Van Bruggen, A. C. & C. C. Appleton. 1977. Studies on the ecology and systematics of the terrestrial molluscs of the Lake Sibaya area of Zululand, South Africa. Zool. Verh. Leiden 154: 1–44.

van Hille, J. C. 1971. Anthicidae (Coleoptera) from Northern Zululand. Trans. R. Soc. S. Afr. 39 (4): 367–391.

Taxonomic Index

(Page numbers in *italics* indicate illustrations.)

Subject index

(Page numbers in *italics* indicate illustrations.)

ABERYSTWYTH
LLYFRGELL